79

Fortschritte der Chemie
organischer Naturstoffe

Progress in the
Chemistry of Organic
Natural Products

Founded by
L. Zechmeister

Edited by
W. Herz, H. Falk,
G. W. Kirby, and R. E. Moore

Authors:
J. C. Braekman, D. Daloze, H. Franzyk,
J. M. Pasteels, and S. Leclercq

SpringerWienNewYork

Prof. W. Herz, Department of Chemistry,
The Florida State University, Tallahassee, Florida, U.S.A.

Prof. H. Falk, Institut für Chemie,
Johannes-Kepler-Universität, Linz, Austria

Prof. G. W. Kirby, Chemistry Department,
The University of Glasgow, Glasgow, Scotland

Prof. R. E. Moore, Department of Chemistry,
University of Hawaii at Manoa, Honolulu, Hawaii, U.S.A.

This work is subject to copyright.
All rights are reserved, whether the whole or part of the material is concerned, specifically those of translation, reprinting, re-use of illustrations, broadcasting, reproduction by photocopying machines or similar means, and storage in data banks.

© 2000 by Springer-Verlag / Wien
Printed in Austria

Library of Congress Catalog Card Number AC 39-1015

Typesetting: Thomson Press (India) Ltd., New Delhi
Printing: Novographic Druck G.m.b.H., A-1230 Wien
Graphic design: Ecke Bonk
Printed on acid-free and chlorine-free bleached paper

SPIN: 10743105

With 11 Figures

ISSN 0071-7886
ISBN 3-211-83361-7 Springer-Verlag Wien New York

Contents

List of Contributors VII

Synthetic Aspects of Iridoid Chemistry
By H. FRANZYK 1

1. Introduction 2
2. Classification of Iridoids 2
   2.1. Iridoid Glycosides 3
   2.2. Non-Glycosidic Iridoids 6
   2.3. Nitrogen-Containing Compounds 7
3. Availability and Production of Iridoids 8
   3.1. Plant Sources 9
   3.2. Cell-Cultures 10
   3.3. Isolation and Purification Methods 11
4. Semi-Synthetic Conversions Between Iridoids 13
   4.1. Hydrolysis and Lactonization 14
   4.2. Decarboxylation of Iridoid Glucosides 15
   4.3. Reduction and Oxidation of Iridoid Glucosides 17
   4.4. Miscellaneous Transformations Used in Structure Elucidation 22
   4.5. Non-Glucosidic Iridoids 25
   4.6. Ring-Cleavage Sequences and Secoiridoid Chemistry 36
5. Monoterpene Alkaloids Structurally Related to Iridoids 45
   5.1. PMTAs – Natural Compounds or Artifacts? 45
   5.2. Semi-Synthesis of Pyridine Monoterpene Alkaloids 48
   5.3. Bacterial Metabolism of Iridoid Glucosides 52
   5.4. Total Synthesis of Pyridine Monoterpene Alkaloids 55
   5.5. Diversity of Bicyclic Cyclopentanoid Piperidines 59
   5.6. Synthesis of Bicyclic Cyclopentanoid Piperidines 59
   5.7. Semi-Synthesis of Glucosidic Secoiridoid Alkaloids 67
6. Syntheses from Iridoids 69
   6.1. Formation of Colored Compounds 69
   6.2. Reactions of Secoiridoids 73
   6.3. Preparation of Marine Diterpenoids 82
   6.4. Building Blocks for Other Types of Cyclopentanoids 87
   6.5. Modifications of the Sugar Moiety in Iridoid Glucosides 103

References 106

The Defensive Chemistry of Ants
By S. LECLERCQ, J. C. BRAEKMAN, D. DALOZE, and J. M. PASTEELS 115

1. Introduction 116
2. Alkaloids 118
   2.1. Structures, Occurrence, and Function 118
      2.1.1. Piperidines and Pyridines 118
      2.1.2. Pyrrolidines and Pyrrolines 120
      2.1.3. Pyrrolizidines 122
      2.1.4. Indolizidines 124
      2.1.5. Tetraponerines 126
      2.1.6. Other Alkaloids 127
   2.2. Synthesis 128
      2.2.1. Piperidines 128
      2.2.2. Pyrrolidines 131
         A. Synthesis of Racemic Pyrrolidines 131
         B. Syntheses of Nonracemic Pyrrolidines 139
      2.2.3. Pyrrolines 149
      2.2.4. Pyrrolizidines 152
         2.2.4.1. 3,5-Dialkylpyrrolizidines 152
            A. Xenovenine 152
            B. (5E,8E)-3-Butyl-5-hexylpyrrolizidine 159
         2.2.4.2. 3-Methyl-5-alkenylpyrrolizidines and
                  3,5-Dialkenylpyrrolizidines 160
      2.2.5. Indolizidines 163
         2.2.5.1. Monomorine I 163
            A. Syntheses of Racemic Monomorine I 163
            B. Syntheses of Nonracemic Monomorine I 173
         2.2.5.2. 3,5-Dialkylindolizidines 190
            A. 3-Butyl-5(4-penten-1-yl)indolizidine 190
            B. 3-Ethyl- and 3-Hexyl-5-methylindolizidines 193
            C. Myrmicarin 237A and 237B 194
            D. Myrmicarin 217 198
      2.2.6. Tetraponerines 200
         A. Syntheses of Racemic Tetraponerines 200
         B. Syntheses of Nonracemic Tetraponerines 205
3. Nonalkaloidal Compounds 211
4. Biosynthesis 217
References 221

Author Index 231

Subject Index 239

List of Contributors

Jean-Claude Braekman, Laboratory of Bio-Organic Chemistry, Department of Organic Chemistry, CP 160/07, University of Brussels, Av. F. D. Roosevelt, 50, B-1050 Brussels, Belgium

Désiré Daloze, Laboratory of Bio-Organic Chemistry, Department of Organic Chemistry, CP 160/07, University of Brussels, Av. F. D. Roosevelt, 50, B-1050 Brussels, Belgium

Henrik Franzyk, Department of Organic Chemistry, The Technical University of Denmark, Building 201, DK-2800 Lyngby, Denmark

Sabine Leclercq, Laboratory of Bio-Organic Chemistry, Department of Organic Chemistry, CP 160/07, University of Brussels, Av. F. D. Roosevelt, 50, B-1050 Brussels, Belgium

Jacques M. Pasteels, Laboratory of Animal and Cellular Biology, CP 160/12, University of Brussels, Av. F. D. Roosevelt, 50, B-1050 Brussels, Belgium

Synthetic Aspects of Iridoid Chemistry

H. Franzyk

Department of Organic Chemistry, The Technical University of
Denmark, Lyngby, Denmark

Contents

1. Introduction	2
2. Classification of Iridoids	2
2.1. Iridoid Glycosides	3
2.2. Non-Glycosidic Iridoids	6
2.3. Nitrogen-Containing Compounds	7
3. Availability and Production of Iridoids	8
3.1. Plant Sources	9
3.2. Cell-Cultures	10
3.3. Isolation and Purification Methods	11
4. Semi-Synthetic Conversions Between Iridoids	13
4.1. Hydrolysis and Lactonization	14
4.2. Decarboxylation of Iridoid Glucosides	15
4.3. Reduction and Oxidation of Iridoid Glucosides	17
4.4. Miscellaneous Transformations Used in Structure Elucidation	22
4.5. Non-Glucosidic Iridoids	25
4.6. Ring-Cleavage Sequences and Secoiridoid Chemistry	36
5. Monoterpene Alkaloids Structurally Related to Iridoids	45
5.1. PMTAs – Natural Compounds or Artifacts?	45
5.2. Semi-Synthesis of Pyridine Monoterpene Alkaloids	48
5.3. Bacterial Metabolism of Iridoid Glucosides	52
5.4. Total Synthesis of Pyridine Monoterpene Alkaloids	55
5.5. Diversity of Bicyclic Cyclopentanoid Piperidines	59
5.6. Synthesis of Bicyclic Cyclopentanoid Piperidines	59
5.7. Semi-Synthesis of Glucosidic Secoiridoid Alkaloids	67
6. Syntheses from Iridoids	69
6.1. Formation of Colored Compounds	69
6.2. Reactions of Secoiridoids	73
6.3. Preparation of Marine Diterpenoids	82

6.4. Building Blocks for Other Types of Cyclopentanoids............... 87
  6.5. Modifications of the Sugar Moiety in Iridoid Glucosides............ 103

References... 106

1. Introduction

The aim of the present review is to cover developments in the chemistry of iridoids and related compounds during the last decade. In order to be able to give a broad, but still comprehensive presentation of this vast area some limitations to the subject were necessary. Accordingly, the chemotaxonomic importance and biosynthesis studies of iridoids will be omitted from the present survey since these subjects have been treated in contemporary reviews (*1–4*). Similarly, biological activity (*5*) and ecological aspects of iridoids (*6, 7*) will not be treated in detail here. A more historical introduction to the chemistry of iridoids has already been given (*8*), and in addition a specialized report on the progress of synthetic iridoid chemistry has appeared (*9*). Here, it is rather intended to provide the reader with an overview of the types of chemical reactions that have recently been applied to iridoids and to present a collection of synthetic possibilities which should give some perspectives for future work in this field.

First a brief introduction to typical structural features of iridoids is presented. It should be emphasized that the subdivision employed here is strictly structural (*10, 11*) – somewhat different subclasses would be the result if biosynthetic relationships were applied (*1, 2*). Both direct isolation from plants and production by cell-cultures can be used for obtaining iridoids, which in turn may be regarded either as starting material for synthesis of other compounds or as synthetic targets themselves. Regarding the latter possibility, only conversions between different iridoids will be discussed in depth in the present review. However, iridoid related monoterpene alkaloids have been included in the present survey and will receive considerable attention, as will the synthetic utility of iridoid-derived building blocks.

2. Classification of Iridoids

The group of monoterpenoid compounds recognized as iridoids comprises at present well above 1200 structures which exhibit a remarkable diversity that will be outlined briefly. First a fundamental

References, pp. 106–114

distinction between iridoid glycosides and non-glycosidic iridoid compounds will be made.

Throughout the present review trivial names (typically having suffixes such as -in or -oside) will be used since the IUPAC nomenclature often is too complex and lengthy even for very simple structures, *e.g.* the general carbocyclic skeleton (R = Glc in Chart 1) would be named (1*S*)-1-β-D-glucopyranosyl-4,7-dimethyl-1,4a,5,6,7,7a-hexahydro-cyclopenta-[*c*]pyran. In many cases the iridoid numbering shown (Chart 1) will be retained, even when another numbering may be equally appropriate, to facilitate recognition of the iridoid skeleton in strongly modified congeners.

2.1. Iridoid Glycosides

The iridoid glycosides themselves may be further subdivided into carbocyclic iridoids and secoiridoids, and their common feature is the dihydropyran ring system. In general a β-D-glucopyranosyl unit is attached at C-1 *via* a β-hemiacetalic bond (Chart 1). The non-glycosidic part of an iridoid glycoside is often referred to as the aglycone or the "genin".

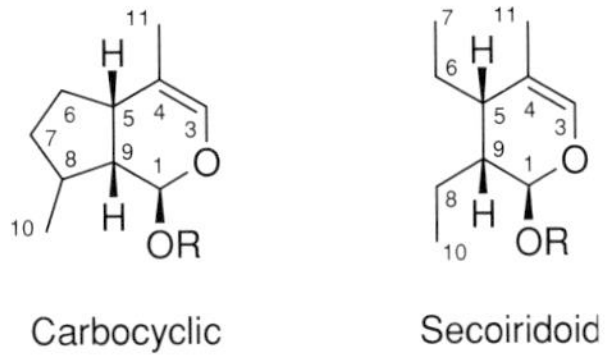

Carbocyclic   Secoiridoid

Chart 1. Basic skeletons of iridoid glycosides

Most carbocyclic iridoids have a *cis*-fused cyclopentane ring with substituents at C-5 and C-9 in β-positions (*i.e.* out of the plane of the paper), but recently a few *trans*-fused compounds have been reported (*12*). A vast structural diversity of carbocyclic iridoids arises from simple modifications of the cyclopentane ring exemplified by some typical functionalities: hydroxy, acyloxy, keto, epoxy, chloro, and olefin. In addition the oxidation state of C-11 may range from methyl to carboxylic acid (and esters thereof). But also the basic skeleton exhibits variation as a result of decarboxylation, thus C-10 and/or C-11 may be absent.

The secoiridoids have a basic skeleton in which the carbon-carbon bond between C-7 and C-8 has been cleaved, but secoiridoids always contain all original ten aglycone carbon atoms. In secoridoids the oxidation state of C-7 and C-8/C-9/C-10 may vary, while C-11 is at the

C₈-Iridoids:

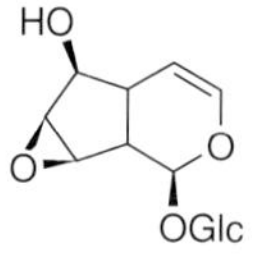

Unedoside (1) 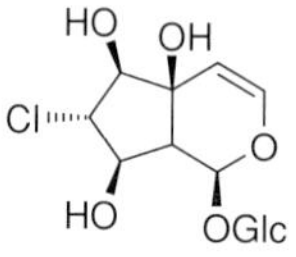 Thunbergioside (2)

C₉-Iridoids with ninth carbon on C-4:

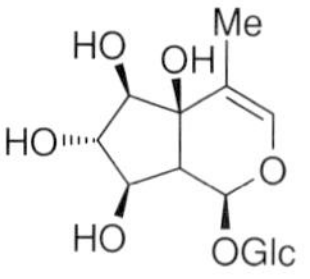

Scabrosidol (3) 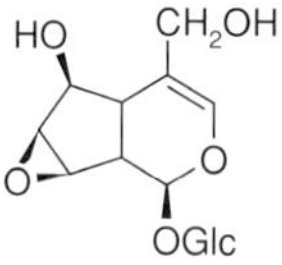 Epoxydecaloside (4) 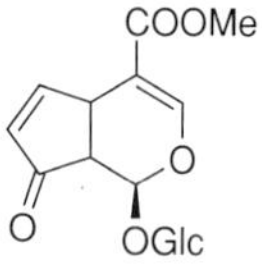 Randioside (5)

C₉-Iridoids with ninth carbon on C-8:

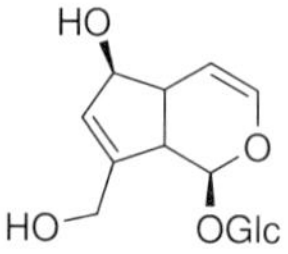

Aucubin (6) 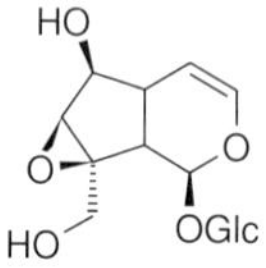 Catalpol (7) 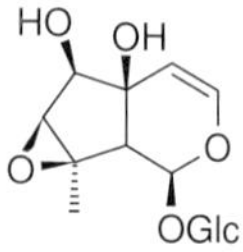 Antirrhinoside (8)

C₁₀-Iridoids:

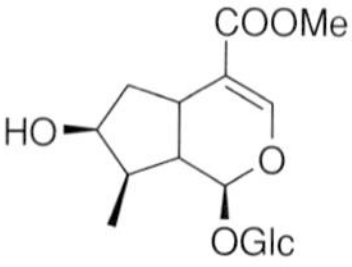

Loganin (9) 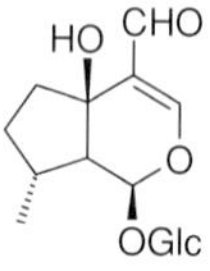 Plantarenaloside (10) 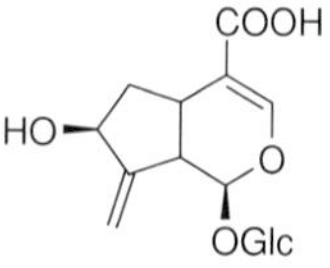 Gardoside (11)

Bisiridoids:

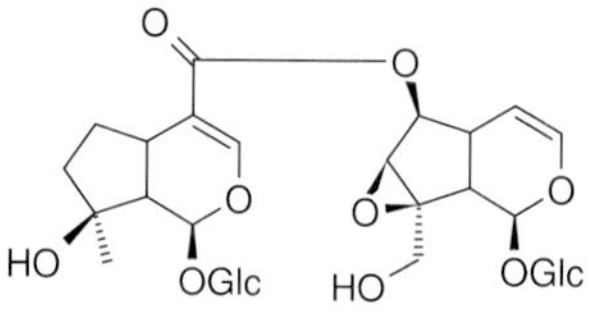

Radiatoside (12)

Chart 2. Carbocyclic iridoid glycosides (*10*, *11*, *13*)

carboxylic stage (either as acid, lactone, methyl ester, or as an ester of a more complex alcohol moiety).

In Chart 2 and Chart 3 typical examples of structures are depicted to illustrate an even more detailed subdivision of iridoid glycosides. In

Simple Secoiridoids:

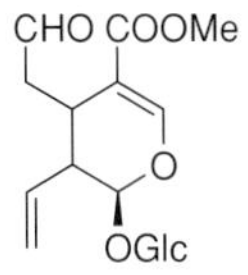

Secologanin (**13**) 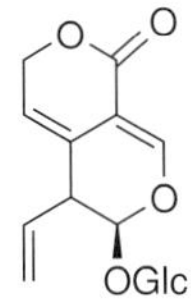 Gentiopicroside (**14**) 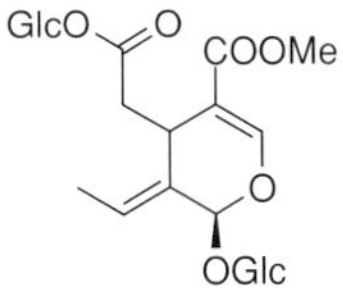 Methyl Gluco-oleoside (**15**)

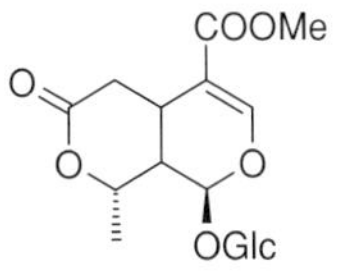

Kingiside (**16**) 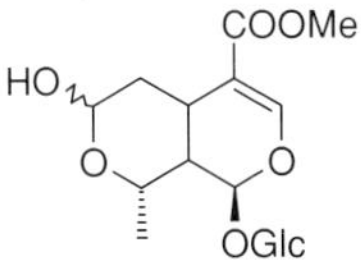 Morroniside (**17**)

Terpene-conjugated Secoiridoids:

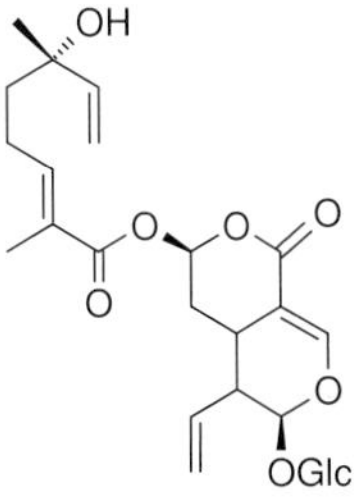

Menthiafolin (**18**) 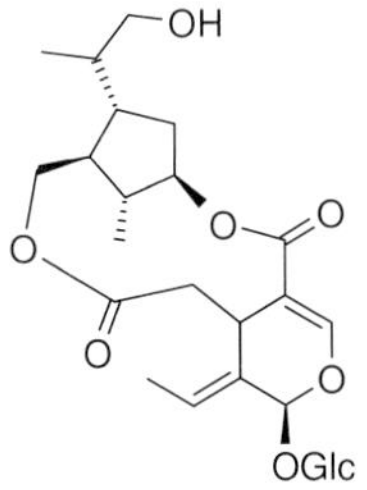 Jasminin (**19**)

Phenolic-conjugated Secoiridoids:

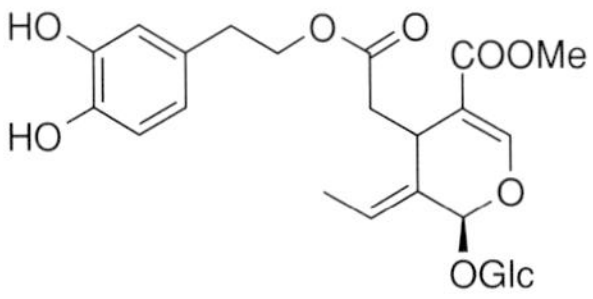

Oleuropein (**20**) 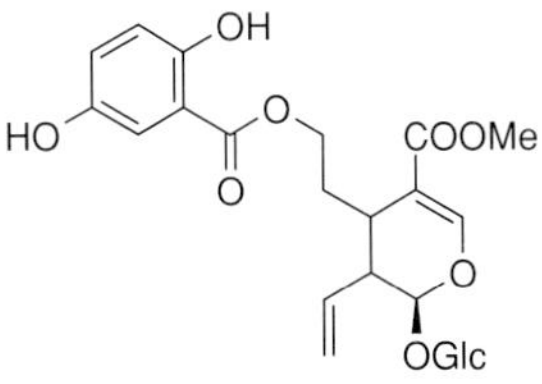 7-*O*-Gentisoylsecologanol (**21**)

Chart 3. Secoiridoids (*10, 11, 13*)

Valeriana Iridoids:

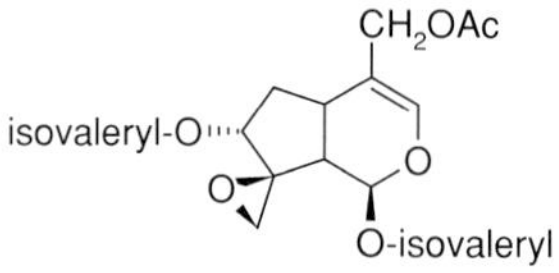

Valtrate (**22**) 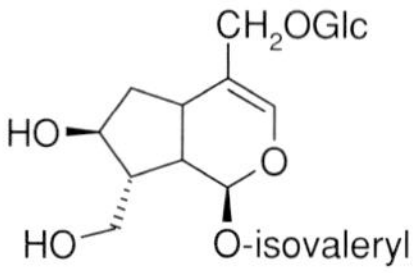 Patrinoside (**23**) 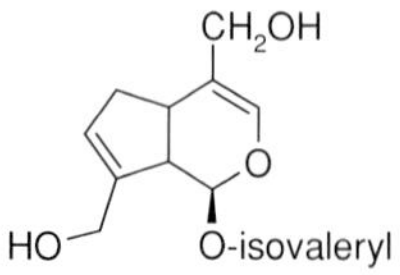 Penstemide Aglycone (**24**)

Plumeria Iridoids:

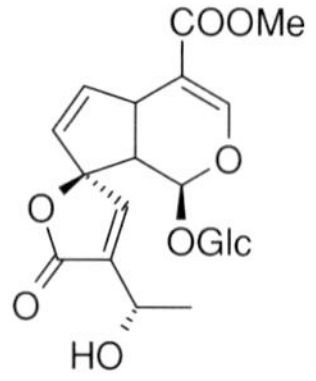

Plumieride (**25**) 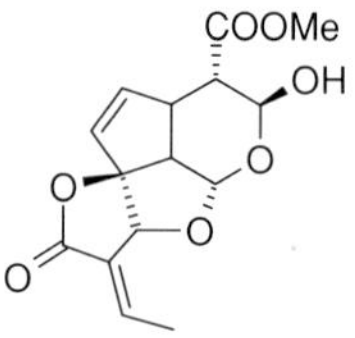 Allamandin (**26**)

Chart 4. *Valeriana* and *Plumeria* iridoids (*10, 11, 13*)

Chart 4 representatives of two groups of special iridoids, namely the *Valeriana* and *Plumeria* iridoids, are shown (*10, 11, 13*). Some of these may be regarded as glycosidic while others may be classified as non-glycosidic, but due to their unusual substitution patterns they are often treated as independent structural assemblies.

2.2. Non-Glycosidic Iridoids

The non-glycosidic iridoids encompass a miscellaneous group of compounds ranging from simple aglycones of ordinary iridoid glycosides, lactones, dehydration products to tricyclic more complex structures. In Chart 5 a few representatives of each subclass are depicted. Many of the simple aglycones may be formed spontaneously during the extraction process due to the liberation of inherent β-glycosidases, which in living plant cells are confined in separate compartments. This tendency may be circumvented by denaturation (brief heating to 60–80°C) of enzymes present in the crude extract. The more apolar compounds, which are extractable by organic solvents are occasionally termed "volatile" iridoids (*14*).

References, pp. 106–114

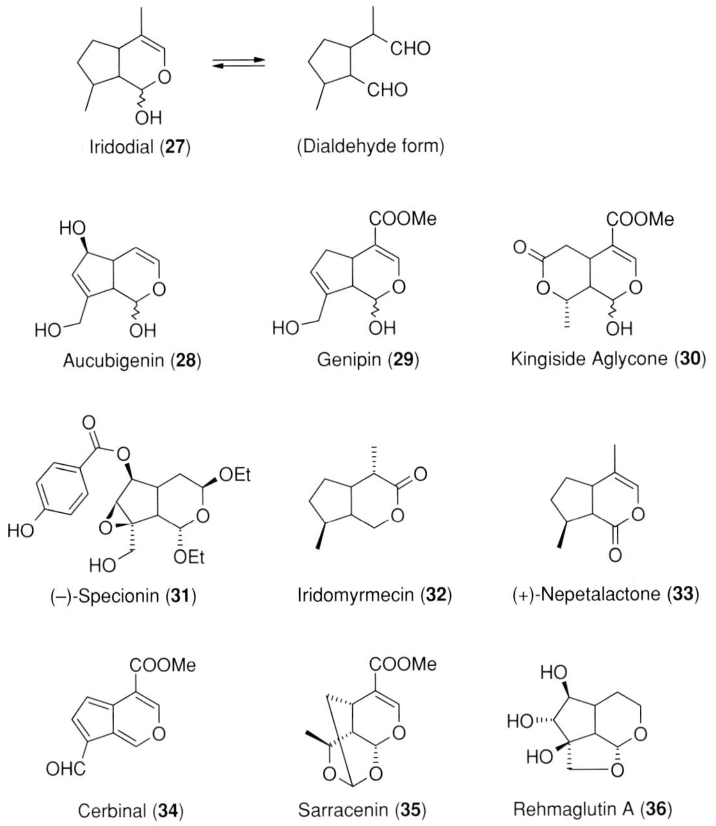

Iridodial (**27**) ⇌ (Dialdehyde form)

Aucubigenin (**28**)   Genipin (**29**)   Kingiside Aglycone (**30**)

(−)-Specionin (**31**)   Iridomyrmecin (**32**)   (+)-Nepetalactone (**33**)

Cerbinal (**34**)   Sarracenin (**35**)   Rehmaglutin A (**36**)

Chart 5. Non-glycosidic iridoids (*10, 11, 13*)

2.3. Nitrogen-Containing Compounds

As pointed out earlier (*15*) the so-called pseudoalkaloids should be regarded as a genuine type of iridoids, since they have proved to be natural constituents and not merely artifacts formed during isolation (*i.e.* when ammonia is applied during extraction), as previously assumed (*16*). A number of quite different alkaloidal iridoids exists: Pyridine monoterpene alkaloids (PMTAs, *e.g.* **37–39**), monoterpenoid piperidines (*e.g.* **40–42**), alkaloidal glucosides (**43–45**), and complex indole alkaloids (*e.g.* **46**) derived from secologanin (**13**) and tryptophan (Chart 6). Emphasis will be given to the first two simple monoterpenoid groups, while the complex indole alkaloids will be omitted in this context.

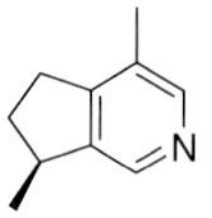

(−)-Actinidine (**37**) 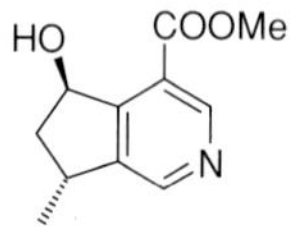   (+)-Rhexifoline (**38**) 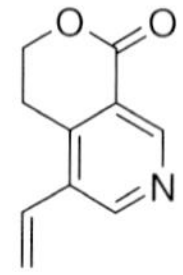   Gentianine (**39**)

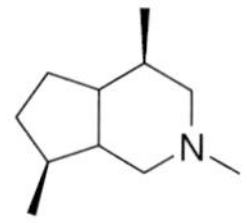

(+)-α-Skytanthine (**40**) 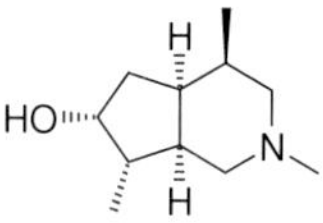   Incarvilline (**41**) 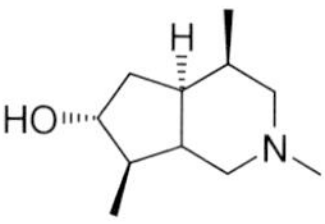   Kinabalurine A (**42**)

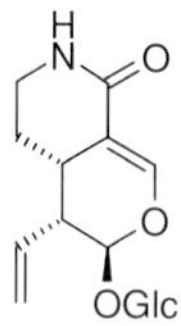

Bankakosin (**43**) 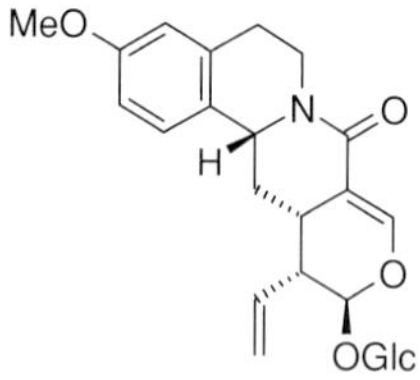   Alangiside (**44**)

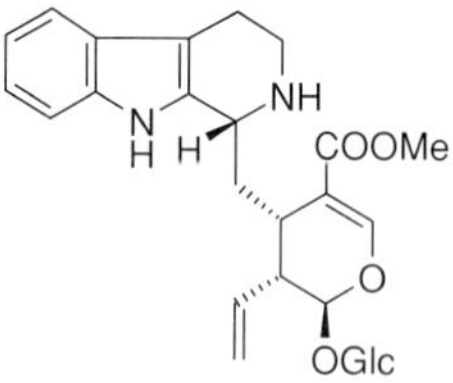

Vincoside (**45**) 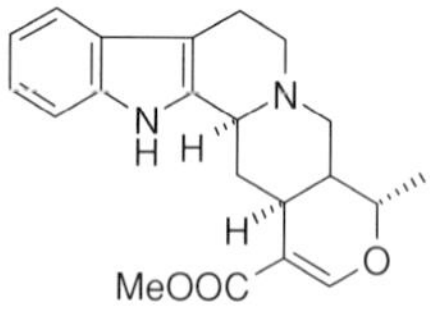   Ajmalicine (**46**)

Chart 6. Iridoid-related alkaloids

3. Availability and Production of Iridoids

As mentioned in the previous section, iridoids are of monoterpenoid origin (*3, 17, 18*) and occur in many orders of the sympetalous angiosperms (*1*). Moreover, both their diversity and biosynthesis seem to be related to the families in which they are found, thus making them useful as chemotaxonomic markers (*2*). During the last decades, one of the main driving forces in iridoid research has been the discovery of novel structures, and this has been facilitated by the appearance of efficient chromatographic methods for the purification of water-soluble compounds. However, when iridoids are employed as starting materials in syntheses of other types of cyclopentanoid compounds, the most important issue becomes the availability of the iridoids. Different approaches to meet this requirement will be discussed in the following sections.

3.1. Plant Sources

Naturally, if an iridoid is to be considered as starting material for a synthesis it must be present in reasonable amounts in a plant species, and occasionally an iridoid content of 0.5–3% of the fresh weight is achievable. But in the selection of a suitable plant species one must also have in mind that it should be easily grown in the appropriate climate. Development of optimal growth conditions and investigation of seasonal variations (*19*) are critical parameters in ensuring an optimal harvest of plant material. Even a species with these characteristics may prove troublesome if compounds with similar adsorption and/or solvent distribution properties are also present.

The requirement for a simple iridoid composition may indeed be crucial for a suitable plant candidate for large-scale production of iridoids. As an example, wild-type *Antirrhinum majus* (Scrophulariaceae) exhibits a diverse content of iridoid glucosides (Chart 7). Thus, antirrhinoside (**8**), 5-glucosylantirrhinoside (**47**), and antirrhide (**48**) co-occur and the compounds are not easily separated on a large scale. In the course of work dealing with use of **8** as starting material for preparation of novel cyclopentanoid compounds (*20, 21*) an investigation of several commercial varieties of *Antirrhinum majus* was performed. A surprisingly high biodiversity with respect to iridoid content and composition was found. The commercially available varieties "Bright Eyes" and "White Wonder" proved superior, since these only contained small amounts of the corresponding chlorohydrin, linarioside (**49**), in addition to the desired antirrhinoside (**8**).

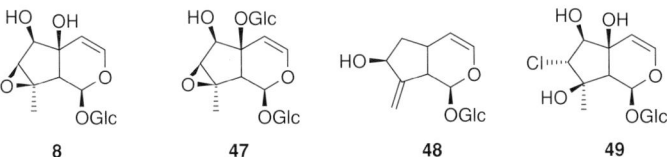

Chart 7. Iridoid glucosides in *Antirrhinum majus* varieties

The plant material from a suitable species may either be extracted directly after harvest or it may be dried prior to extraction. Alternatively, fresh plant may be frozen immediately after harvest and subsequently kept (below −20°C) for several years if the stability of the iridoid is sufficient (*e.g.* **6** may decompose under such conditions). Fresh material has the advantage of being readily macerated using a blending device, and the compounds are transferred immediately to the solvent, usually methanol, ethanol or acetone (sometimes in admixture with water). Nevertheless, when working with multi-kilograms of plant, it may prove more convenient to employ dry material as it reduces the weight to 1/5, and then grinding may be performed without solvent. Furthermore, no water is released during extraction so it allows for fast concentration of the extract at lower temperatures. Yet, in the case of very polar substances, it may in fact be necessary to add water to the extraction solvent to enhance their solubility (*e.g.* 90% aqueous ethanol is suitable for most iridoids in dry material), and frequently an extended extraction time is necessary.

3.2. Cell-Cultures

A few recent examples of production of iridoids by tissue cultures will be given here (Chart 8), but this method still seems to need further refinement to be exploitable. Suspension cultured cells of *Penstemon serrulatus* showed an ability to produce iridoids comparable to that of the intact plant (*22, 23*). Thus, the two *Valeriana*-type iridoids, penstemide (**50**) and serrulatoloside (**51**) were formed in up to 4% and 2% yields (based on dry weight), respectively, and could be isolated using normal extraction and chromatographic procedures. After selection of iridoid-producing cell lines, callus and cell suspension cultures of *Genipa americana* allowed production of tarennoside (**52**), geniposidic acid (**53**) and gardenoside (**54**) in 0.3%, 0.04% and 0.15% isolated yields,

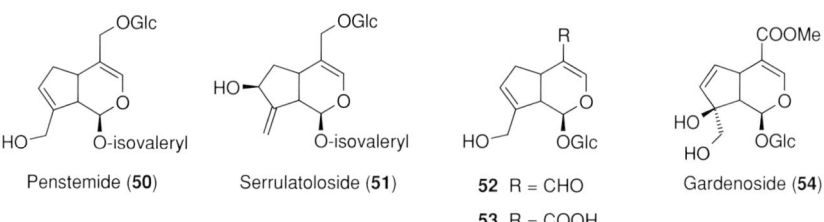

Chart 8. Iridoids produced by cell-cultures

respectively (*24*). Occasionally, callus cultures express an altered iridoid composition or may entirely lack the capability to produce iridoids; *e.g.* even though the crude alcoholic extract of *Nyctanthes arbor-tristis* callus tissue cultures gave a positive iridoid test on TLC plates, the spots did not correlate with any known iridoid isolated from the intact plant (*25*).

3.3. Isolation and Purification Methods

The inherent instability towards acidic conditions seen for iridoid glycosides is mainly due to the presence of a hydrolyzable glycosidic linkage, but other functionalities such as epoxides in the cyclopentane ring may also undergo acid-catalyzed reactions. The sensitivity towards acids is most pronounced for compounds with either a 5-hydroxyl substitution or a double bond in the cyclopentane ring, and several rearrangements and partial decompositions have been reviewed previously (*8*). However, these tendencies do not normally pose any problems during standard isolation and chromatographic procedures. Actually, acetic acid is frequently added to plant extracts prior to purification on reverse phase columns in order to avoid that iridoid acids elute in partially ionized form (with tailing or double-peaks as the result). Moreover, iridoids having either a lactone or an ester functionality are susceptible to hydrolysis in basic media. On the other hand, the sensitivity to heat seems to have been overstated, since most iridoid compounds, including the otherwise sensitive aucubin (**6**), survive prolonged heating to at least 50–60°C in neutral solutions.

Solvent partitioning of the crude plant extract is often the method of choice as a first means of fractionation. Commonly, an initial partitioning between water and diethyl ether separates the more lipophilic compounds. The iridoid glycosides left in the aqueous layer may then to some degree be extracted selectively into chloroform, ethyl acetate, and *n*-butanol by successive extractions. Only the most polar substances, such as sugars, salts, and iridoid glycosides with a highly oxidized skeleton, will still remain in the aqueous phase. Depending on the structure of the iridoid of interest, the next step of purification may be treatment with basic alumina, which irreversibly removes all *ortho-*dihydroxylated phenolics such as caffeic esters (*e.g.* verbascoside) and flavonoids. Among the procedures for obtaining highly iridoid-enriched fractions, the charcoal method still seems the most versatile. It is rapid even when working on a 100 g scale due to its simplicity: an aqueous solution of the extract is stirred with an appropriate amount of charcoal for 30 minutes. The suspension is filtered and the cake is washed with water to remove non-adsorbed compounds (sugars, salts *etc.*). When the sugars are eluted completely or the level of iridoid in the washings becomes significant, the adsorbed iridoid is subsequently eluted with methanol or ethanol. Separation of an iridoid mixture may be performed by using a gradient elution with increasing alcohol:water ratio.

Traditional column chromatographic methods applicable for large-scale work include medium pressure liquid chromatography (MPLC) using either normal or reverse phase packing material, but due to the low capacity and high flow rate it is both time- and solvent-consuming. Flash column chromatography (CC) on conventional silica types also suffers from these drawbacks, whereas the use of TLC-mesh silica gel as packing material in either flash CC or vacuum liquid chromatography (VLC) allows a relatively fast separation of large quantities on easy-handled column sizes. More comprehensive surveys of the above-mentioned and other more advanced techniques for chromatographic separation of iridoids have been presented previously (*3*, *8*, *15*, *26*).

A number of large-scale isolations of iridoid glucosides have been reported and a few examples will be reviewed here (Chart 9). Aucubin (**6**) was obtained from the ornamental plant *Aucuba japonica* (Aucubaceae) in 1% of the fresh weight (*27*) in an amount of 500 g: extraction was performed with hot water and chromatography was carried out on Celite with aqueous *n*-butanol as eluent. The hexaacetate of **6** (75 g) was obtained from the same source by another procedure: the water-soluble part of an ethanolic extract was purified by solvent partitioning in water:butanon–ethyl acetate. The aqueous phase was

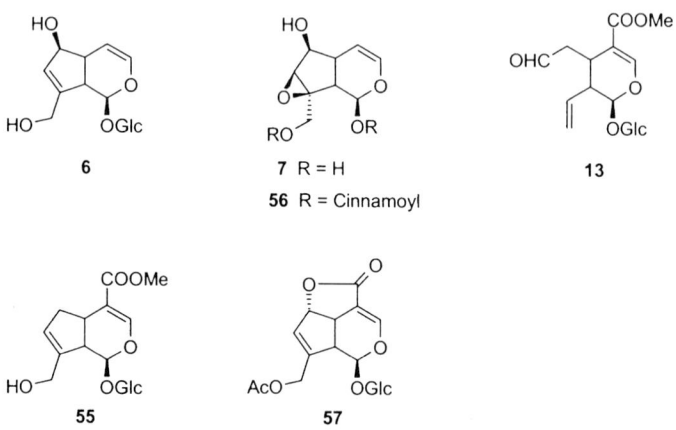

Chart 9. Readily accessible iridoid glucosides

concentrated, and the residue was acetylated and then crystallized to afford 1.25% (based on fresh weight) of hexaacetate (**28**). Catalpol (**7**) was isolated in crystalline form (800 g) from a commercial drug (50 kg) of *Picrorhiza kurrooa* (Scrophulariaceae) by a combination of the charcoal method and final purification by chromatography on alumina (*29, 30*). In addition, catalpol (**7**) itself is commercially available form Merck AG, Darmstadt, Germany, while geniposide (**55**) is available from Glico Foods Corporation, Osaka, Japan (*31*). A large batch (178 g, 0.17% yield) of secologanin (**13**) was obtained from berries of *Symphoricarpos albus* (Caprifoliaceae) by successive extraction with acetone:chloroform, filtration, freeze-drying, adsorption on silica, and elution with ethyl acetate:isopropanol (*32*). Among several species of *Scutellaria* (Lamiaceae) investigated, *S. subvelutina* seemed promising as scutellarioside I (**56**) was predominant in the water-soluble part of the crude ethanolic extract (*21*). This cinnamoyl ester even proved sufficiently apolar to allow its semi-continuous extraction from an aqueous solution into ethyl acetate (1.3% of the fresh weight). Subsequent acetylation and crystallization yielded the pentaacetate of **56** (37% yield; motherliquor: ∼80% pure). Similarly, the Soxhlet method has been employed in the isolation of crystalline asperuloside (**57**) (*33*).

Several spray reagents exist for the detection of iridoid spots on TLC plates by promoting the acid-catalyzed formation of colored products (*3, 8*), but in most cases spraying with dilute sulfuric acid followed by heating gives adequate results. The enol-carbonyl conjugated iridoids can be detected in UV-light (*e.g.* 254 nm), while simultaneous detection of enol-carbonyl conjugated (UV-max ∼235–245 nm) and decarboxylated iridoid glycosides (UV-max ∼195–210 nm) may be achieved by an analytical HPLC setup with a two-channel UV-detector.

4. Semi-Synthetic Conversions Between Iridoids

In early iridoid chemistry, a number of standard reactions were developed for structure elucidation purposes, and the most common include hydrolysis and methanolysis of complex iridoid esters to yield more simple identifiable subunits. Likewise simple reduction, especially hydrogenation, but also oxidation have been employed widely as proof of basic stereochemical correlation to known compounds. In the last three decades where biosynthetic studies have been undertaken, longer and slightly more advanced reaction sequences have evolved to enable synthesis of labelled precursors for feeding experiments. Due to the scarcity of material, a few syntheses that will be discussed have only been reported on labelled compounds (but labelling will be omitted from the schemes presented here).

4.1. Hydrolysis and Lactonization

As already mentioned, alkaline hydrolysis of iridoid esters or bisiridoids is a common way of obtaining structural knowledge that otherwise would require sophisticated NMR experiments to obtain. Fortunately, most iridoids are stable towards the alkaline conditions necessary for hydrolysis since a low excess of hydroxide usually is sufficient for a reasonably fast (typically 2–5 hours) and quantitative conversion into salts of iridoid acids.

However, in certain cases, *e.g.* in the hydrolysis of the *p*-hydroxybenzoic acid derivative, fontanesioside (**58**), a large excess of hydroxide is necessary, as is a prolonged reaction time. After the initial ionization of the phenol, further reaction then requires attack of the similarly charged hydroxide ion, which is sluggish probably due to electronic

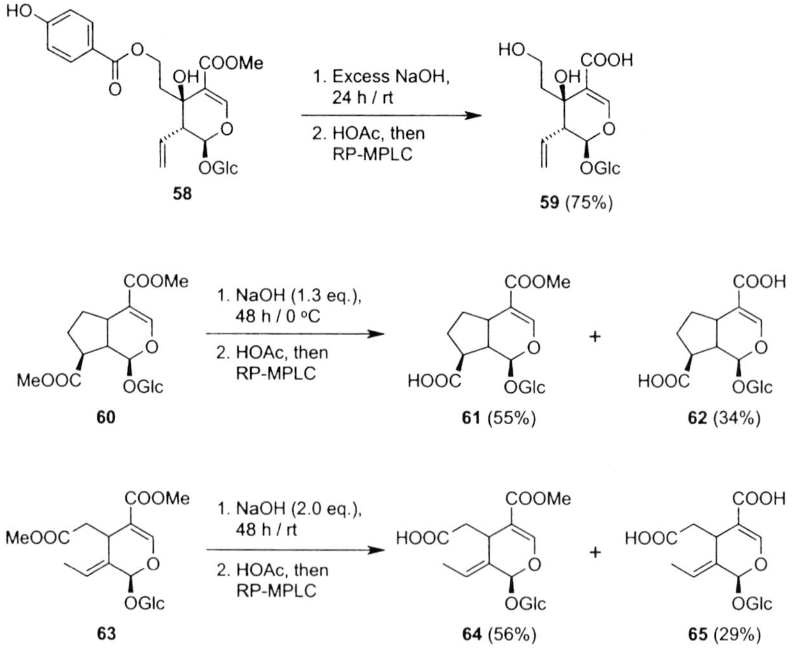

Scheme 1. Hydrolysis of iridoid methyl esters (*34–36*)

repulsion (*34*). A similar effect is also seen for iridoids with two ester moieties where hydrolysis of the second ester is more slow. Thus selective hydrolysis of the more reactive ester in *e.g.* forsythide dimethyl ester (**60**) and oleoside dimethyl ester (**63**) may be performed in moderate yields (*35, 36*). The iridoid acids may then be obtained after addition of an excess of acetic acid followed by chromatography (Scheme 1).

Lactonizations are occasionally brought about under mild conditions such as dilute weak base or acid, or simply by contact with silica gel. Examples are the conversions between alboside A and B (**66** and **67**) (*37*), between secologanol (**68**) and sweroside (**69**), and between 5-hydroxysecologanol (**70**) and swertiamarin (**71**) (*34*) (Scheme 2). Accordingly, iridoid lactones isolated from plants may in some cases be artifacts formed from the corresponding hydroxy acids during the purification process.

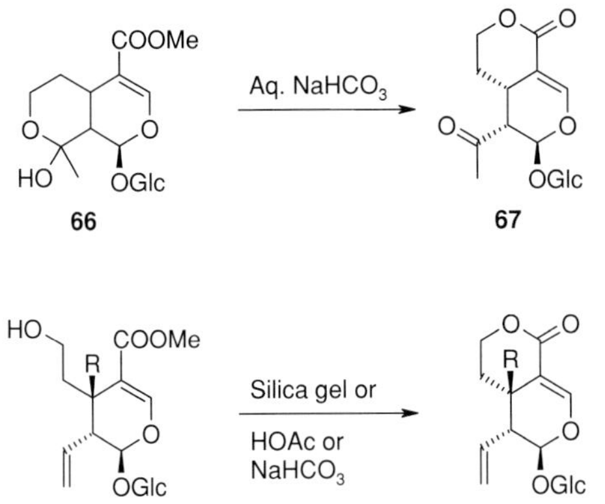

Scheme 2. Lactonizations of secoiridoids (*34, 37*)

4.2. Decarboxylation of Iridoid Glucosides

A feasible conversion of 4-carboxylated iridoids (*i.e.* the C_{10}-subgroup) into decarboxylated iridoids (*i.e.* C_9-iridoids with ninth carbon on C-8) was originally developed by MURAI and TAGAWA (*38*) for the conversion of deoxyloganic acid tetraacetate (**72a**) into 8β-aucuban tetraacetate (**73a**). Since then, similar transformations have been performed on various iridoid substrates (Scheme 3). In general the

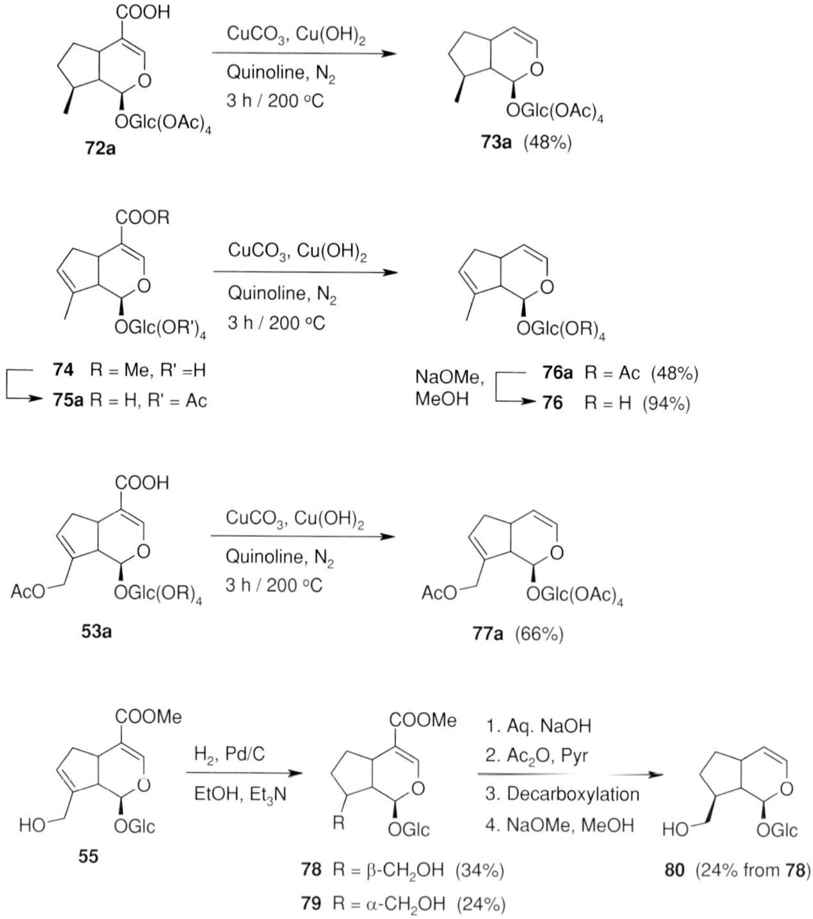

Scheme 3. Decarboxylation of iridoid glucosides (*38–40*)

acetate of an iridoid 11-carboxylic acid undergoes loss of C-11 when heated to 200°C in quinoline in the presence of a basic copper catalyst (copper(II) carbonate and copper(II) hydroxide). Deoxygeniposide (**74**) was (via **75a**) converted into 6,10-dideoxyaucubin (**76**), while geniposidic acid tetraacetate (**53a**) similarly afforded bartsioside tetraacetate (**77a**) (*39*). Hydrogenation of geniposide (**55**) gave only a moderate yield of a mixture of adoxoside (**78**) and *epi*-adoxoside (**79**), and decarboxylation of the former gave capensioside (**80**) (*40*).

Recently, an olefinic C_8-iridoid was prepared by a route starting from gardenoside (**54**) (*41*) (Scheme 4). Oxidation with one equivalent of

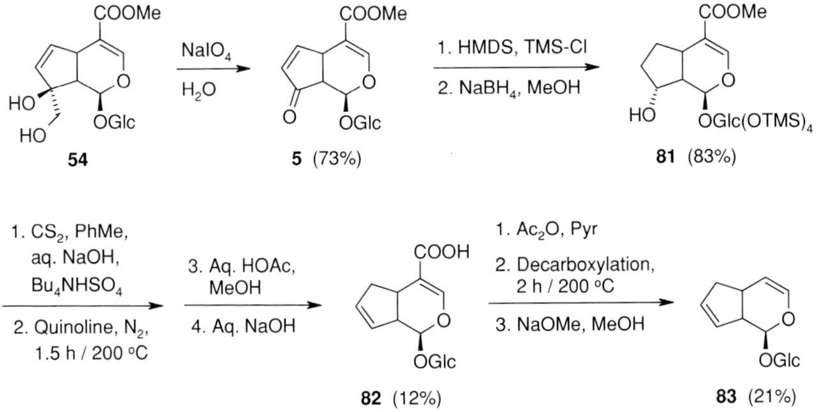

Scheme 4. Preparation of a decarboxylated C_8-iridoid glucoside (41)

sodium periodate gave selectively randioside (**5**). Silylation and subsequent 1,4-reduction afforded alcohol **81** which was subjected to a Chugaev reaction *via* its methylxanthate. After deprotection and ester hydrolysis, the elimination product (**82**) was in turn decarboxylated to give olefin **83**.

4.3. Reduction and Oxidation of Iridoid Glucosides

Syntheses of the simple 8α- and 8β-methyl substituted carbocyclic iridoid glucosides started from gardenoside hexaacetate (**54a**) (*42–44*), which by transfer hydrogenolysis and subsequent deacetylation was converted into deoxygeniposide (**74**). At this point the two epimer-enriched series arose from the different selectivities obtained in the hydrogenation of **74** with Pd/C and Rh/C catalysts (Scheme 5). When 5% palladium on carbon was used as catalyst, the 8β-epimer (**85**) was predominant (∼85%) in the mixture **84/85** (86% total yield) (*42*). By contrast, hydrogenation of **74** with 5% rhodium on carbon as catalyst afforded mainly the 8α-epimer (**84**) (95% epimeric purity; 71% total yield) (*43*). Reduction of the 11-methyl esters in **84** and **85** with *in situ* formed LiAl(OMe)$_2$H$_2$ afforded the 11-hydroxymethyl analogues **86** and **87** in 41% and 48% yield, respectively (and with the same epimeric purity as the starting material). Oxidation of **86** and **87** using Pt and O$_2$ allowed preparation of the corresponding 11-aldehydes **88** (95% epimeric purity; 80% yield) and **89** (85% epimeric purity; 96% yield),

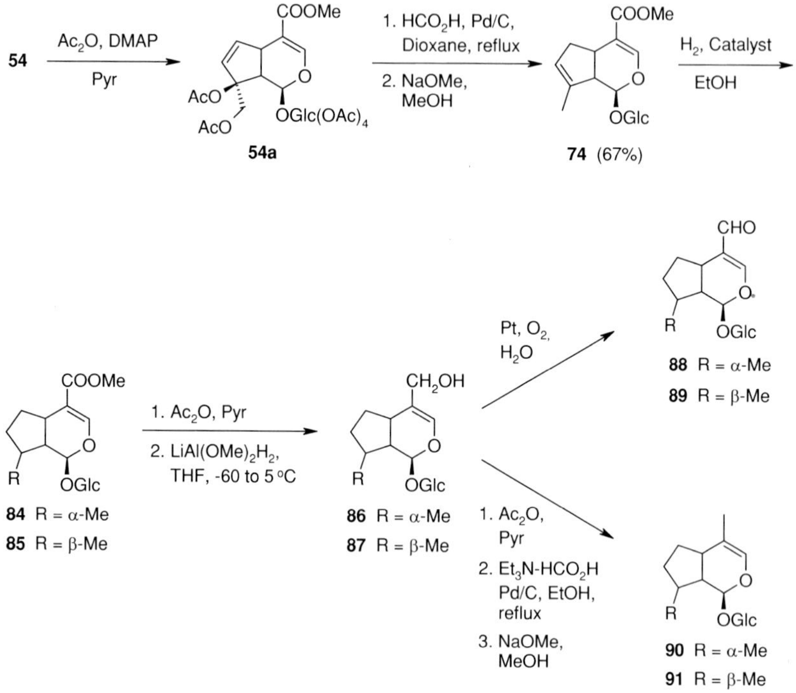

Scheme 5. Reductive pathways to highly deoxygenated iridoid glucosides (42–44)

whereas mild transfer hydrogenolysis of the corresponding acetates with triethylamine-formic acid complex as hydride donor gave *epi*-iridodial glucoside (**90**, 73%, crystallized to 99% epimeric purity) and iridodial glucoside (**91**, 69% with 85% epimeric purity), respectively (*42, 44*).

In order to obtain antibodies specific to deoxyloganin (**85**) and 8-*epi*-deoxyloganin (**84**), these iridoid glucosides were prepared in optically pure form starting from naturally occurring cornin (**92**) and boschnaloside (**88**), respectively (*45*) (Scheme 6). Cornin tetraacetate (**92a**) was converted into the corresponding ethylene dithioketal (**93**), which was subjected to desulfurization with Raney-Nickel to provide tetraacetate **85a**. Optically pure deoxyloganin (**85**) was then obtained after Zemplén deacetylation of **85a**. On the other hand, oxidation of the 4-formyl functionality in tetraacetate **88a** led to an acid which upon methylation and Zemplén deacetylation afforded optically pure 8-*epi*-deoxyloganin (**84**). Further processing of **84** and **85** towards haptens and protein conjugates will be discussed in Section 6.5.

References, pp. 106–114

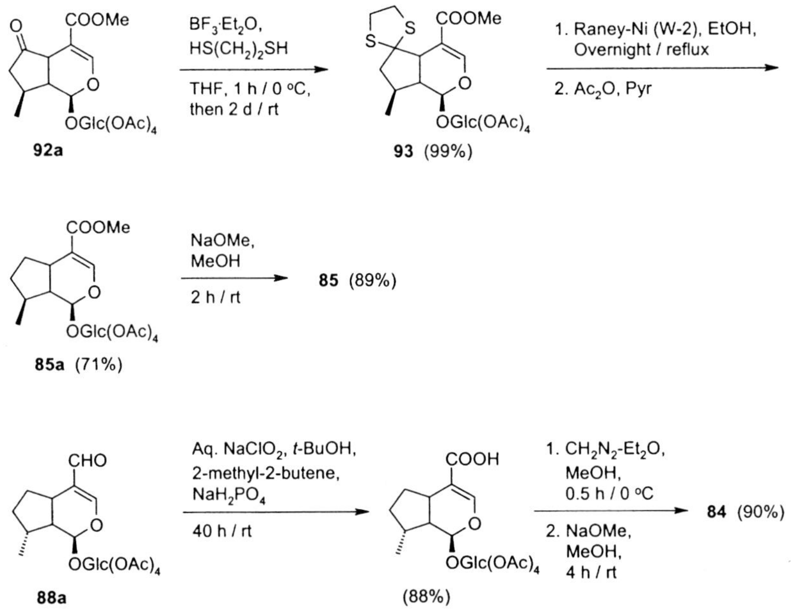

Scheme 6. Semi-synthesis of epimeric pure deoxyloganin (**85**) and of its 8-epimer (**84**)(*45*)

In an approach towards 7-oxidized 8β-methyl carbocyclic iridoid glucosides, deoxygeniposide tetraacetate (**74a**) was epoxidized with *m*-chloroperbenzoic acid to give the isomeric epoxides **94** and **95** (*46, 47*) (Scheme 7). Next, the boron trifluoride etherate catalyzed rearrangement (*46*) to the 7-keto compound was optimized for both epoxides (*47*). The α-epoxy compound (**95**) proved significantly less reactive, but at the same time its reaction proceeded more cleanly, and after concurrent deacetylation and equilibration of the stereochemistry at C-8, ketologanin (**96**) was obtained in good yield (*47*). Stereoselective reduction of the keto group in **96a** followed by deacetylation and saponification allowed preparation of 7-*epi*-loganic acid (**97**) in good yield. Also, the 7-epimeric loganic acid (**98**) and loganin (**9**) were obtained from **96a** by successive reduction and Mitsunobu inversion of the resulting free 7α-hydroxyl substituent followed by a one-pot deacetylation and partial hydrolysis (*40, 48*).

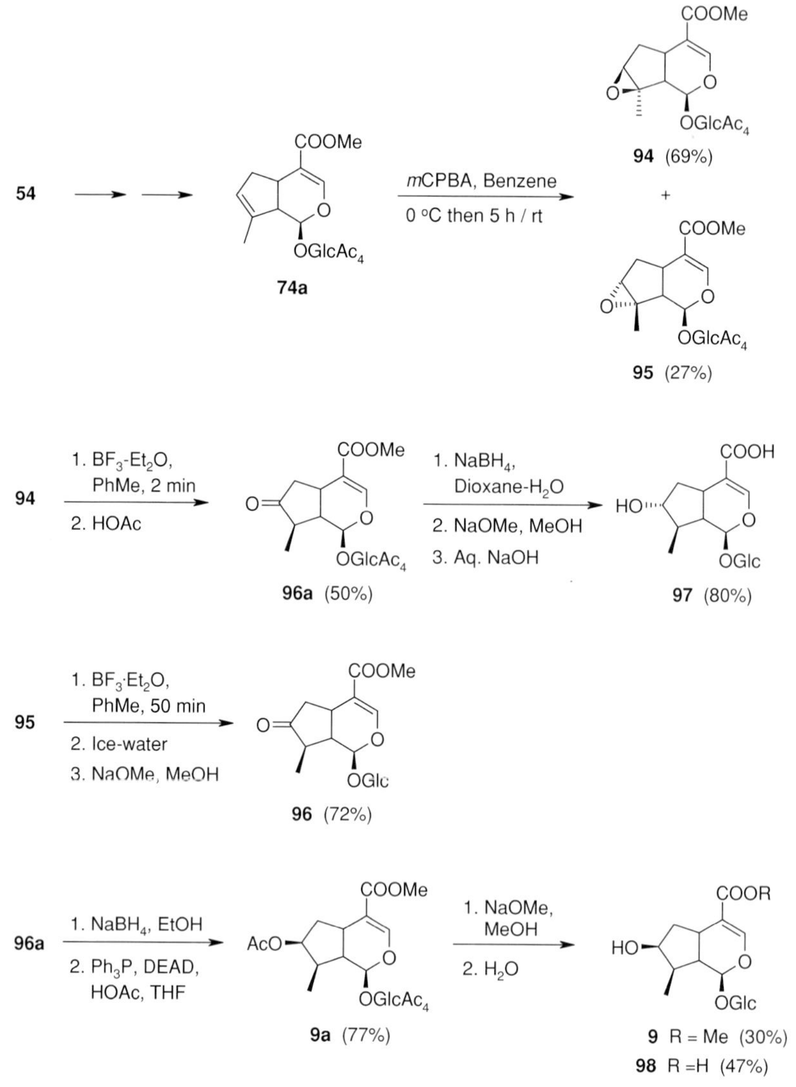

Scheme 7. Simple reduction and oxidation routes (40, 46–48)

The difference in stereospecificity observed for Pd/C and Rh/C catalysts in the above-mentioned hydrogenation of **74** was almost completely lost when the substrate was gardoside methyl ester (**99**) featuring an *exo*-methylene group. In this case, 8-*epi*-loganin (**100**) was the major product with both catalysts (*39, 47*) (Scheme 8).

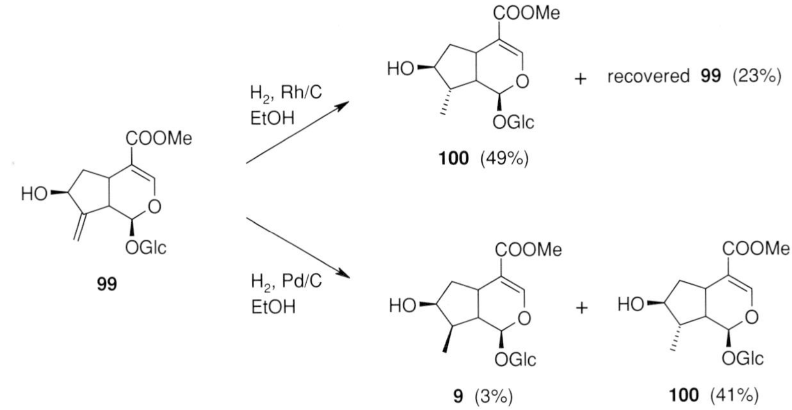

Scheme 8. Hydrogenation of gardoside methyl ester (**99**) (*39, 47*)

For use in biosynthetic studies a short route from aucubin hexaacetate (**6a**) to 10-deoxycatalpol (**103**) was developed (*49*) (Scheme 9). First, transfer hydrogenolysis of **6a** followed by deacetylation afforded a mixture of partially deoxygenated products: 6,10-dideoxyaucubin (**76**), 10-deoxyaucubin (**101**), linarioloside (**102**), and recovered aucubin (**6**). The epoxidation of **101** with *m*-chloroperbenzoic acid proved unsuccessful, whereas 10-deoxycatalpol (**103**) could be obtained in low yield after oxidation of **101** with hydrogen peroxide and tungsten(VI) oxide.

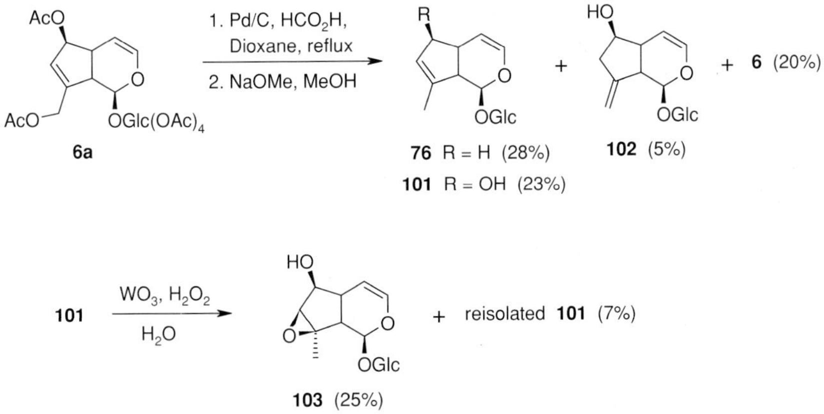

Scheme 9. Preparation of 10-deoxycatalpol (**103**) (*49*)

4.4. Miscellaneous Transformations Used in Structure Elucidation

The conversion of catalpol (**7**) into aucubin (**6**) has been performed in high yield starting from acetate **7a** (*50*) (Scheme 10). Reduction with lithium aluminum hydride, re-acetylation, and phosphorus oxychloride mediated dehydration of the resulting **104a** yielded hexaacetate **6a**. Transformation of decarboxylated iridoid glucosides into the corresponding 4-formylated analogues was originally developed by JENSEN *et al.* (*51*) and involved a Vilsmeier formylation of a peracetylated iridoid derivative. This procedure was employed for the conversion of **6a** into **105a**, which subsequently was oxidized further into the 4-carboxylic acid, scandoside hexaacetate (**106a**) (*50*).

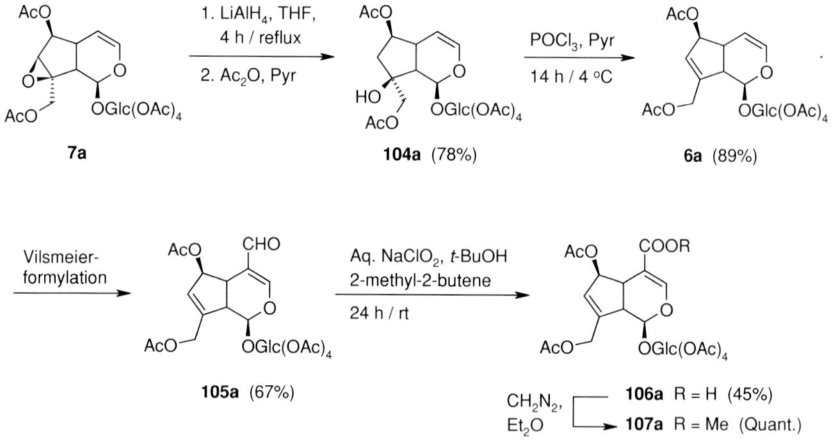

Scheme 10. Vilsmeier formylation of a decarboxylated iridoid glucoside (*50*)

The *trans*-diaxial selectivity already observed in the elimination reaction **104a**→**6a** was also utilized in the structure elucidation of an iridoid, nepetanudoside (**108**), with an unusual 1,5,9-tri-*epi*-stereochemistry so far found only in the genus *Nepeta*. Hence, its tetraacetate (**108a**) afforded under similar conditions the corresponding protected 7,8-olefin (**109**) (*52*) (Scheme 11).

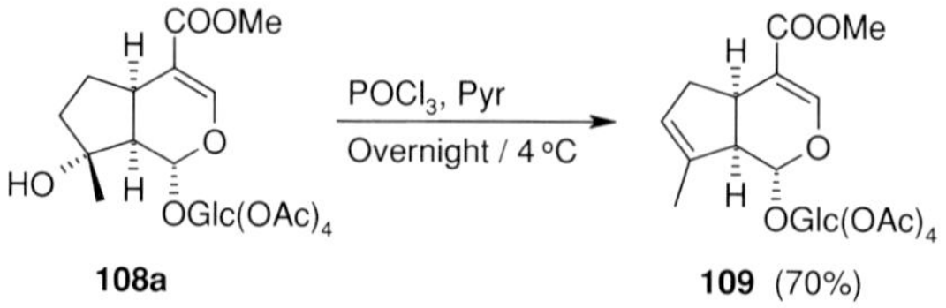

Scheme 11. Dehydration of nepetanudoside tetraacetate (**108a**) (*52*)

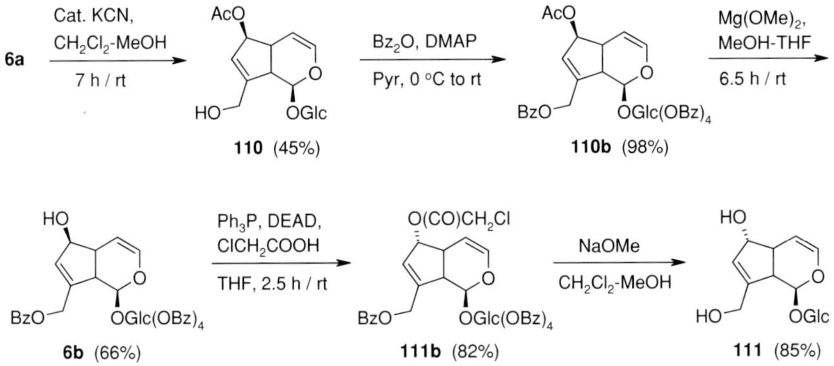

Scheme 12. Inversion of C-6 in aucubin (**6**) (*53*)

Recently, an efficient protocol for the selective inversion of the 6-position in aucubin (**6**) has been reported (*53*) (Scheme 12). Thus, hexaacetate **6a** was selectively deacetylated in dichloromethane-methanol to 6-monoacetyl derivative **110** by using a catalytic amount of potassium cyanide. The remaining hydroxyl groups were benzoylated, and subsequently the acetyl group was removed in a magnesium methoxide solution. Mitsunobu inversion at C-6 of the resulting pentabenzoate (**6b**) was performed with monochloroacetic acid as nucleophile to give **111b**, from which 6-*epi*-aucubin (**111**) was obtained by deacylation.

Occasionally, NMR spectroscopic methods are insufficient for unequivocal determination of the absolute stereochemistry in iridoid glucosides (*54, 55*). For example nOe experiments proved inadequate for determining the configuration of C-8 in 10-hydroxyhastatoside (**113**) since the originally proposed structure for this compound in fact corresponded to its 8-epimer. However, careful comparison of the ^{13}C NMR data for several pairs of similar 5-hydroxy/5-deoxy compounds indicated that **113** and the known 10-hydroxycornin (**112**) constituted such a pair with identical stereochemistry at C-8, for which chemical evidence was found (*54*) (Scheme 13). Hence, **112** was via its enol acetate **112a** oxidized in the 5-position to yield hexaacetate **113a**, which proved identical with one of the acetates obtained from prolonged acetylation of the natural compound (**113**). At the same time, the configuration of 5,6β-dihydroxyadoxoside (**114**) was settled, since sodium borohydride reduction of **113a** yielded **114a** as the sole product (after full migration of the 5-*O*-acetyl group to the 6-position), and this was identical with the acetate obtained from the natural compound (**114**).

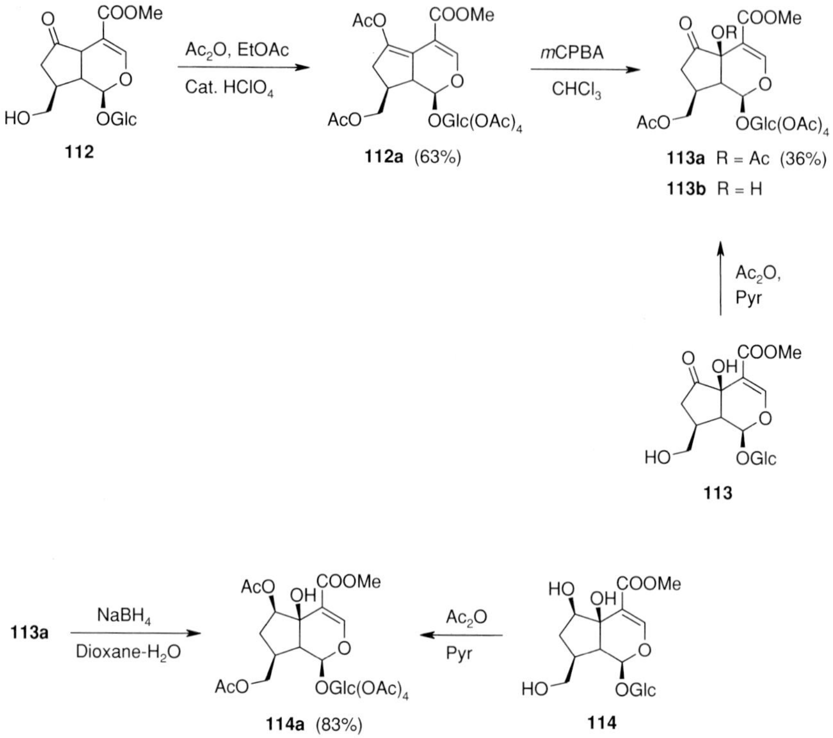

Scheme 13. Chemical correlation of iridoids (**113** and **114**) from *Penstemon* ssp. with the known **112** (*54*)

Another example illustrating the need for supporting chemical evidence was reported recently for some iridoids related to antirrhinoside (**8**) (*55*) (Scheme 14). During a study of the chemistry of antirrhinoside (**8**), epoxide opening reactions with pyridinium halides in dimethylformamide were examined. When **8** was treated with pyridinium chloride for one day both chlorohydrin isomers (**49** and **115**) were isolated together with recovered educt (23%). Similarly, treatment of **8** with pyridinium bromide for six days afforded bromohydrins **116** and **117**. All four halohydrins (**49** and **115**–**117**) were readily converted back to the starting **8** by treatment with aqueous sodium hydroxide. Incidentally, it was found that isolinarioside (**115**) and the earlier reported *epi*-muralioside (**118**) and *epi*-antirrhinoside (**119**) all had identical NMR data. The last two structures can now be considered redundant (*55*) (Scheme 14).

References, pp. 106–114

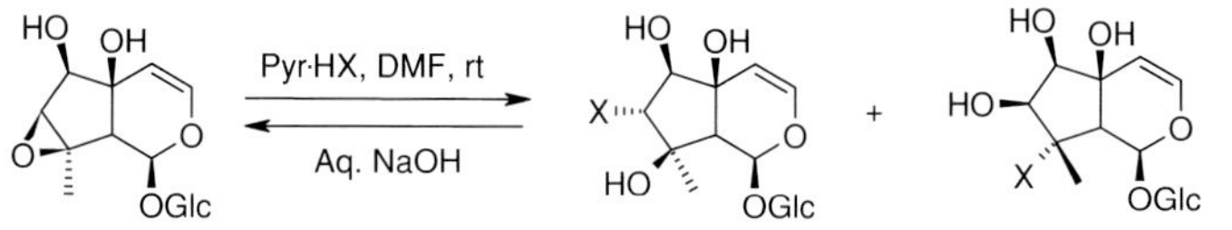

8 **49** X = Cl (45%) **116** X = Br (55%) **115** X = Cl (11%) **117** X = Br (10%)

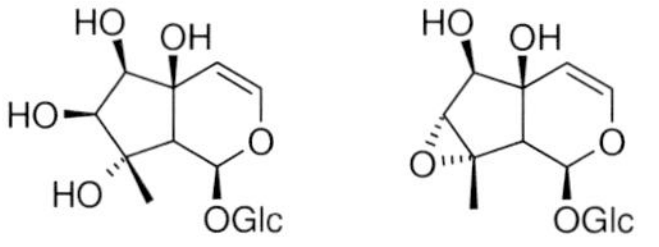

118 **119**

Scheme 14. Formation of halohydrins from antirrhinoside (**8**) (*55*)

4.5. Non-Glucosidic Iridoids

In this section some transformations of iridoid glucosides or their aglycones into non-sugar iridoids such as naturally occurring lactones and simple intramolecular acetals will be discussed but conversions regarding methyl and ethyl acetals as well as 1-*O*-silyl ethers of iridoid aglucones will be reviewed as well.

As a final proof for the assumed structure of a novel iridolactone, 7,8-dehydro-6β,10-dihydroxy-11-*nor*-iridomyrmecin (**123**), its partial synthesis from aucubin (**6**) was undertaken (*56*) (Scheme 15). The

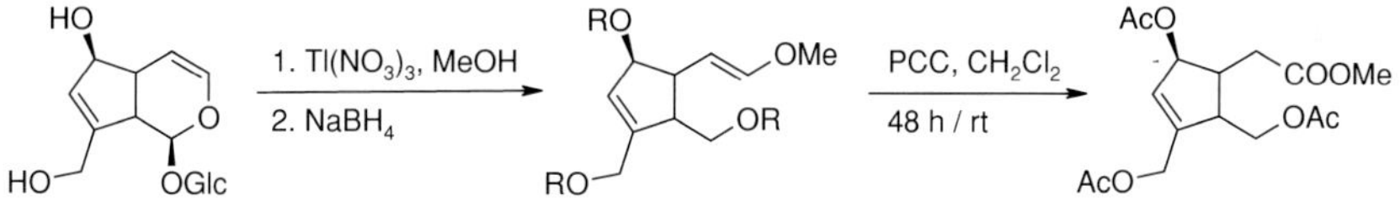

6 **120** R = H (43%) **120a** R = Ac (74%) **121** (89%)

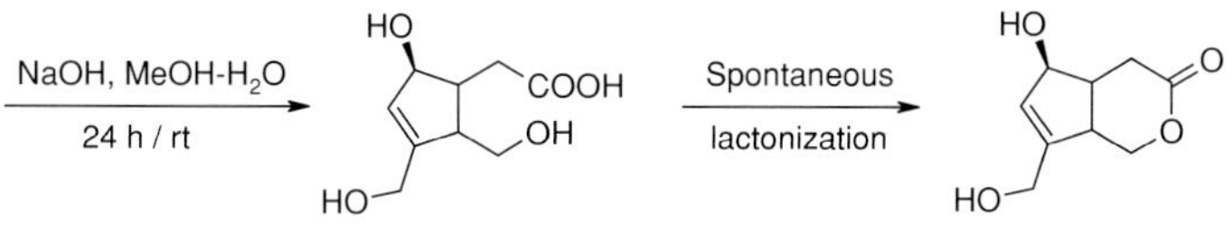

122 (Quant.) **123**

Scheme 15. Deglucosylation of aucubin (**6**). Preparation of an iridolactone (**123**) (*56*)

dihydropyran ring in **6** was opened with thallium(III) nitrate and sodium borohydride in methanol to give enol ether **120**, which subsequently was oxidized with pyridinium chlorochromate to methyl ester **121**. Alkaline hydrolysis followed by neutralization gave an acid (**122**) that cyclized spontaneously to the corresponding lactone (**123**).

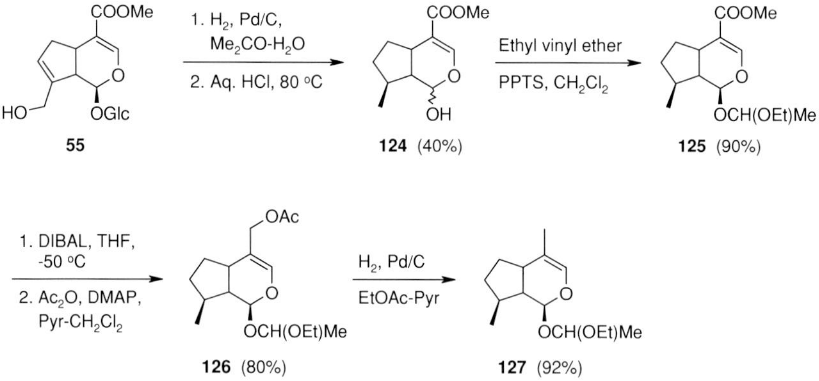

Scheme 16. Synthesis of iridolactone precursors from geniposide (**55**) (*57*)

Geniposide (**55**) was employed as the starting material for the preparation of three iridolactones isolated from *Actinidia polygama* (silver vine; Actinidaceae) (*57*) (Schemes 16 and 17). Initially, hydrogenolysis of the allylic alcohol in **55** followed by acid-catalyzed cleavage of the glycosidic bond afforded deoxyloganin aglucone (**124**) in moderate yield. After protection of the hemiacetal as the ethoxyethyl ether, the resulting single epimer (**125**) was subjected to reduction with diisobutylaluminum hydride followed by acetylation. This afforded **126**, which upon hydrogenolysis gave the deoxygenated product (**127**) in high yield (Scheme 16). These intermediates were further processed to yield the three different iridolactones (Scheme 17). Hydrolysis of **127** gave iridodial (**27**) that was oxidized with pyridinium dichromate (PDC) to yield the desired (+)-nepetalactone (**33**). Similarly, isodihydronepetalactone (**130**) was obtained *via* hydrogenation of **127** to **128** and subsequent hydrolysis and oxidation. Reductive ring-opening of **124** with diisobutylaluminum hydride gave allylic diol **131**, which in turn was converted into α-methylenelactone **132** by a manganese dioxide mediated oxidation. Finally, hydrogenation of the exocyclic double bond led stereoselectively to (+)-iridomyrmecin (**32**) (*57*).

References, pp. 106–114

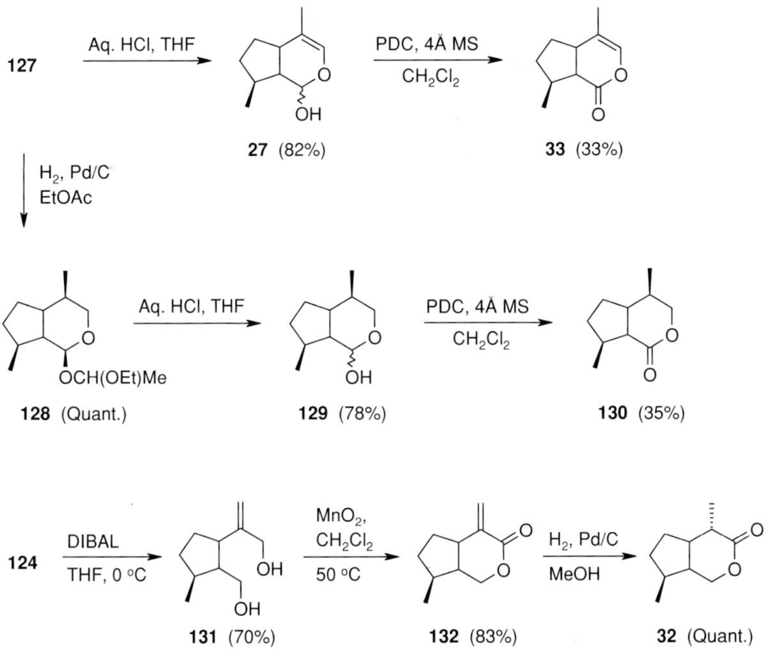

Scheme 17. Preparation of iridolactones (57)

Enantiomerically pure (+)-Mitsugashiwa-lactone (**135**) has been prepared in four steps from aucubin hexaacetate (**6a**) (58) (Scheme 18). Regio- and stereoselective transfer hydrogenolysis of **6a** afforded 8β-aucuban tetraacetate (**73a**) (51, 58) which was further hydrogenated to

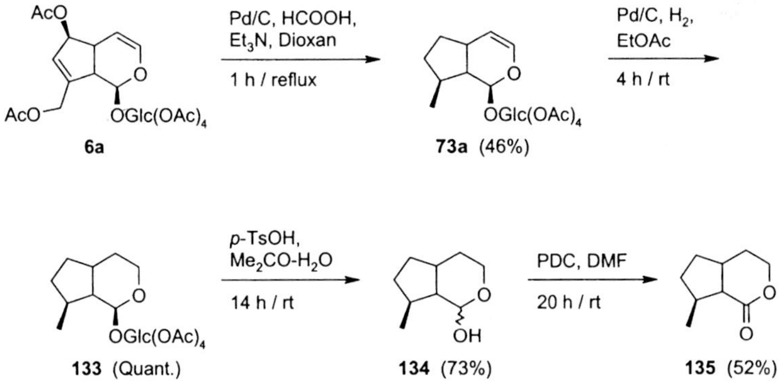

Scheme 18. Conversion of aucubin hexaacetate (**6a**) into (+)-Mitsugashiwa-lactone (**135**) (58)

the corresponding 3,4-dihydro compound (**133**). Acid-catalyzed hydrolysis of the glycosidic bond in **133** gave aglucone **134** that upon oxidation furnished lactone **135** (*58*).

Another iridolactone, antirrhinolide (**139**), was isolated as a minor constituent from *Antirrhinum majus* (Scrophulariaceae) and to prove its stereochemical relationship with antirrhinoside (**8**), a partial synthesis was investigated (*59*) (Scheme 19). In a previous study (*20*) it was shown that **8** readily could be converted into the aglucone of 5,6-*O*-isopropylidene-3,4-dihydroantirrhinoside (**136**). Oxidation of **136** with *in situ* formed ruthenium(VIII) oxide gave the intermediary β,γ-epoxylactone **137**, which by treatment with triethylamine during work-up was rearranged to give protected lactone **138** in high yield. Deprotection was accomplished with *p*-toluenesulfonic acid monohydrate in chloroform to yield the desired lactone (**139**) together with recovered **138**. Standard methods for ketal hydrolysis led to extensive decomposition of **138** (*59*).

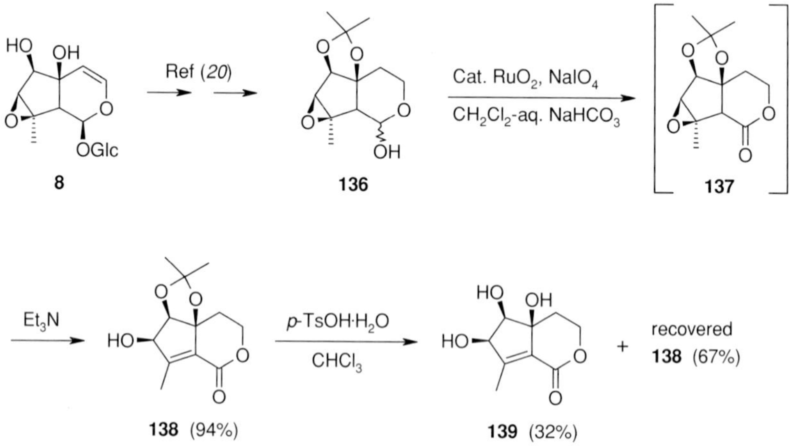

Scheme 19. Semi-synthesis of antirrhinolide (**139**) (*59*)

The spruce budworm antifeedant (−)-specionin (**31**) and some analogues were prepared from aucubin hexaacetate (**6a**) in several steps (*60*) (Scheme 20). First, a triphenylphosphine hydrobromide promoted acetal formation followed by deacetylation furnished the corresponding ethyl bisacetal (**141**) as a mixture of isomers. Epoxidation with *m*-chloroperbenzoic acid occurred selectively from the convex side; subsequent protection of the primary hydroxyl as its di-(*p*-methoxy)trityl ether gave **143**, which then was acylated with *p*-(benzyloxy)benzoyl

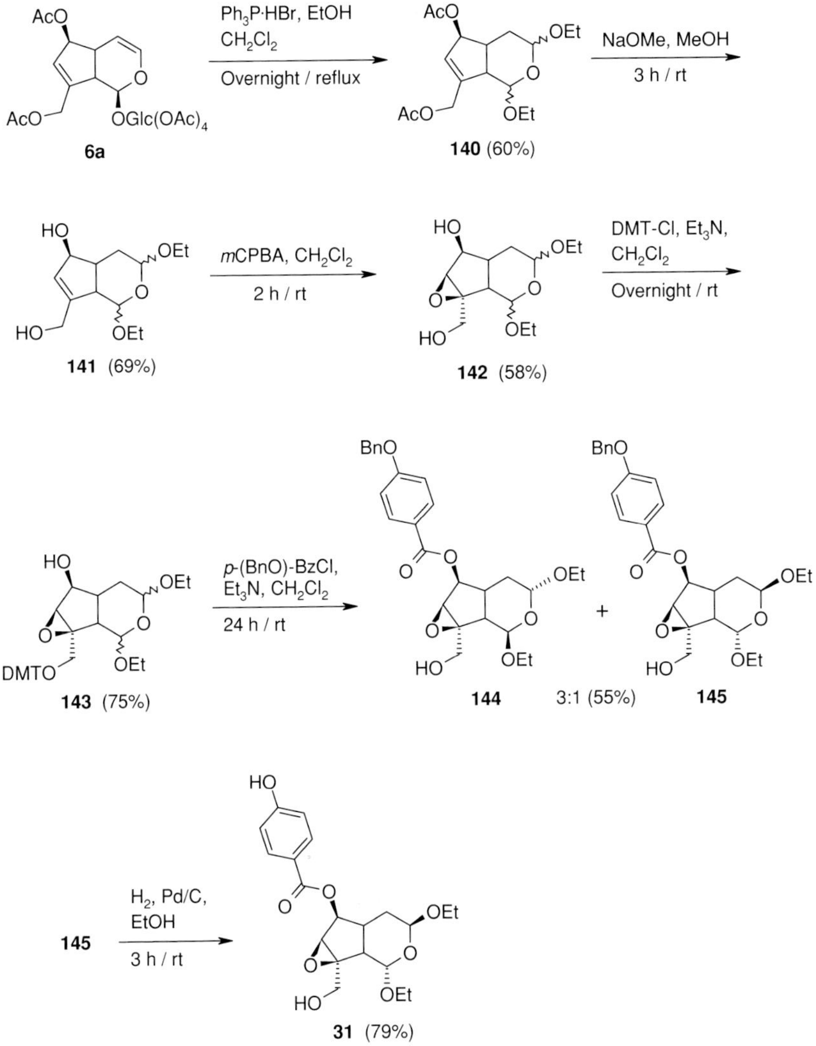

Scheme 20. Semi-synthesis of (−)-specionin (**31**) from aucubin hexaacetate (**6a**) (*60*)

chloride. This produced in a 3:1 ratio two distinct protected benzoates (**144** and **145**) of which the minor component upon debenzylation yielded (−)-specionin (**31**) (*60*).

The unusual aromatic iridoids cerbinal (**34**) and baldrinal (**149**) have been synthesized from genipin (**29**) by two similar routes (*9*, *61*), of which the shortest is outlined in Scheme 21. Selective silylation of

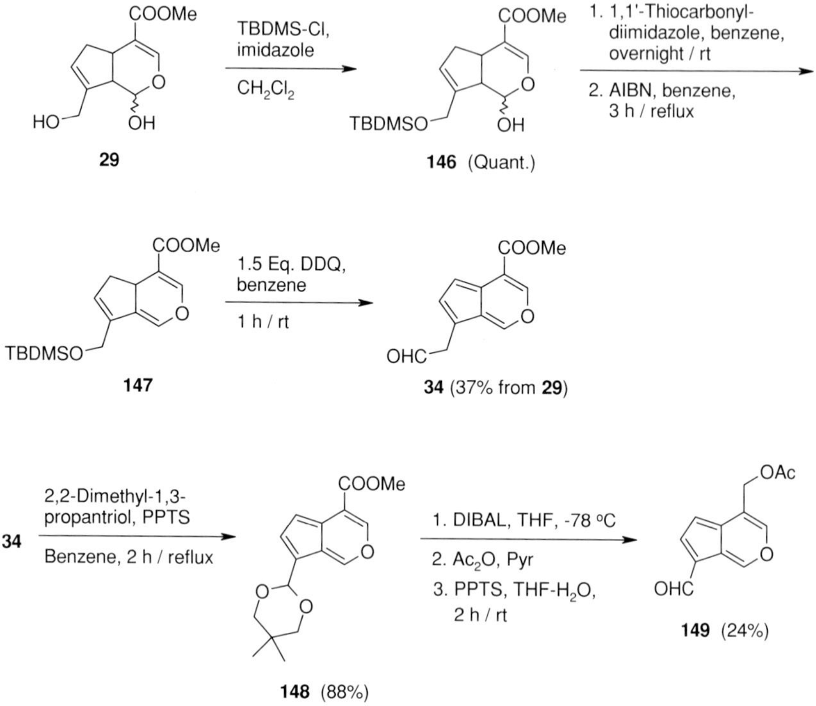

Scheme 21. Preparation of 10π-aromatic iridoids from genipin (**29**) (*9*)

genipin (**29**) at the primary position gave monosilyl ether **146** in quantitative yield. For the subsequent dehydration to **147**, conversion of **146** into the corresponding thioimidazolide followed by heating in benzene in the presence of azobisisobutyronitrile (AIBN) proved most efficient. Treatment of **147** with 2,3-dichloro-5,6-dicyano-1,4-benzoquinone (DDQ) led to oxidation at the 5,6-position and to concomitant oxidation of the allylic 10-position to give cerbinal (**34**) (*9*). The aldehyde moiety in **34** was protected as a 1,3-dioxane derivative, and subsequent reduction of the methyl ester followed by acetylation and acetal hydrolysis under mild conditions afforded baldrinal (**149**) (*9*).

Furthermore, the *Valeriana*-type compound didrovaltrate (**150**) has been prepared from **29** by a multi-step sequence (Scheme 22) (*9*). After a two-step selective protection of the hemiacetalic hydroxyl group in **29**, a thioether at C-10 was formed with diphenyl disulfide in the presence of tributylphosphine. This gave **151** in which the methyl ester functionality was reduced to the hydroxymethyl stage and then re-oxidized with barium manganate to yield aldehyde **152**. After desilylation, an

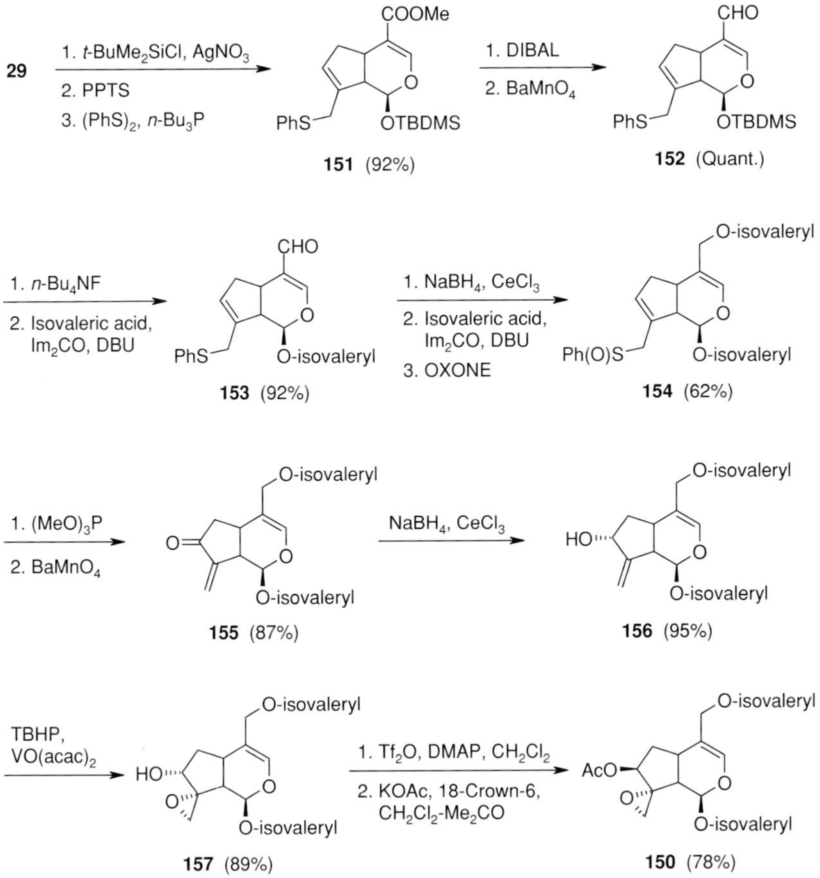

Scheme 22. Semi-synthesis of a *Valeriana*-type iridoid from genipin (**29**) (*9*)

isovaleryl group was introduced at the hemiacetalic position by a N,N'-carbonyldiimidazole mediated condensation. The aldehyde in the resulting **153** was reduced; the resulting primary alcohol was acylated to yield a diisovalerate in which the phenyl sulfide was then oxidized with potassium monopersulfate triple salt (OXONE). Evans rearrangement of the obtained phenylsulfoxide **154** and subsequent oxidation of the allylic alcohol at C-7 gave *exo*-methylene ketone **155**. Reduction of ketone **155** was followed by epoxidation of the resulting **156** to give epoxy-alcohol **157**. Inversion of the 7α-hydroxyl was performed *via* displacement of the corresponding triflate with potassium acetate to yield didrovaltrate (**150**) (*9*).

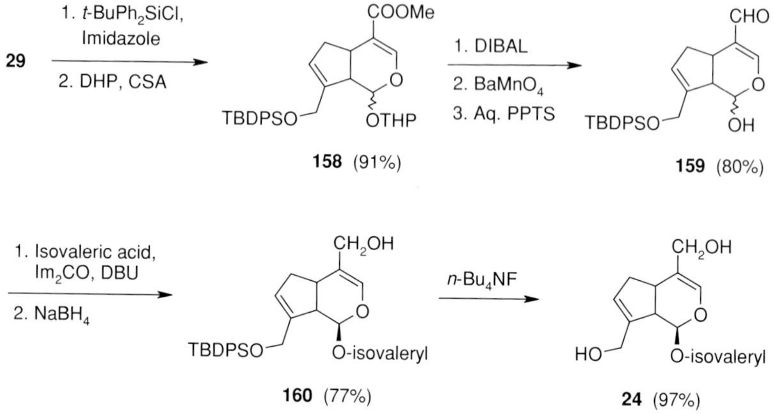

Scheme 23. Semi-synthesis of penstemide aglycone (**24**) (*9*)

Likewise, penstemide aglycone (**24**) was prepared from **29** using a modified protection strategy (Scheme 23). In this case, after protection of the primary hydroxyl group, the hemiacetalic position was protected as a tetrahydropyranyl ether (**158**). Again, the 11-methyl ester was converted into the corresponding aldehyde, and subsequently the THP ether was removed to yield **159**. The isovaleryl moiety was introduced, then reduction of the aldehyde functionality and desilylation furnished the desired **24** (*9*).

Also *Plumeria* iridoids have been prepared with genipin 1-*O*-TBDMS ether (**161**) as starting material (Schemes 24 and 25) (*9*). Acetylation afforded allylic acetate **162** which was coupled smoothly with the sodium salt of methyl acetoacetate in a palladium π-allyl complex mediated reaction. The ketone in **163** was protected as a cyclic 1,3-dioxane ketal and subsequent *cis*-dihydroxylation with osmium tetroxide and *N*-methylmorpholine-*N*-oxide (NMO) yielded diol **164**. Lactonization was accomplished by treatment with sodium methoxide and the following dehydration was performed *via* the corresponding triflate. Deprotection of the ketal in **165** proceeded well on treatment with trityl tetrafluoroborate. Then a phenylselenyl group was introduced between the carbonyls, and upon oxidation a selenoxide elimination afforded the unsaturated keto lactone **166**. Desilylation of **166** was accompanied by an intramolecular Michael addition of the hemiacetalic hydroxyl group, and this produced tetracyclic ether **167** as an epimer mixture (α:β ratio 3.5:1) (*9*). The next step was stereoselective reduction of the keto group in **167** which most conveniently was performed with triethylsilane in trifluoroacetic acid at 0°C (Scheme 25). In addition to

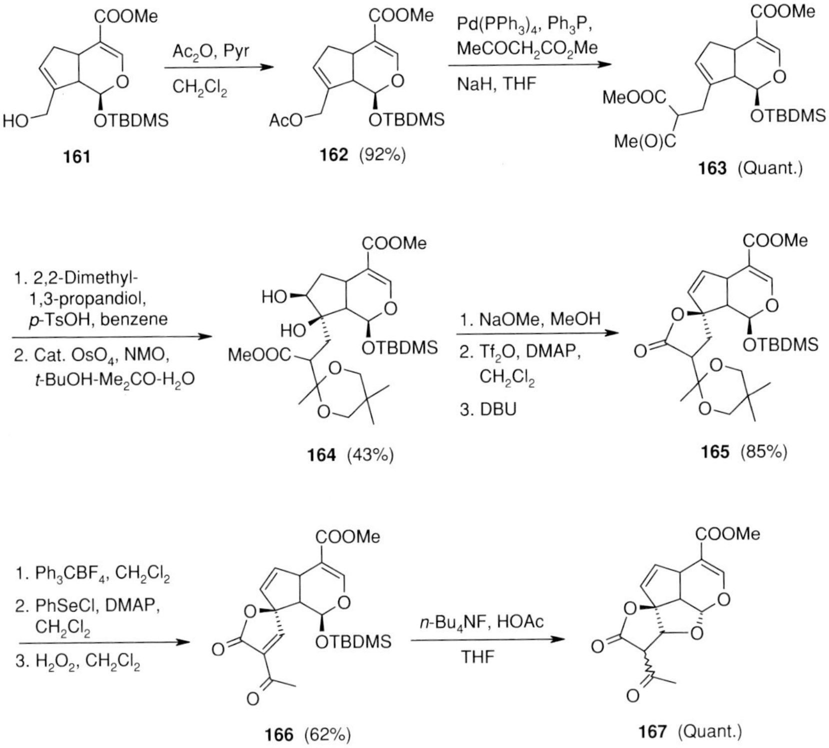

Scheme 24. Preparation of precursors (**166** and **167**) of *Plumeria* iridoids (*9*)

(+)-allamandicin (**168**), (+)-*epi*-allamandicin (**169**) and (+)-*iso*-allamandicin (**170**) were also produced. When this reduction was carried out at room temperature, a mixture of (+)-plumericin (**171**) and (+)-*iso*-plumericin (**172**) was produced. Treatment of the mixture **168/169** (3:1) with phosphorus oxychloride afforded **171/172** in the same ratio. Reduction of **166** with triethylsilane furnished a mixture of plumeiride aglycone 1-*O*-TBDMS ether (**173**) and its epimer (**174**) (*9*).

Preparations of gardenoside aglycone 1-*O*-TBDMS ether (**175**), asperuloside aglycone 1-*O*-TBDMS ether (**176**) and garjasmine (**177**) have also been accomplished with genipin (**29**) as starting material (*9, 31, 62*) (Scheme 26). The TBDMS ether **178** was dihydroxylated selectively on the convex face to yield diol **179**, which was subjected to selective elimination of the secondary hydroxyl group *via* its triflate to give the disilyl ether of gardenoside aglycone (**180**). Hydrolysis of **180** under mild conditions for a shorter period allowed isolation of gardenoside aglycone 1-*O*-TBDMS ether (**175**) in low yield, but with

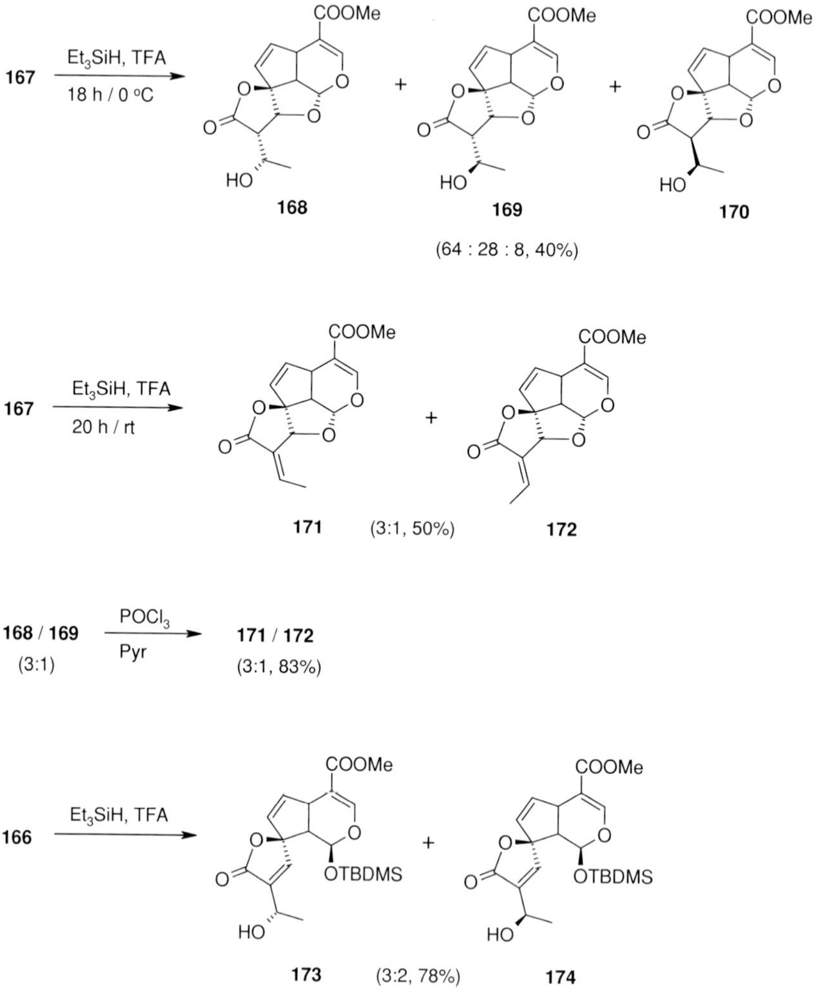

Scheme 25. Preparation of *Plumeria* iridoids (9)

a high recovery of educt. Prolonged reaction time under the same conditions resulted in transposition of the tertiary hydroxyl group to yield the desired 6α-hydroxy compound **181** in moderate yield. Selective silylation at the primary position gave silyl ether **182**. Conversion of **182** into the corresponding acid was performed with potassium hydride in tetrahydrofuran. Subsequent lactonization mediated by dicyclohexylcarbodiimide (DCC) afforded **183**. Finally, selective desilylation at the primary position in **183** followed by acetylation furnished the desired

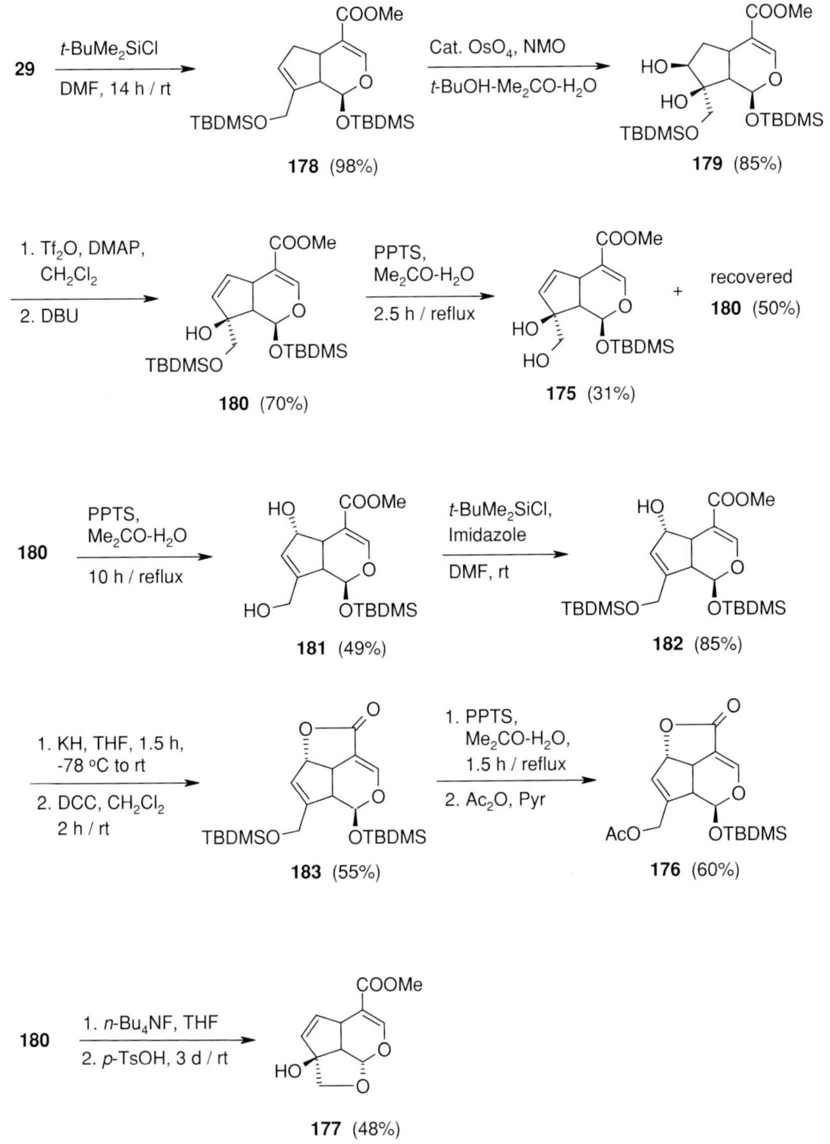

Scheme 26. Syntheses of aglycone 1-*O*-TBDMS ethers **175** and **176**, and of garjasmine (**177**) (*9, 31, 62*)

asperuloside aglycone 1-*O*-TBDMS ether (**176**). Garjasmine (**177**) was obtained from the common intermediate **180** by successive desilylation and acidification (*9, 31, 62*).

4.6. Ring-Cleavage Sequences and Secoiridoid Chemistry

Conversion of carbocyclic iridoid glucosides into secoiridoids obviously involves cleavage of the carbon-carbon bond between C-7 and C-8, and so far oxidation of 7,8-diols (*63–65*) or a Baeyer-Villiger oxidation of a 7-keto compound (*65*) have been employed to effect this transformation. Here, two recent slightly modified ring-cleavage sequences will be outlined.

The key intermediate for the preparation of suitable 7,8-diols is deoxygeniposide (**74**), which after protection of the sugar moiety may be hydroxylated either directly (*63, 66*) or by a two-step procedure (*36, 65*). As already described, **74** may be obtained from transfer hydrogenolysis of gardenoside (**54**) (Section 4.3), but two other similar routes to this compound have been devised (Scheme 27). Asperuloside (**57**) or asperulosidic acids **184** and **185** were converted into methyl ester **186** by successive treatment with sodium methoxide and acetylation, or by methylation and subsequent acetylation, respectively (*66*). Transfer hydrogenolysis of **186** gave a 9:1-mixture of the desired **74a** and the partially reduced 6α-acetoxy compound (**187**) (*66*). In a different approach, the pentaacetate of shanzhiside methyl ester (**188**) was dehydrated with thionyl chloride in pyridine to give 6β-acetoxy compound **189** together with a small amount of the 8,9-olefin (**190**). Thus, elimination occurs, as expected, most favourably when the hydroxyl group and the abstractable proton are *anti* to each other. Again, **74a** was readily obtained from **189** upon transfer hydrogenolysis (*66*). After the protecting groups on the sugar moiety were changed to benzyl groups, **74b** was selectively hydroxylated at the 7,8-double bond with osmium tetroxide/trimethylamine-*N*-oxide (Scheme 28). The pronounced predominance of **191** over **192** reflects the lower steric hindrance at the β-face of the molecule. Both diols were cleaved oxidatively with sodium periodate to provide directly 7-formyl-8-keto secoiridoid derivative **193b** in high yield. Jones oxidation afforded the 7-carboxylic acid **194b** which was esterified with 3,4-(di-*O*-benzyl)dopaol in a dicyclohexylcarbodiimide mediated reaction. The 8-keto functionality in **195** was reduced with sodium borohydride to give **196** in almost complete diastereomeric excess. Subsequent dehydration with thionyl chloride afforded benzyl protected oleuropein (**20b**), which was

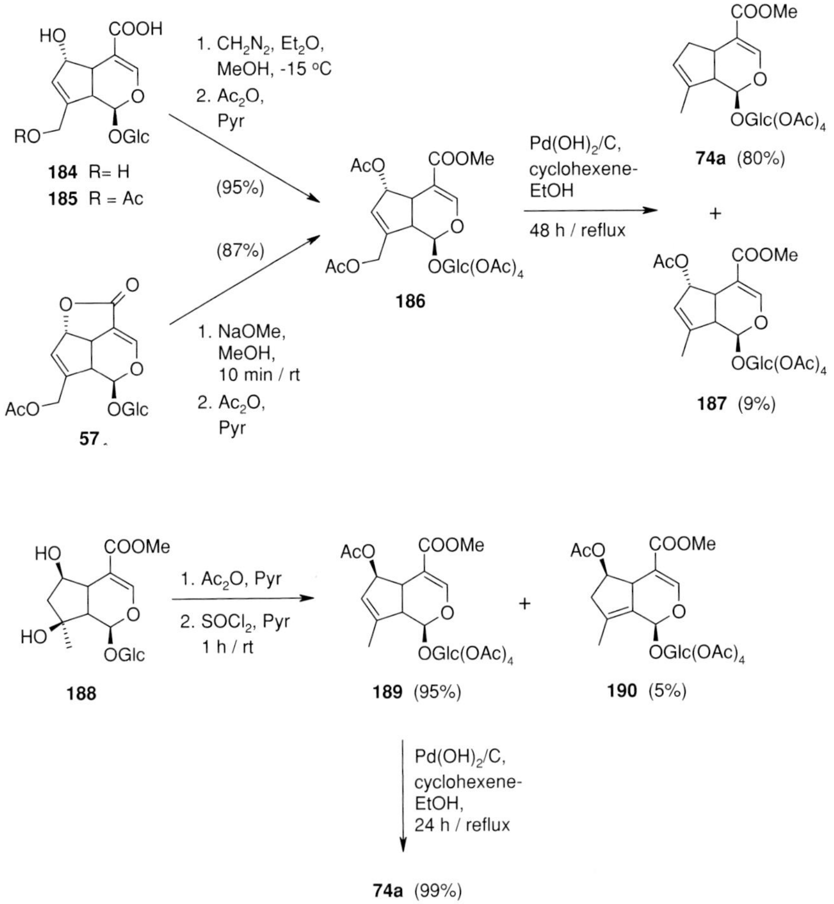

Scheme 27. Alternative routes to deoxygeniposide tetraacetate (**74a**) (*66*)

deprotected to give **20** in 17–23% overall yield depending on the starting material (*66*).

Another procedure involved the analogous *trans*-diols (**197** and **198**). These were obtained by an acid-catalyzed hydrolysis of a crude mixture of epoxides **94** and **95** prepared from **74a** (*36*). Oxidative cleavage of these diols had previously been effected with lead tetraacetate (*65*), but in order to avoid this reagent sodium periodate was examined here (*36*). This turned out to be a sluggish but selective reagent since only **197** was cleaved to the expected 7-formyl-8-keto product (**193a**), whereas **198** was left unchanged (Scheme 29).

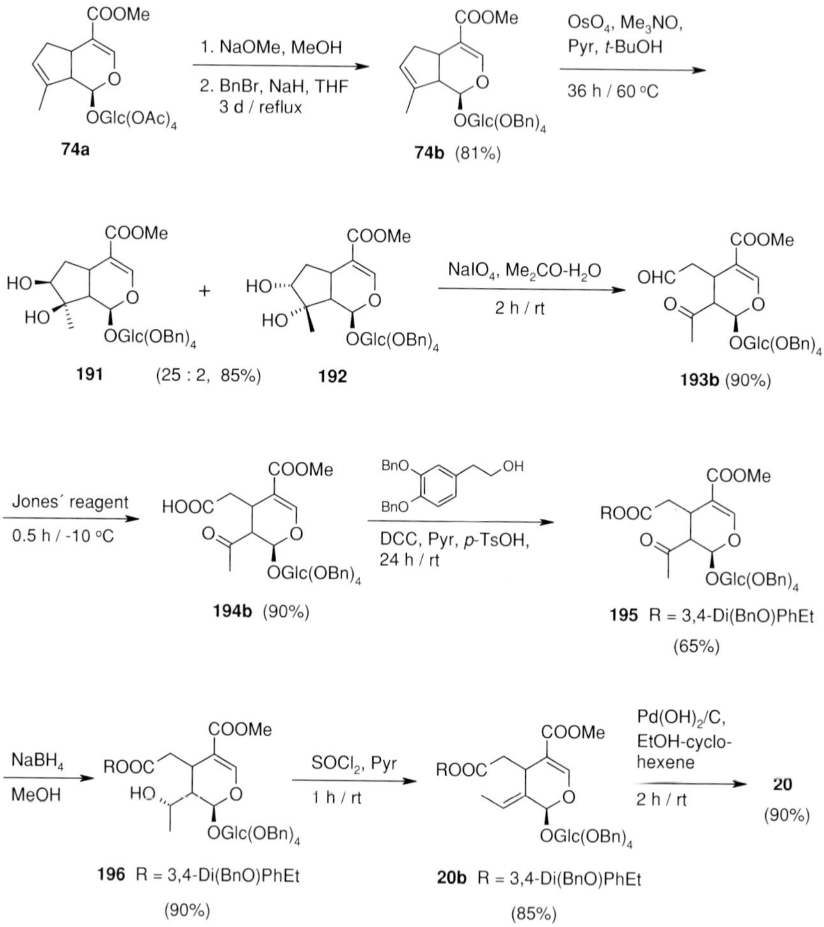

Scheme 28. Partial synthesis of oleuropein (**20**) (66)

Probably the 8α-hydroxyl in **198** is too hindered to participate in the formation of a cyclic ester intermediate. However, Jones oxidation of **198** followed by methylation produced smoothly 8-keto methyl ester **199**, which was also accessible *via* Jones oxidation of diol **197** or aldehyde **193a** followed by methylation. Reduction of **199** with sodium borohydride gave a mixture of 8-epimeric kingiside tetraacetates (**16a/200a**) and the corresponding hemiacetals (**17a/201**) (Scheme 30). Hence, the reaction apparently proceeds by a relatively slow initial reduction of the keto group followed by a fast lactonization, and the resulting lactones are then further reduced to their hemiacetal counter-

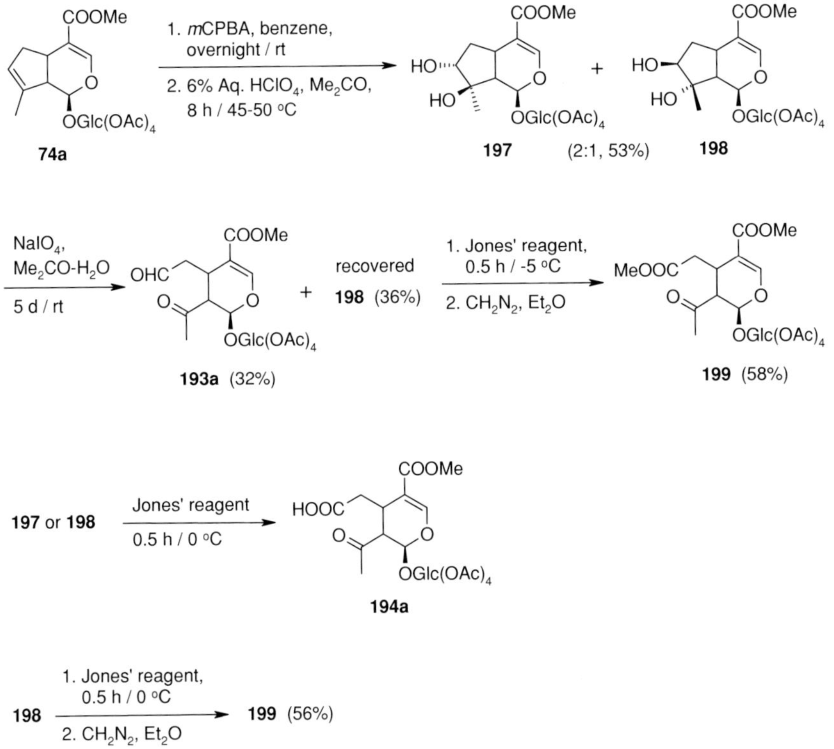

Scheme 29. Ring-cleavage sequence to secoiridoid intermediate **199** (*36*)

parts. But after a prolonged reaction time and repeated Jones oxidation of the crude product mixture, kingiside tetraacetate (**16a**) and 8-*epi*-kingiside tetraacetate (**200a**) were isolated in reasonable yields. Deacetylation afforded the unprotected secoiridoid glucosides **16** and **200** (*36*).

For use in biosynthetic studies, 8,10-epoxysecologanin and 8,10-epoxysecoxyloganin derivatives were prepared from secologanin tetraacetate (**13a**) (*67*) (Scheme 31). Due to the presence of an acid-sensitive epoxy functionality in the target compounds, the photo-sensitive *o*-nitrophenylethylene acetal was chosen as protection for the aldehyde group in **13a**. But this also implied that diastereomeric mixtures would be obtained in the following series of reactions. Acetal **202** was dihydroxylated with osmium tetroxide/*N*-methylmorpholine-*N*-oxide to give diol mixture **203**, which then *via* tosylation was converted into the separable epoxides **204** and **205**. Acetal deprotection by photolysis and

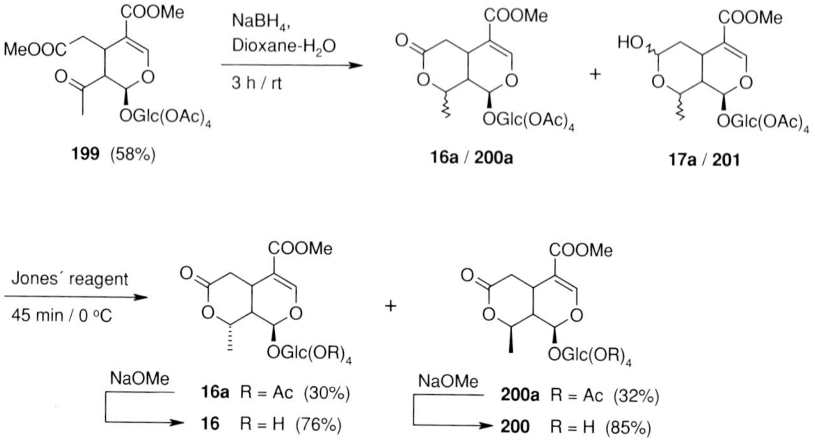

Scheme 30. Preparation of kingiside (**16**) and 8-*epi*-kingiside (**200**) (*36*)

subsequent deacetylation afforded 8,10-epoxysecologanins **206** and **207** in moderate yields. Jones oxidation of aldehydes **206a** and **207a** followed by deacetylation gave 8,10-epoxysecoxyloganins **208** and **209**, respectively (*67*).

The assignments of the structures for epoxides **206a** and **207a** were performed by chemical correlation with the established structures for 8,10-epoxyswerosides **212** and **213** (Scheme 32). Dihydroxylation of sweroside tetraacetate (**69a**) gave diols **210** and **211**, which after separation were converted into authentic samples of **212** and **213**. Alternatively, **212** and **213** were obtained from **206a** and **207a**, the structures of which then were settled (*67*).

In addition to the classical chemical conversions encountered so far, a remarkable enzymatic transformation of 7-*O*-alkyl-10-hydroxyoleoside secoiridoid glucosides into jasmolactones has been reported (*68*). It was found that β-glucosidase catalyzed the production of jasmolactones B and D (**216** and **217**) from 10-hydroxyoleosides **214** and **215**, respectively, *via* a stereospecific rearrangement (Scheme 33). Lactonization between the 7-carbonyl and 10-hydroxyl groups and a stereospecific alkoxyl transfer to C-8 appear to be concomitant steps involved in this conversion. Moreover, both steps might well be concerted with the observed stereospecific 9β-protonation. In the same reaction, the 1-β-D-glucosyl moiety was hydrolyzed also (*68*).

In a recent study, cell suspension cultures of *Lonicera japonica* (Caprifoliaceae) that otherwise did not produce any iridoid or secoiridoid glycosides were shown to possess a latent capability to transform

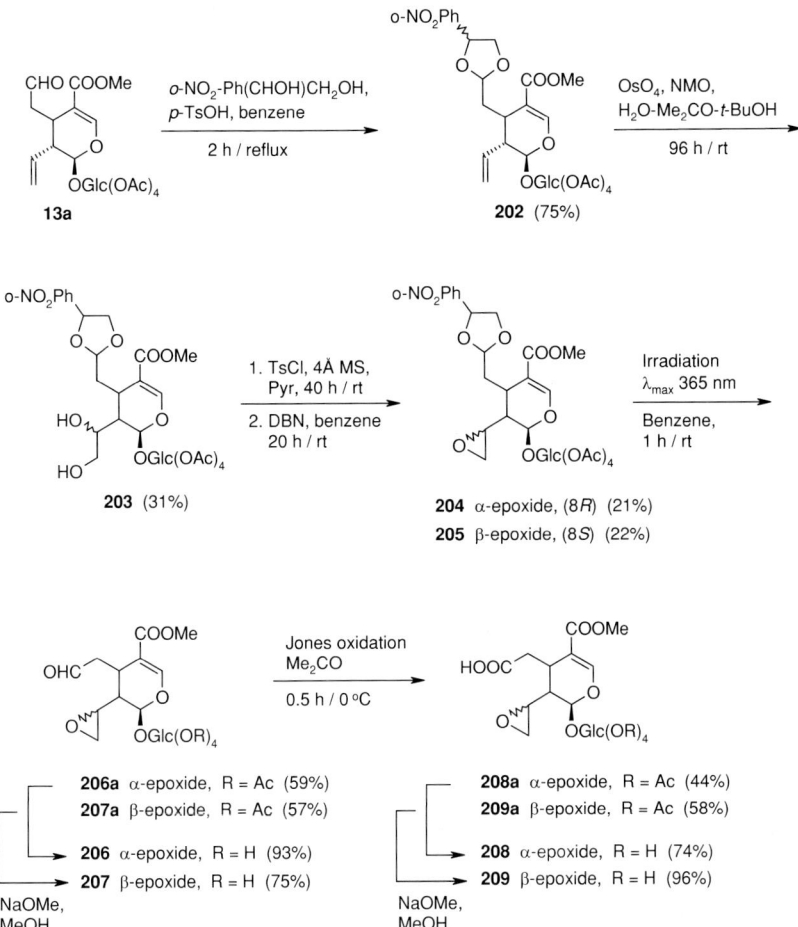

Scheme 31. Semi-synthesis of epoxysecologanins and epoxysecoxyloganins (67)

deoxyloganin (**85**) sequentially into loganin (**9**) and secologanin (**13**), whereas administration of **9** produced **13** (Scheme 34). However, no extracellularly sequestered **13** could be detected, and isolation of biotransformed compounds thus involved chromatography of whole-cell extracts (69).

Several secoiridoid aglycone silyl ethers have been prepared from genipin monosilyl ether **161** (9). After conversion of **161** into a 10-thioether and subsequent oxidation to the corresponding sulfoxide, Evans rearrangement yielded *exo*-methylene alcohol **218**, which was oxidized to the corresponding ketone (**219**). Next, (trimethylsilyl)tributylstannane

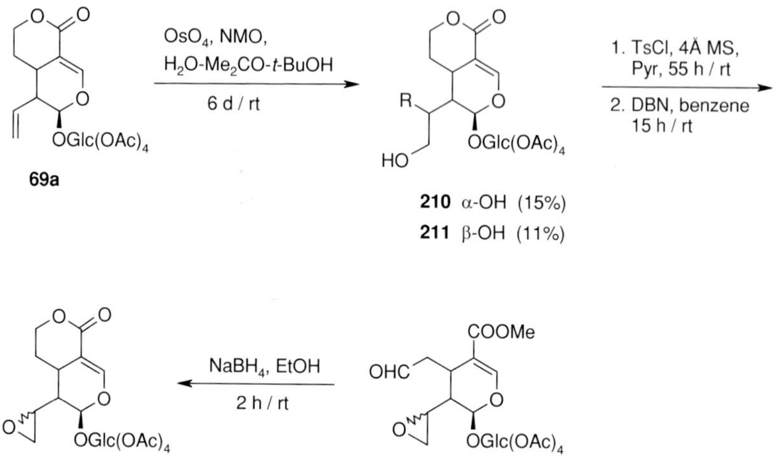

Scheme 32. Chemical correlation of epoxysecologanins and epoxyswerosides (67)

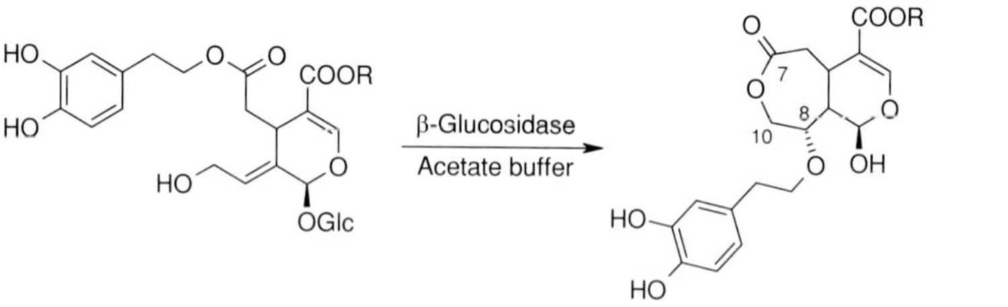

210 α-OH (15%)
211 β-OH (11%)

212 α-epoxide (37%) (71%) **206a** α-epoxide
213 β-epoxide (42%) (65%) **207a** β-epoxide

214 R = Me
215 R = (3,4-di-OH-Ph)Et

216 R = Me (10%)
217 R = (3,4-di-OH-Ph)Et (20%)

Scheme 33. Enzymatic conversion of 10-hydroxyoleosides into jasmolactones (68)

proved to be an excellent reagent for 1,4-addition to the enone functionality in **219**. Selective removal of the formed 7-O-TMS group afforded keto stannane **220** in good yield. Reduction of the keto group followed by oxidative fragmentation with lead tetraacetate afforded secologanin aglycone 1-O-TBDMS ether (**221**), which by reduction with sodium borohydride was transformed into sweroside aglycone 1-O-TBDMS ether (**222**) (Scheme 35).

References, pp. 106–114

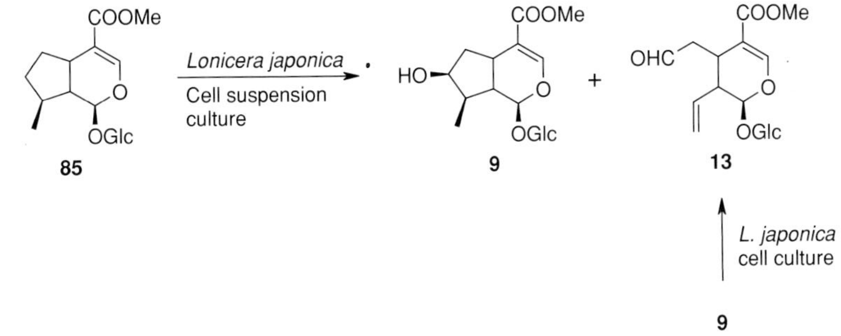

Scheme 34. Biotransformation of iridoid glucosides (69)

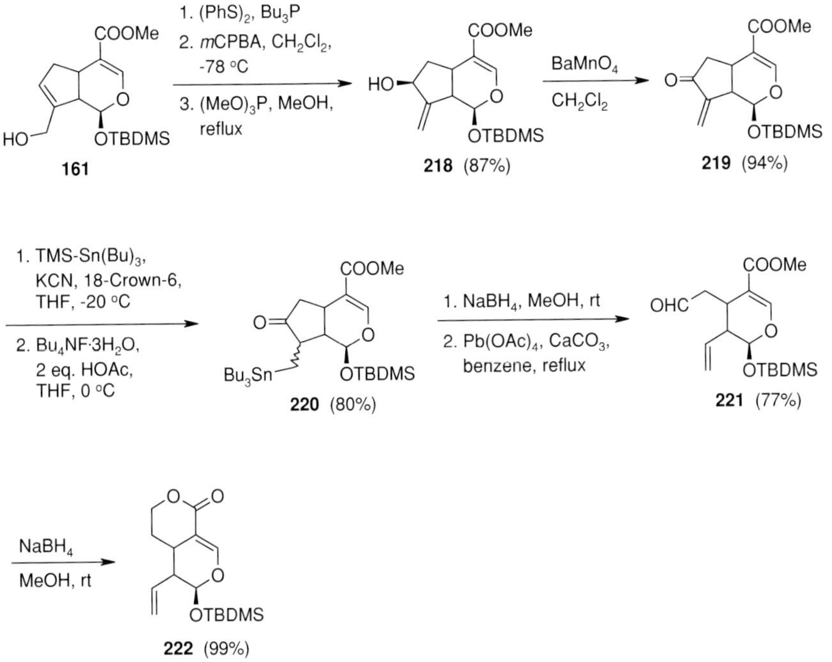

Scheme 35. Preparation of secoiridoid aglycone silyl ethers (9)

Secologanin aglycone silyl ether **221** was converted into an enamine and then a phenylselenyl group was introduced next to the aldehyde to give **223**. Oxidation of the phenylselenide led to concomitant elimination thus yielding α,β-unsaturated aldehyde **224**. A mixture of the open-chain compound **225** (38%) and the desired lactone **226** (20%) was obtained upon reduction of **224**. Irradiation of **225** in the presence of a catalytic amount of sodium methoxide and 2-acetonaphthone as sensitizer allowed its complete conversion into gentiopicroside aglycone 1-*O*-TBDMS ether (**226**) (Scheme 36).

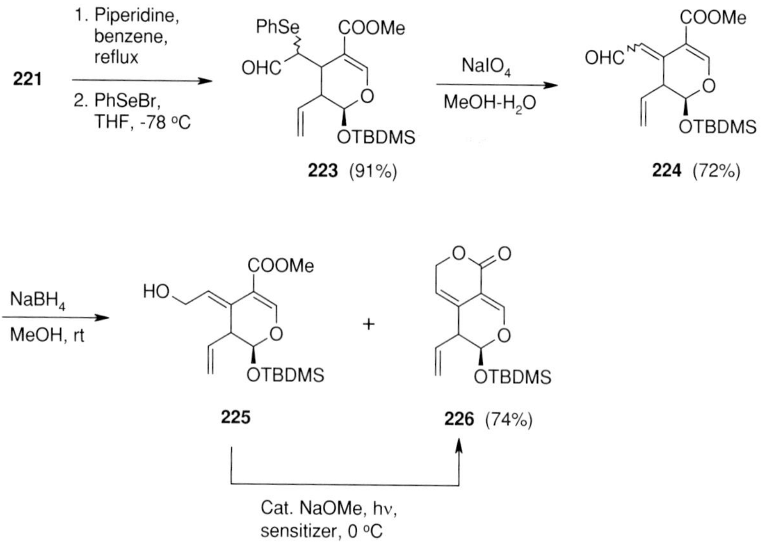

Scheme 36. Preparation of gentiopicroside aglycone silyl ether **226** (*9*)

Hydrogenation of *exo*-methylene ketone **219** with Pd/C as catalyst afforded selectively the 8α-epimer (**227**), which readily was equilibrated with diazabicyclo[5.4.0]undec-7-ene (DBU) to give the more stable 8β-epimer (**228**). Baeyer-Villiger oxidation of the more labile **227** afforded the corresponding lactone, kingiside aglycone 1-*O*-TBDMS ether (**229**). Morroniside aglycone 1-*O*-TBDMS ether (**231**) was produced by chemoselective diborane reduction of **229**. Upon desilylation of **231** and acid-catalyzed cyclization (−)-sarracenin (**35**) was obtained. Starting from **228** the corresponding 8β-epimers 8-*epi*-kingiside aglycone 1-*O*-TBDMS ether (**230**) and 8-*epi*-morroniside aglycone 1-*O*-TBDMS ether (**232**) were synthesized by the same protocol and the latter was converted further into 8-*epi*-sarracenin (**233**) (*9*) (Scheme 37).

References, pp. 106–114

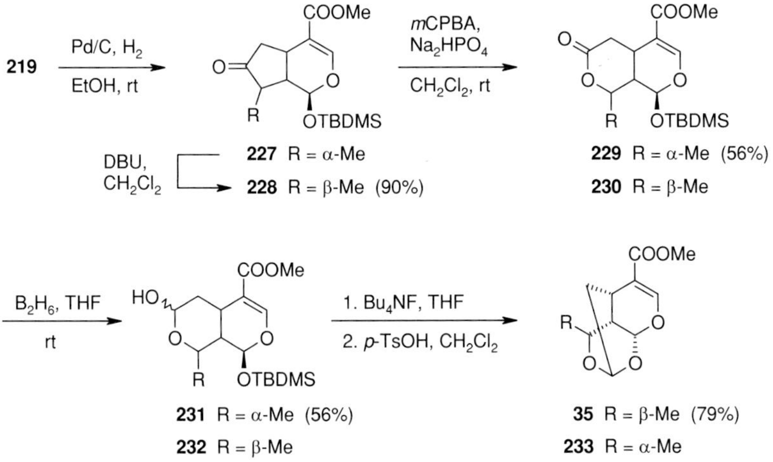

Scheme 37. Preparation of kingiside aglycone silyl ethers and sarracenins *(9)*

5. Monoterpene Alkaloids Structurally Related to Iridoids

In the present survey of the chemistry of iridoid related monoterpene alkaloids, two major subgroups are recognized, namely the pyridine monoterpene alkaloids (PMTAs) and the bicyclic cyclopentanoid piperidine alkaloids. Regarding the PMTAs, a discussion of whether they are true natural metabolites or merely artifacts formed during work-up and isolation will be presented. Next, three different pathways to PMTAs are evaluated: (i) chemical semi-synthesis from iridoid glucosides, (ii) metabolism of iridoid glucosides by intestinal bacteria, and (iii) total synthesis. A brief introduction to the bicyclic cyclopentanoid piperidines is provided by reviewing some novel compounds reported in the last decade. Some enantioselective total synthetic routes and one recent semi-synthetic pathway to these compounds are outlined. Finally, reference is given to a few conversions of secoiridoids into glucosidic monoterpenoid alkaloids.

5.1. PMTAs – Natural Compounds or Artifacts?

For several decades the origin of monoterpene alkaloids has been subject to dispute; however, because of their structural resemblance to the iridoids they have frequently been treated as related to, or derived from, iridoid glucosides. Whether pyridine monoterpenes were genuine natural compounds was especially questionable since in many cases these alkaloids could not be obtained when aqueous ammonia was omitted from the extraction procedure (*16, 70*). Accordingly, such pyridines have often been considered as artifacts, but in more recent reports on novel substances this appears not always to be the case.

A possible example of pyridine artifacts was the isolation of the racemic plectrodorine (**234**) and isoplectrodorine (**235**) (from *Plectronia odorata*) using classical acid-ammonia extractions (*71*) (Scheme 38). Chemical correlation of **234/235** with racemic deoxyrhexifoline (±)-(**236**) was performed by a tributyltin hydride reduction of dibenzoate

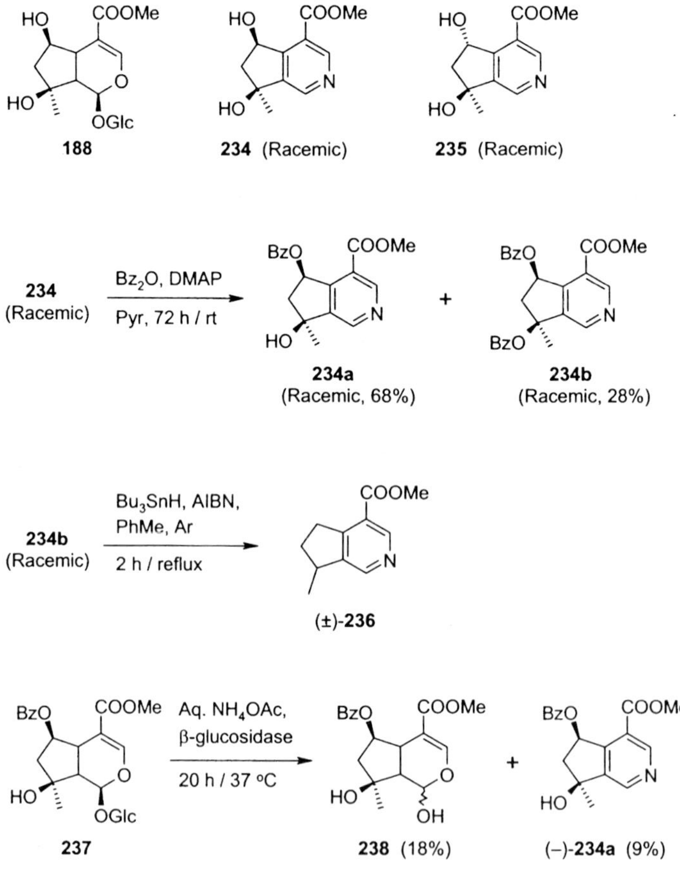

Scheme 38. Racemic PMTAs from *Plectronia odorata*. Correlation with known compounds (*71*)

234b. Optically active monobenzoate (−)-**234a** was prepared from the co-occurring iridoid glucoside 6-O-benzoylshanzhiside methyl ester (**237**) by treatment with β-glucosidase in an ammonium acetate solution; also genin **238** was obtained as a likely intermediate in this reaction. Although no obvious explanation for the racemization was given, **234/235** were considered to be artifacts derived from shanzhiside methyl ester (**188**) also present in the plant (*71*).

Two PMTAs, namely boschniakine (**239**) and the novel euphrosine (**240**) were isolated from *Orthocarpus* species (Scrophulariaceae) in which euphroside (**241**) was the main iridoid component whereas plantarenaloside (**10**) was a minor constituent (*72*) (Scheme 39). In this case, both **239** and **240** could be isolated on basification with sodium hydroxide/sodium bicarbonate rather than with aqueous ammonia although the yields were ten-fold lower. Attempts at a direct conversion of **241** into **240** similar to that described above for **237** failed; however, when **242** was prepared from **241** first, successive treatments with strong aqueous acid and ammonia allowed isolation of **240** (*72*).

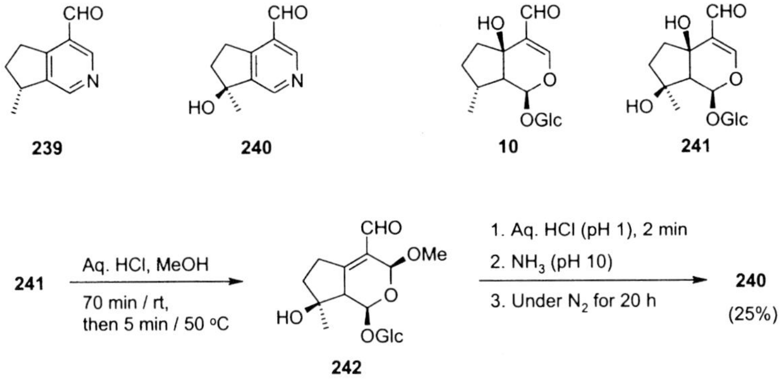

Scheme 39. Possible pairs of iridoid glucosides and PMTAs from *Orthocarpus* species (*72*)

Likewise, (−)-oxerine (−)-(**243**) has been isolated together with harpagide (**244**) (*73*), the latter of which might be considered the parent iridoid glucoside of the former (Chart 10). Indeed **244** could be transformed into **243** by a one-pot procedure using β-glucosidase and ammonium acetate. Nevertheless, that this PMTA might be an artifact seems unlikely, since ammonia was not employed during work-up, isolation, or purification (*73*).

Furthermore, two alkaloids (**245** and **246**) related to the co-occurring pyridine jasminine (**247**) were obtained from *Osmanthus austrocaledo-*

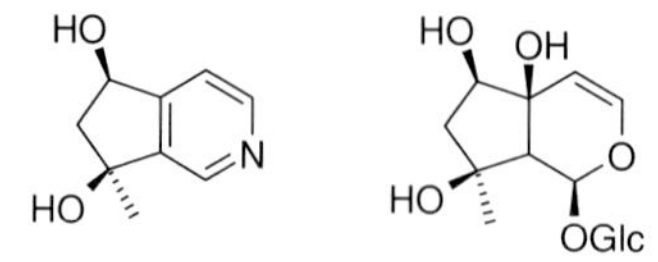

Chart 10. (−)-Oxerine (−)-(**243**) and harpagide (**244**)

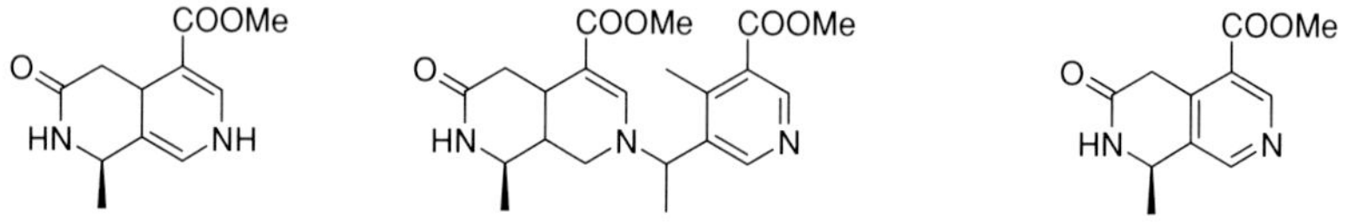

Chart 11. Iridoid glucosides and alkaloids from *Osmanthus austrocaledonica* (*74*)

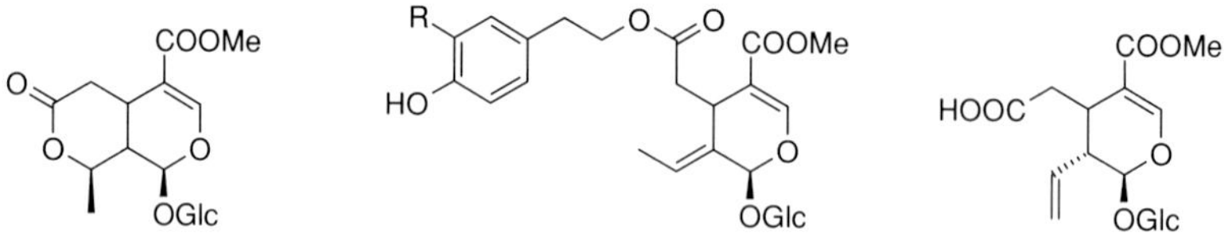

nica (Oleaceae) (*74*) (Chart 11). Both traditional ammoniacal extraction and a procedure using sodium carbonate as the alkaline agent afforded dihydrojasminine (**245**), austrodimerine (**246**) and **247**, thus ruling out that the alkaloids were artifacts (*74*). Besides, the structures of the secoiridoids also isolated from this species 8-*epi*-kingiside (**200**), ligstroside (**248**), oleuropein (**20**) and secoxyloganin (**249**), do not implicate them as obvious parent iridoids, which under acid-base extraction might yield alkaloids **245–247** even though they seem related to **200**.

In conclusion PMTAs appear to be genuine natural alkaloids, but whenever possible a comparative isolation conducted in the absence of ammonia seems appropriate to rule out artifacts.

5.2. Semi-Synthesis of Pyridine Monoterpene Alkaloids

As already outlined briefly in the previous section, so-called biomimetic syntheses of PMTAs from iridoid glucosides are possible under relatively mild conditions, although the yields are often low to

References, pp. 106–114

moderate. Here, additional recent examples of preparations of some known PMTAs and a few purely synthetic analogues will be presented.

Enzymatic hydrolysis of loganin (**9**) followed by amination of the resulting aglycone and subsequent acid treatment allowed isolation of cantleyine (**250**) and tetrahydrocantleyine (**251**), the former of which was the starting point for the synthesis of three (8S)-(−)-forms of naturally occurring PMTAs (*75, 76*) (Scheme 40). Radical deoxygenation of the 7-hydroxyl group in **250** was performed *via* its thioimidazolide **252** by tributyltin reaction to give (−)-deoxyrhexifoline (−)-(**236**) in good yield. Sodium borohydride reduction of (−)-**236** afforded (−)-tecostidine (**253**) (from *Tecoma stans*, Bignoniaceae). Successive acetylation and hydrogenolysis led to (−)-actinidine (−)-(**37**) (from Actinidiaceae and Valerianaceae) in high yield (*76*).

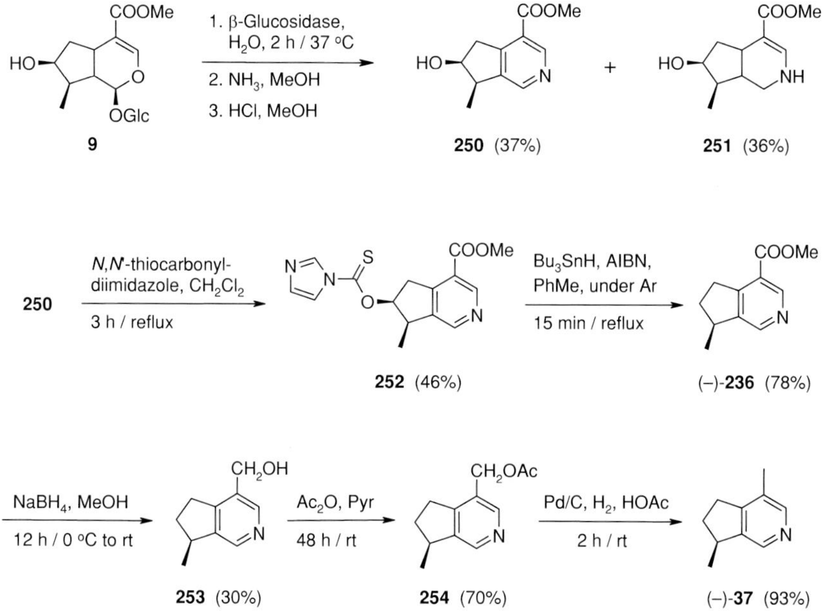

Scheme 40. Semi-syntheses of PMTAs from loganin (**9**) (*75, 76*)

Four iridoid glucosides were transformed into PMTAs in one-pot procedures using β-glucosidase for aglucone formation and ammonium acetate as the ammonia source (*77*) (Scheme 41). Enzyme treatment of 8-*epi*-loganin (**100**) in the presence of ammonium acetate afforded a low yield of the expected pyridine (**255**), whereas cornin (**92**) under these conditions produced corninine (**256**), which underwent degradation

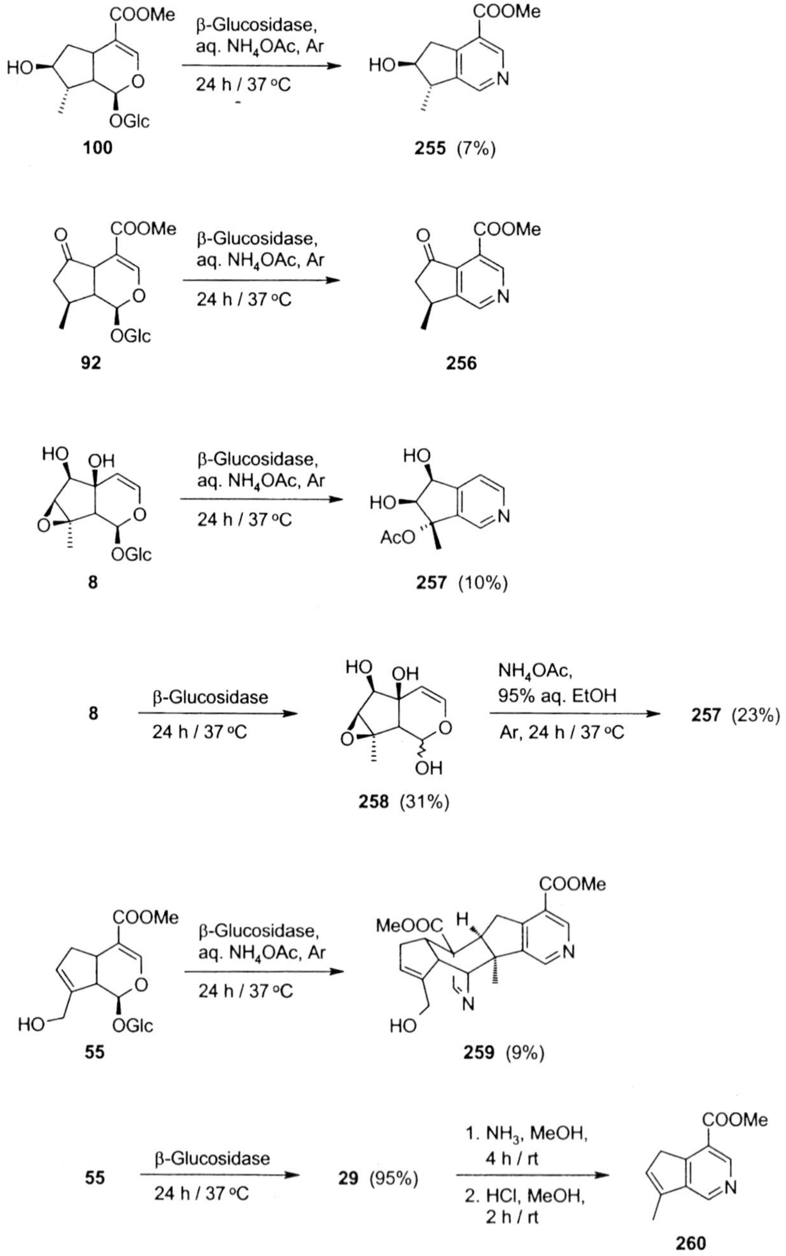

Scheme 41. Semi-syntheses of PMTAs from iridoid glucosides (77)

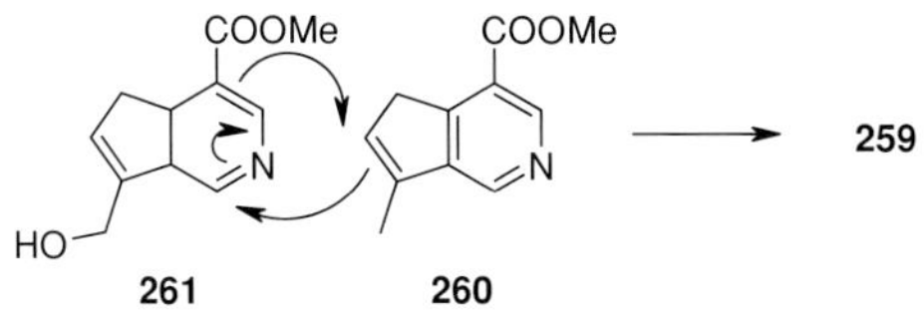

Scheme 42. Possible mechanism for the formation of dimer **259** (*77*)

during purification attempts. On the other hand, antirrhinoside (**8**) gave a somewhat unexpected pyridine (**257**) in which the epoxide had undergone acid-catalyzed opening due to the ammonium acetate present in the mixture. But these conditions did not affect **8** itself, so the oxirane ring must have opened at a later stage. Alternatively, the aglucone (**258**) of **8** was prepared prior to the amination step, which likewise afforded **257** in a modest yield. Surprisingly, a dimeric product (**259**) was obtained when geniposide (**55**) was treated with β-glucosidase in an ammonium acetate solution. It seemed most likely that dimer **259** arose from a Diels-Alder type dimerization between racemigerine (**260**) and a putative dihydropyridine (**261**) (Scheme 42). In contrast, if the aglucone genipine (**29**) was subjected to successive treatments with ammonia and hydrochloric acid in dry methanol racemigerine (**260**) was the predominant alkaloidal product (*77*).

Treatment of the individual iridoid glucosides harpagide (**244**), harpagoside (**262**) or 8-*O*-*p*-coumaroylharpagide (**263**) first with

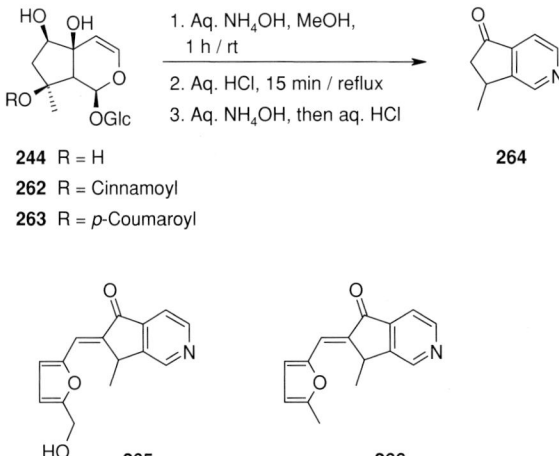

Scheme 43. Preparation of PMTAs from an extract of *Harpagophytum procumbens* (*78*)

ammonia and then with hydrochloric acid afforded aucubinine B (**264**), but only in a low yield (~1%) (*78*). Compounds **244**, **262** and **263** are the main iridoids present in *Harpagophytum procumbens*, and a commercial extract of this species was similarly treated successively with ammonia and hydrochloric acid (Scheme 43). But in addition to **264**, two novel PMTAs beatrine A and B (**265** and **266**) were obtained. The attached furan substituents appeared to originate from sugars present in the crude plant extract, since **265** was formed when a mixture of *e.g.* fructose and **262** was subjected to the above-mentioned conditions for PMTA preparation (*78*).

5.3. Bacterial Metabolism of Iridoid Glucosides

The first report on bacterial metabolism of an iridoid glucoside dealt with swertiamarin (**71**), and both its metabolism by intact human fecal flora and by defined strains of intestinal bacteria were examined (*79*). For comparison **71** was hydrolyzed by β-glucosidase, and this afforded a mixture of gentiopicral (**267**) and erythrocentaurin (**268**) (Scheme 44).

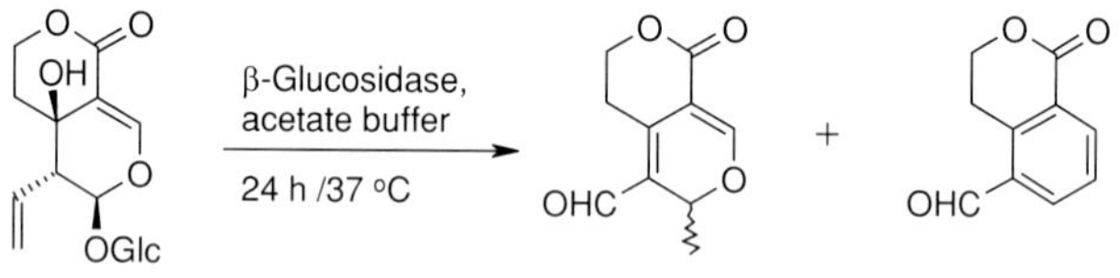

Scheme 44. Enzymatic transformation of swertiamarin (**71**) (*79*)

Anaerobic incubation with a bacterial mixture from human feces converted **71** into three metabolites, which were identified as the known gentianine (**39**), **268**, and 5-hydroxymethylisochroman-1-one (**271**). Most of the individually tested intestinal bacteria (25 species) consumed **71** and produced **268** and **271** along with small amounts of **39**. Since gentianine (**39**) was formed even in a phosphate buffer without a source of ammonia, the nitrogen incorporated into **39** might well originate from the bacteria. Formation of the benzene ring in **271** may be rationalized as follows: the dialdehyde form of the aglucone (**269**) is dehydrated; rotation around the C-5/C-9 bond would give the intermediate **270** which then is cyclized and further dehydrated (Scheme 45) (*79*).

References, pp. 106–114

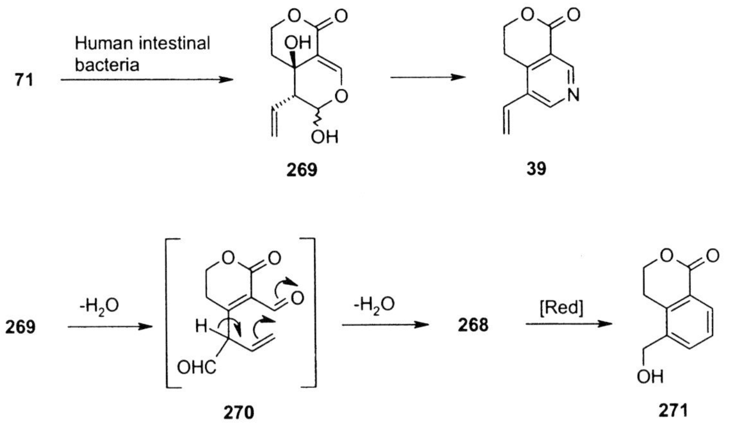

Scheme 45. Formation of aromatic compounds from swertiamarin (**71**) (*79*)

Likewise, anaerobic incubation with defined strains of human intestinal bacteria, or with fecal flora of humans, transformed aucubin (**6**) into the nitrogen-containing aucubinines A and B (**273** and **264**) (*80*). The aglucone aucubigenin (**28**) might react with an ammonia equivalent (possibly by an enzymatic process) to give the putative intermediate **272**, which by ensuing oxidative and reductive processes may be converted into **273** and **264** (Scheme 46).

In a similar manner, geniposide (**55**) and gardenoside (**54**) were converted into nitrogen-containing compounds both by enzymatic hydrolysis in an ammonium chloride solution and by incubation with

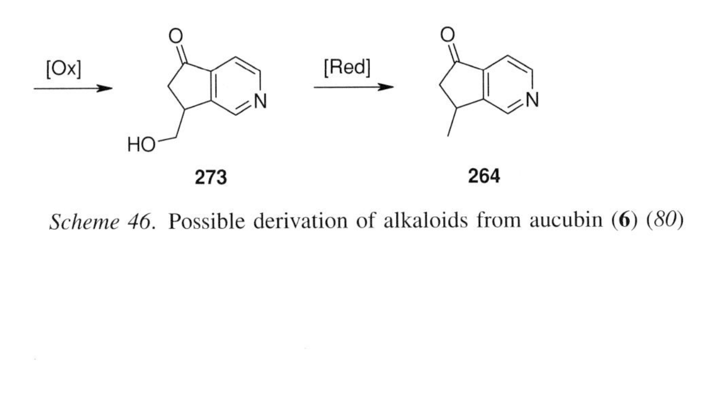

Scheme 46. Possible derivation of alkaloids from aucubin (**6**) (*80*)

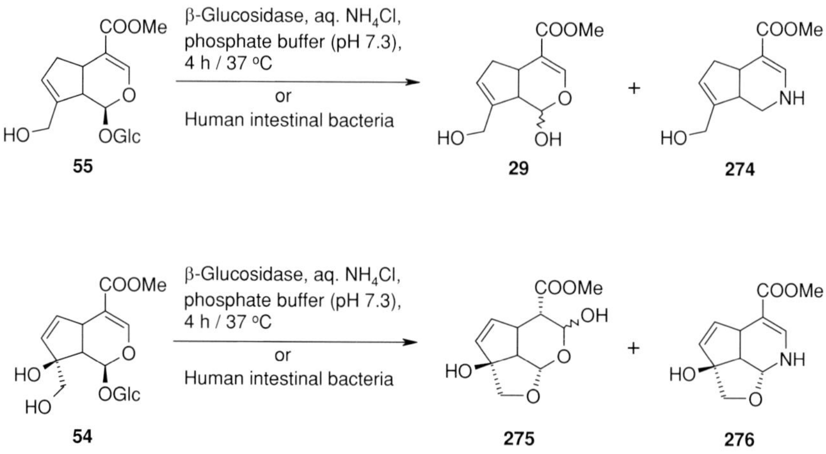

Scheme 47. Formation of alkaloids from gardenoside (**54**) and geniposide (**55**) (*81*)

human intestinal bacteria (*81*) (Scheme 47). Both methods transformed **55** into genipin (**29**) and genipinine (**274**), the latter of which appears simply to be the product from a selective reductive amination of the former. Apparently, the presence of an 8α-hydroxymethyl substituent favoured formation of tricyclic metabolites (**275** and **276**) when gardenoside (**54**) was subjected to these conditions. In gardiol (**275**) water is added to the 3,4-double bond, whereas the substitution of the 2-oxygen for a nitrogen seems to stabilize the α,β-unsaturated system in gardenine (**276**) (*81*).

Finally, the previously discussed harpagide (**244**), harpagoside (**262**) and 8-*O*-*p*-coumaroylharpagide (**263**) from *Harpagophytum* species were shown to give relatively high conversions (1.5–12.5%) into aucubinine B (**264**) when incubated with human intestinal bacteria (*82*) (Scheme 48).

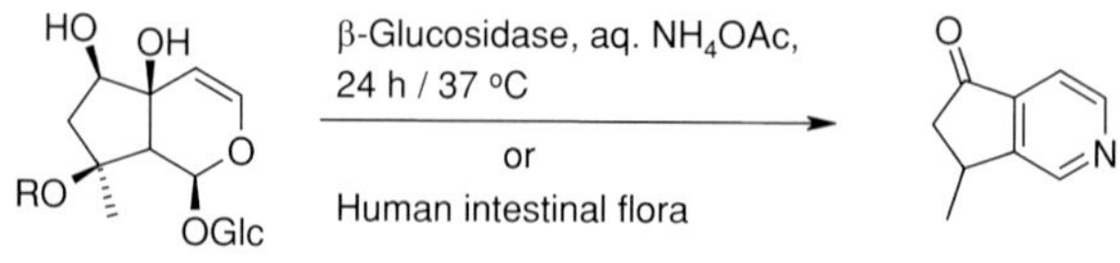

244 R = H
262 R = Cinnamoyl
263 R = *p*-coumaroyl

264

Scheme 48. Formation of aucubinine B (**264**) (*82*)

References, pp. 106–114

5.4. Total Synthesis of Pyridine Monoterpene Alkaloids

Only a few recent papers have dealt with the total synthesis of PMTAs with a cyclopentano[c]pyridine skeleton. In one approach the key step was a photoreductive cyclization of N,N-diallyl-2-oxocyclopentanecarboxamide **277** (*83*) (Scheme 49). Treatment of 4-methylcyclohexane-1,3-dione (**278**) with tosyl azide in the presence of triethylamine led to diazo diketone **279**, and subsequent Wolff rearrangement of **279** afforded **277**. Irradiation of **277** in acetonitrile–triethylamine produced a mixture of racemic 2-oxo-3-azabicyclo[4.3.0]-nonan-6-ols **280** and **281**. This mixture was reduced to the corresponding bicyclic piperidines **282/283**, which by heating in a 1:1-mixture of nitrobenzene and xylene in the presence of palladium catalyst were transformed into (±)-**37**.

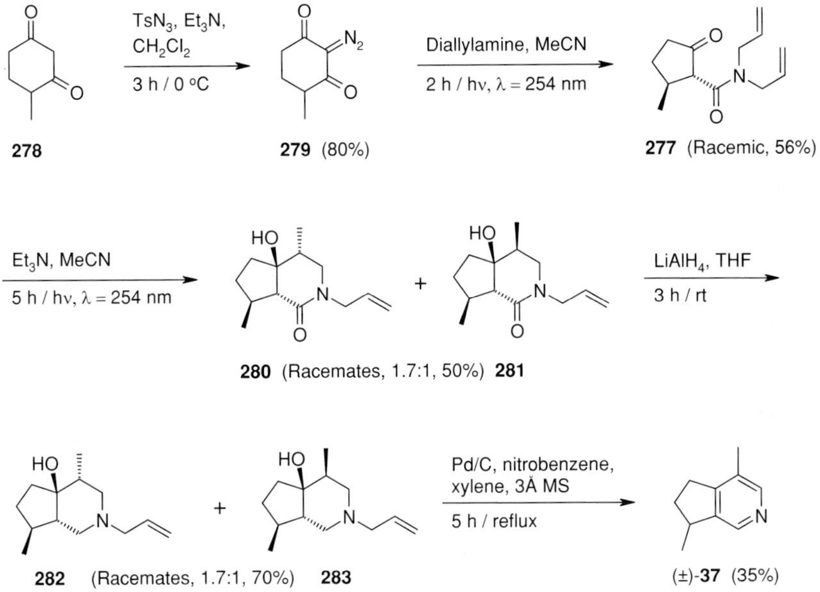

Scheme 49. Total synthesis of racemic actinidine (±)-(**37**) (*83*)

Racemic actinidine (±)-(**37**) was also obtained by a short elegant route starting from methyl 5-methylnicotinate (**284**) (*84*) (Scheme 50). Conversion of **284** into the corresponding pyridinium salt **285** was followed by a regioselective addition of a mixed copper, zinc alkyl organometallic nucleophile (**286**). The labile intermediary 1,4-dihydro-

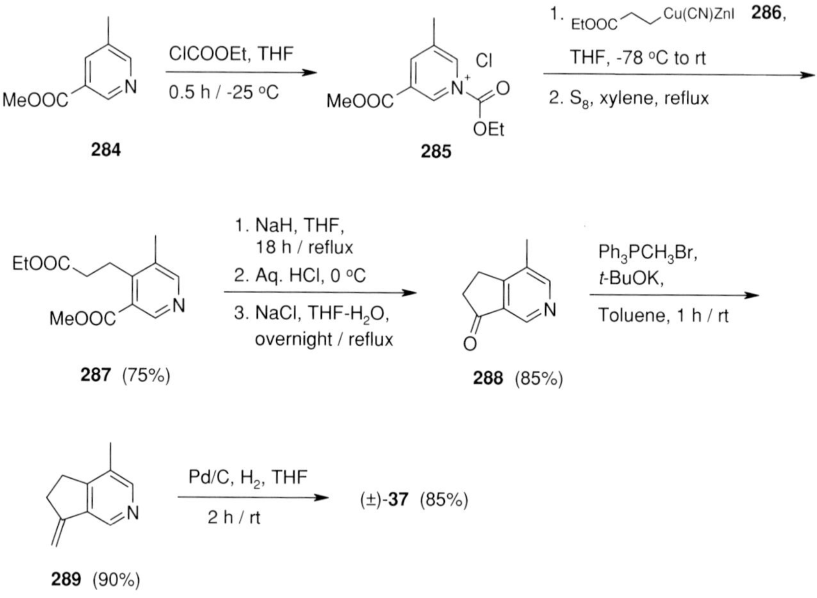

Scheme 50. A pyridine-based route to racemic actinidine (±)-(**37**) (*84*)

pyridine product was immediately oxidized with sulfur in boiling xylene to give 3,4,5-substituted pyridine **287** in good yield. Consecutive anionic cyclization and sodium chloride promoted decarboxylation afforded bicylic 8-keto pyridine **288**, which by a Wittig olefination and subsequent hydrogenolysis was transformed into (±)-(**37**).

A different protocol employed the commercially available 3-bromopyridine (**290**) as starting material for the preparation of racemic oxerine (±)-(**243**) through a multi-step sequence (*85*) (Scheme 51). First, γ-ethynyl bromide **291** was prepared in seven steps, and then a samarium(II) iodide mediated intramolecular cyclization produced the expected cyclopentano[*c*]pyridine **292** with an exocyclic methylene group. Ozonolysis of **292** afforded the corresponding ketone (**293**) which was subjected to a Grignard reaction with methylmagnesium bromide. This yielded exclusively a *syn*-configurated product that upon debenzylation furnished (±)-(**243**).

Later, the intermediate **293** was prepared by a shorter route (*86*) (Scheme 52). In this case, tributyltin hydride/AIBN promoted radical cyclization conditions were employed for the construction of the cyclopentano[*c*]pyridine skeleton from **294** in which funtionalization was the reverse of that in **291**.

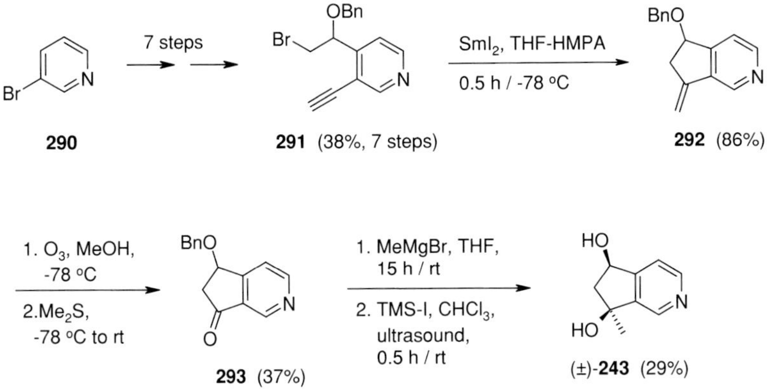

Scheme 51. Total synthesis of racemic oxerine (±)-(**243**) (*85*)

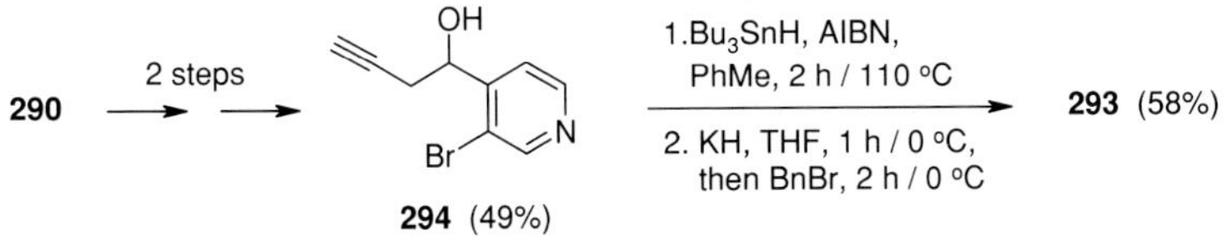

Scheme 52. Improved synthesis of bicyclic pyridine intermediate **293** (*86*)

Racemic actinidine (±)-(**37**) has also been prepared in a sequence using pyridine radical chemistry (*87*) (Scheme 53). First 3,4-lutidine (**295**) was converted into homoallylic alcohol **296** in three steps, and from this intermediate two routes to (±)-(**37**) were investigated.

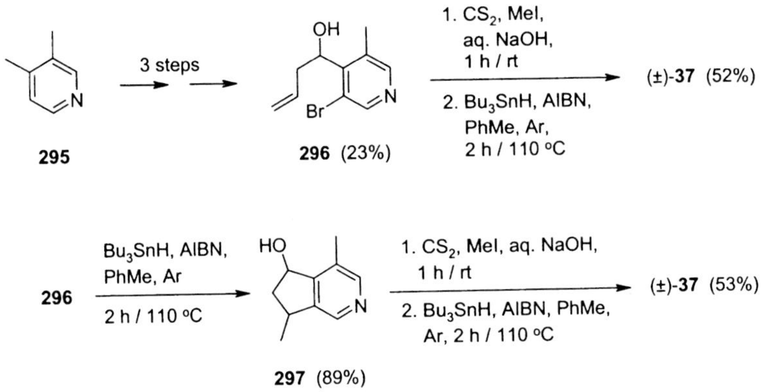

Scheme 53. PMTA synthesis using pyridine radical chemistry (*87*)

Conversion of **296** into its methyl xanthate followed by concurrent deoxygenation and reductive radical cyclization afforded directly (±)-**37**, whereas the other path involved cyclization (to **297**) prior to the deoxygenation step.

Interestingly, a novel enantioselective synthesis of (−)-actinidine (**37**) has recently emerged (*88*). By this method, chirality was introduced by using the commercially available (−)-(*S*)-citronellol (**298**) as starting material (Scheme 54). Vinyl sulfides **299/300** were obtained from **298** in four steps. They were separately subjected to intramolecular [3+2] dipolar cycloaddition *via* the corresponding silyl nitronates (**301**) generated *in situ* by treatment with *N,O*-bis(trimethylsilyl)acetamide (BSA) in the presence of triethylamine. Treatment of the resulting cyclopentano-isoxazolidines **302/303** with potassium fluoride brought about the following possible sequence. Desilylation with concomitant *N,O*-ring opening to **304**, loss of phenylsulfide to liberate the aldehyde, and rearrangement of the nitroso-olefinic part to give the α,β-unsaturated oxime (**305**), which after cyclization and dehydration afforded pyridine-

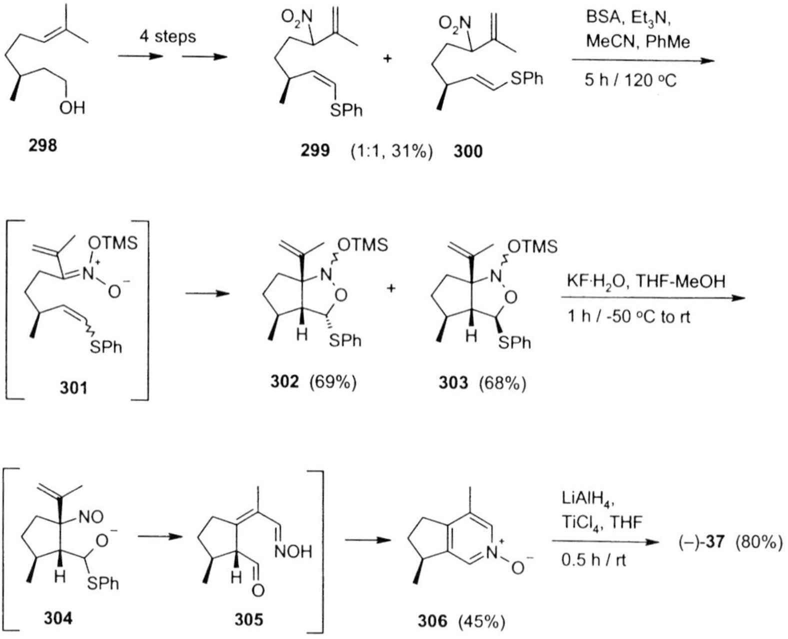

Scheme 54. Enantioselective route from (−)-citronellol (**298**) to (−)-actinidine (−)-(**37**) (*88*)

N-oxide **306**. Reduction of the *N*-oxide proceeded smoothly with the titanium tetrachloride/lithium aluminum hydride system to yield enantiopure (−)-**37**.

5.5. Diversity of Bicyclic Cyclopentanoid Piperidines

Two partially overlapping subgroups of bicyclic piperidines with a 3-azabicyclo[4.3.0]nonane skeleton have been recognized for many years, namely the *Tecoma* and *Skytanthus* alkaloids (Chart 12). The alkaloids from *Skytanthus* (Apocynaceae) comprise compounds with three different types of stereochemistry, *e.g.* α-, β-, and δ-skytanthine (**40**, **307** and **308**) as well as hydroxylated and dehydrated derivatives thereof (*16, 70*). Recently, *Tecoma stans* (Bignoniaceae) was reinvestigated, and the major alkaloids isolated were 7-hydroxydehydroskytanthine (**309**), 4-hydroxytecomanine (**310**), 5-hydroxyskytanthine (**311**) and (−)-tecomanine (**312**) (*89*). A commonly seen feature in *Tecoma* alkaloids is the 5,6-double bond.

Most of the related *Incarvillea* alkaloids found so far have the 4β-Me,5α-H,8α-Me,9α-H configuration and appear to be derived from incarvilline (**41**) (*90, 91*). Several kinds of esters of **41** have been found in *Incarvillea sinensis* (Bignoniaceae) *e.g.* incarvine C (**313**) (*92*), monomeric and dimeric esters with open-chain monoterpenoid acids (*92–94*), and more complex dimeric compounds like incarvillateine (**314**) (*94, 95*). In addition hydroxyincarvilline (**315**) has been found in this thoroughly investigated species (*91*). It may be noted that most compounds presently obtained from *I. sinensis* contain a 7α-hydroxy substitution. More recently, similar classes of monoterpene alkaloids have been reported from *Kopsia* species (Apocynaceae) (*96–98*). Trans-fused compounds of the 5α-H,8β-Me,9β-H series: *e.g.* kinabalurines A, B and C (**42**, **316**, and **317**) as well as compounds of the 5β-H,8β-Me,9α-H series: *e.g.* kinabalurines D, E and F (**318–320**) appear to occur equally frequently in these species. Moreover, 7-hydroxy and 7-keto substitutions seem common in both series (*97, 98*).

5.6. Synthesis of Bicyclic Cyclopentanoid Piperidines

Several syntheses of racemates corresponding to naturally occurring azabicyclo[4.3.0]nonane alkaloids have appeared during the past decade (*83, 99–102*). In addition four different stereoselective approaches towards the present type of piperidine monoterpene alkaloids have been

60   H. Franzyk

Skytanthus alkaloids:

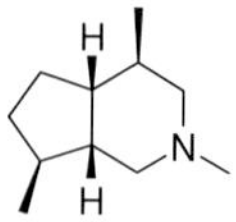

            40 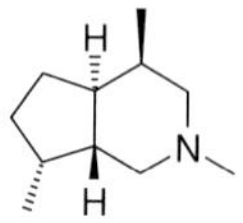   307 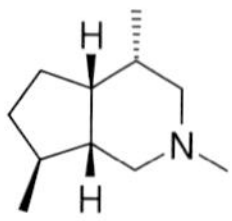   308

Tecoma alkaloids:

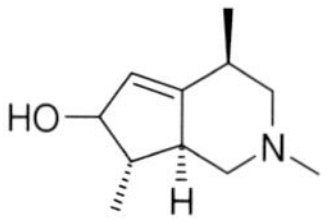

      309 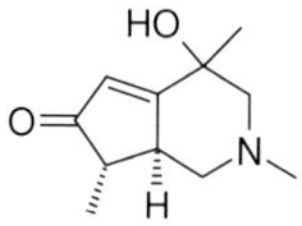   310 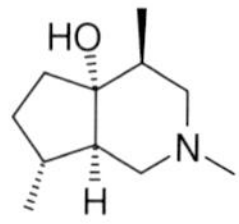   311 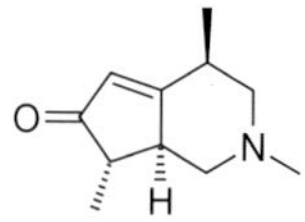   312

Incarvillea alkaloids:

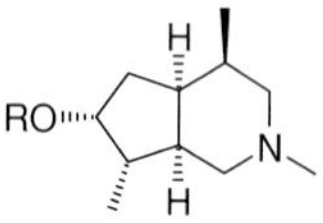

    41 R = H
    313 R = feruloyl
                        314 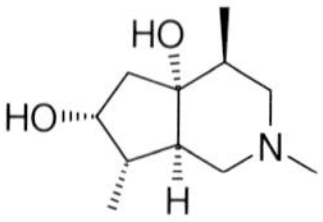   315 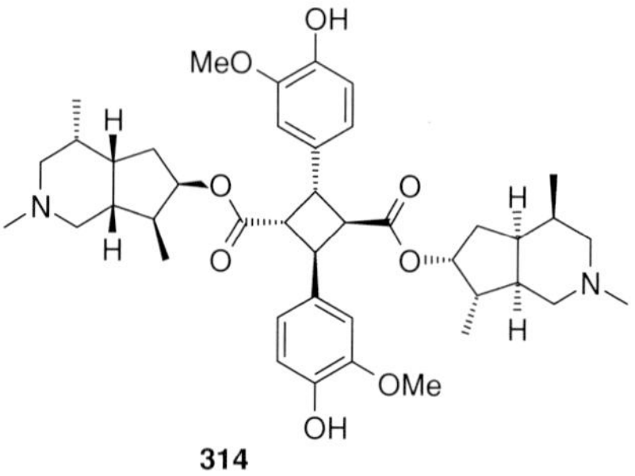

Kopsia alkaloids:

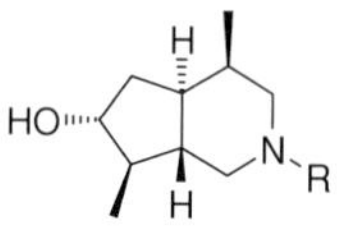

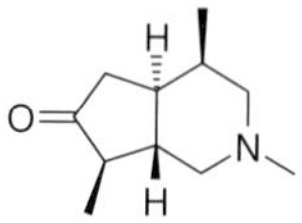

    42 R = Me   316 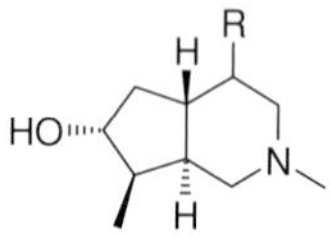   318 R = α-Me   319
    317 R = H   320 R = β-Me

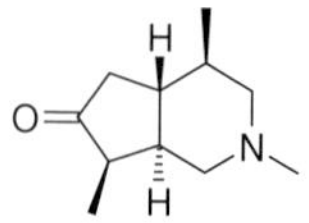

Chart 12. Diversity of bicyclic cyclopentanoid piperidines (*16, 70, 89–98*)

reported (*103–110*) and here attention will be given to these methods. Also, a semi-synthesis of related bicyclic 11-*nor*-piperidines from the iridoid glucoside antirrhinoside (**8**) has been described (*20*).

In the first asymmetric synthesis, (+)-α-skytanthine (**40**) and (+)-δ-skytanthine (**308**) were prepared from alcohol **321**, accessible in high

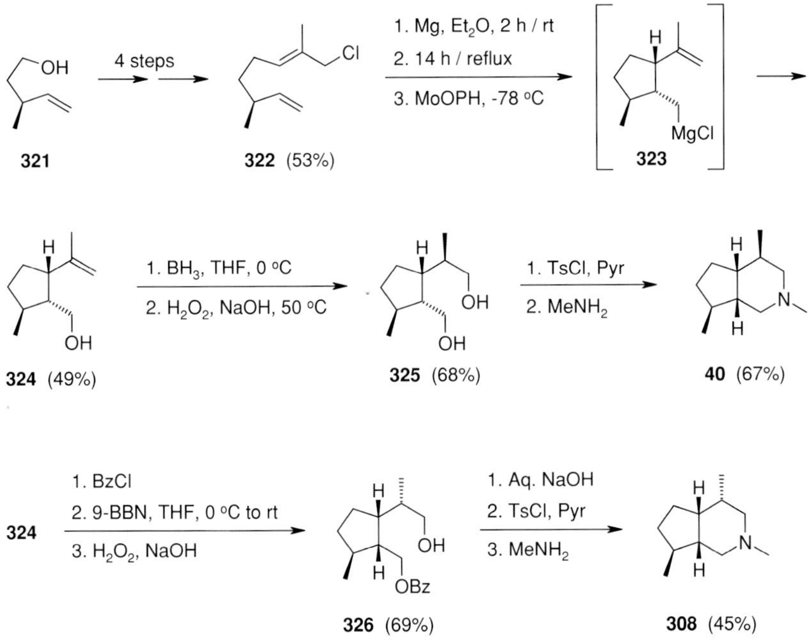

Scheme 55. Use of a magnesium-ene reaction in the synthesis of azabicyclo[4.3.0]nonane alkaloids (*104*)

enantiopurity (*103*), and the crucial step was an intramolecular magnesium-ene reaction with high diastereoselectivity (*104*) (Scheme 55). Four steps were required to convert **321** into the allylic chloride **322**. Cyclization of the magnesium-metalated **322** produced intermediate **323** which by oxidative trapping with molybdenum pentoxide pyridine-hexamethylphosphoramide complex (MoOPH) at low temperature furnished an isomer mixture from which the major cyclopentane (**324**) was obtained. Hydroboration and subsequent oxidation of **324** gave (+)-α-iridodiol (**325**) which by successive tosylation and ring-closure with methylamine was converted to (+)-α-skytanthine (**40**). Interestingly, when the benzoate of **324** was subjected to hydroboration with 9-borabicyclo[3.3.1]nonane (9-BBN) the reaction proceeded with reversed stereoselectivity to give alcohol **326** which after saponification was transformed into (+)-δ-skytanthine (**308**).

(−)-Carvone monoepoxide (**327**) was employed as a chiral starting material for the synthesis of (+)-tecomanine (**332**), the antipode of the naturally occurring compound (*105*) (Scheme 56). The essential step in this sequence was an initial Favorskii-type rearrangement of **327** which

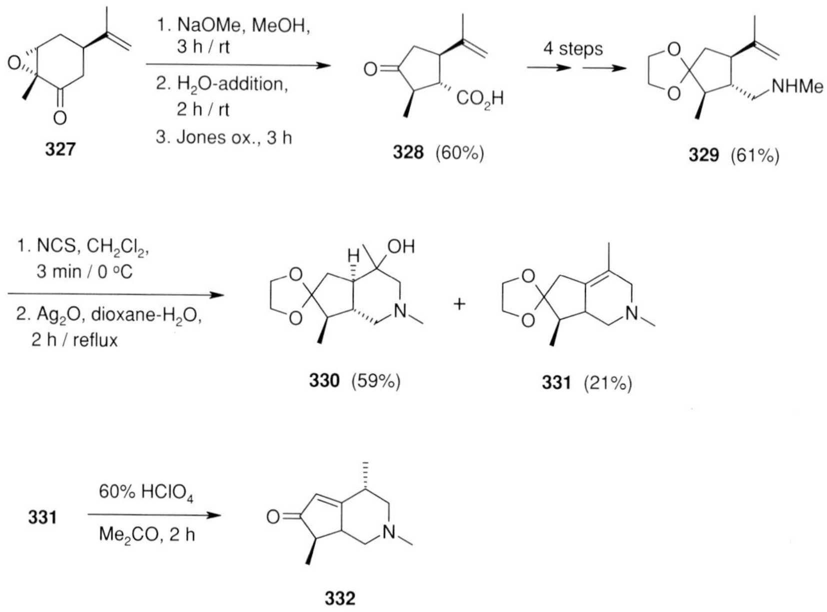

Scheme 56. Enantioselective route to (+)-tecomanine (**332**) (*105*)

upon subsequent hydrolysis and Jones oxidation afforded keto acid **328**. After the keto group was protected as a 1,3-dioxolane ketal three additional steps were employed to reach *N*-methylamino compound **329**. Activation of **329** by *N*-chlorination with *N*-chlorosuccinimide (NCS) followed by cyclization, promoted by silver(I) oxide in dioxane-water, produced a mixture of bicyclic tertiary alcohol **330** and olefin **331**. Ultimately, deprotection of the minor product (**331**) induced a concomitant migration of the double bond to the more favourable α,β-position to yield (+)-tecomanine (**332**).

Another stereocontrolled route to 3-azabicyclo[4.3.0]nonane alkaloids that exploited (*R*)-1-phenylethylamine (**333**) as an inducer of chirality was designed for the preparation of (−)-*N*-demethyl-δ-skytanthine (**334**), (−)-α-skytanthine (**335**), and (+)-epidihydrotecomanine (**336**) (Schemes 57 and 58) (*106, 107*). The key step in this synthesis was the initial ketene aza-Claisen rearrangement which was carried out by treating optically active 2-(1-phenylethyl)-2-azabicyclo[2.2.1]hept-5-ene **338** with *in situ* generated dichloroketene (Scheme 57). This provided the *cis*-fused α,α-dichloro 3-azabicyclo[4.3.0]nonenone **339**, which was reduced to lactam **340**. Monomethylation followed by epoxidation yielded epimer mixture **341** which under basic conditions

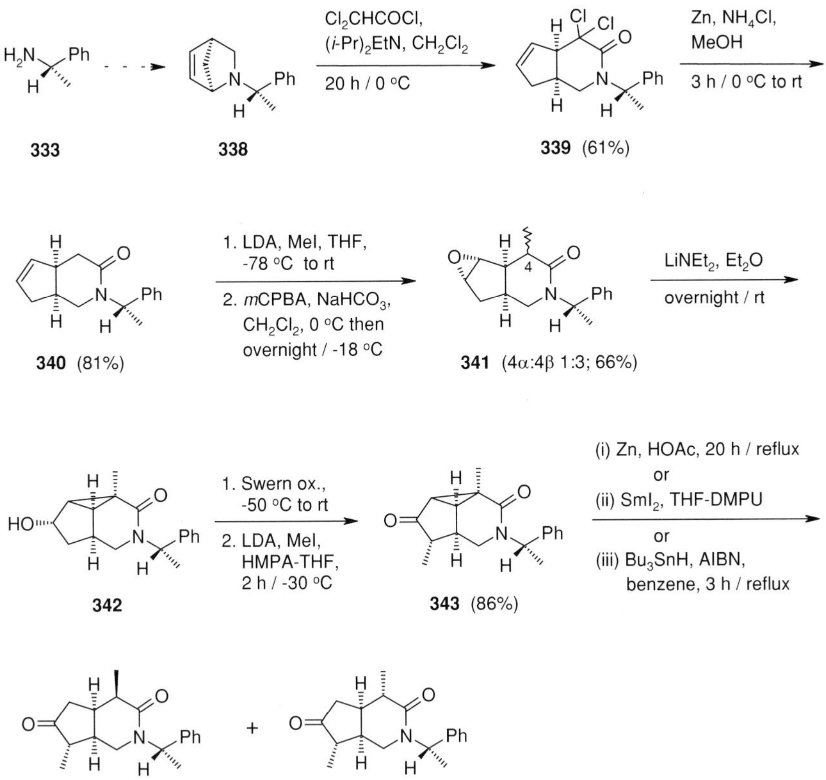

(i)	**344**	(1:2, 73%)	**345**
(ii)	**344**	(1:1, 67%)	**345**
(iii)	**344**	(5:1, 65%)	**345**

Scheme 57. Synthesis of bicyclic lactam intermediates for bicyclic cyclopentanoid piperidines (*106*, *107*)

via an intramolecular epoxide opening was transformed into tricycle **342**. Swern oxidation of **342** and stereoselective monomethylation of the resulting ketone produced keto lactam **343**. Three different methods for reductive cleavage of the cyclopropane moiety in **343** were examined. With samarium(II) iodide equal amounts of keto lactams **344** and **345** were obtained, while zinc/acetic acid and tributyltin hydride yielded ratios of 1:2 and 5:1, respectively.

The keto group in **344** was converted to the corresponding dithiane (**346**) from which the lactam carbonyl, the dithiane moiety and the *N*-benzylic group were removed by successive reductions (*via* **347**) to give (−)-*N*-demethyl-δ-skytanthine (**334**). Alternatively, the functionalities in

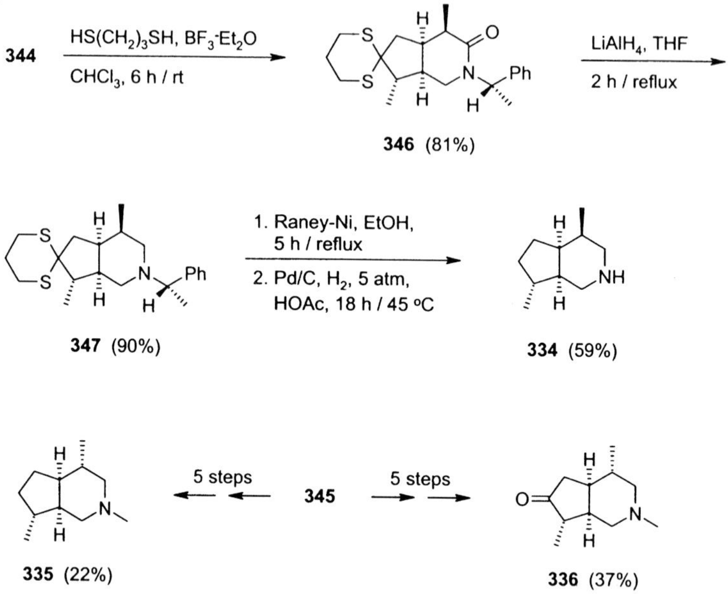

Scheme 58. Preparation of azabicyclo[4.3.0]nonane alkaloids **334–336** (*106, 107*)

345 were manipulated to allow its conversion into (−)-α-skytanthine (**335**) and (+)-epidihydrotecomanine (**336**) (*107*) (Scheme 58).

In the latest total synthesis of (+)-α-skytanthine (**40**) a similar protocol for the introduction of chirality was employed. Cyclopentene alcohol **348** was converted into *N*-(*S*)-1-phenylethylcarboxylamide **349**, and then an asymmetric aza-Claisen rearrangement afforded *exo*-methylene cyclopentanoid amide **350** (*108*). Reduction of the amido group in **350** and subsequent hydroboration/oxidation of the *exo*-methylene group led to amino alcohols **351** which by a Mitsunobu-type reaction with cyanomethylenetrimethylphosphorane (CMMP) (*109*) was cyclized to piperidines **352/353**. The major piperidine isomer (**353**) was debenzylated by hydrogenolysis and *N*-methylated by the Eschweiler-Clarke procedure to give (+)-α-skytanthine (**40**) (*110*).

The preparation of 3-azabicyclo[4.3.0]nonane alkaloids from the decarboxylated iridoid glucoside antirrhinoside (**8**) has also been investigated (*20*). In the first attempted strategy, the aim was to obtain a cyclopentanoid diol similar to **325** which could be cyclized *via* its ditosylate. Hence, **8** was converted into the corresponding diacetonide (**354**) which in turn was subjected to a selective partial deketalization, and then the 3,4-enol ether was hydrogenated to yield monoacetonide

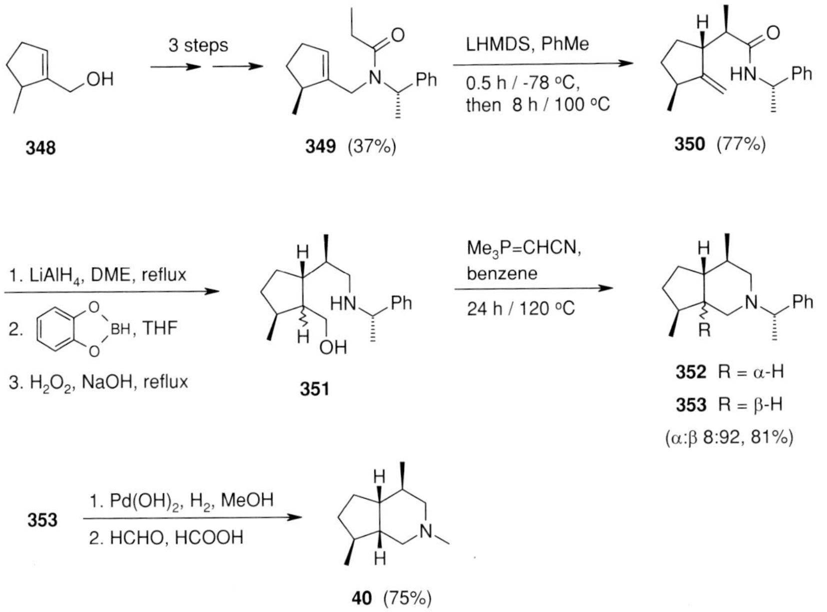

Scheme 59. Total synthesis of (+)-α-skytanthine (**40**) (*110*)

355. Upon enzymatic cleavage of the glycosidic bond aglycone acetonide **136** was obtained in high yield. However, sodium borohydride reduction of **136** proved troublesome, since the expected diol **356** was very prone to undergo intramolecular epoxide opening to give tricyclic ether **357** during work-up and purification (Scheme 60).

Alternatively, aglucone **258**, prepared from **8**, was allowed to react with benzylamine hydrochloride under slow addition of sodium cyanoborohydride which directly yielded bicyclic piperidine **358** by a double reductive amination. Reduction and azidolysis of the epoxide functionality afforded piperidines **359** and **360**, respectively. Hydrogenolysis of the benzyl group in **358/359** gave directly the unprotected piperidines (**361/362**) (*20*) (Scheme 61).

Enantioselective total synthetic routes to the known types of PMTAs and cyclopentanoid bicyclic piperidines are not straightforward and usually require advanced synthetic techniques in comparison with their rather simple preparation from iridoid glucosides (Sections 5.2 and 5.4). Further details on this subject are available in a recent review (*111*) on monoterpene alkaloids in general which covers the literature from 1974 to 1997.

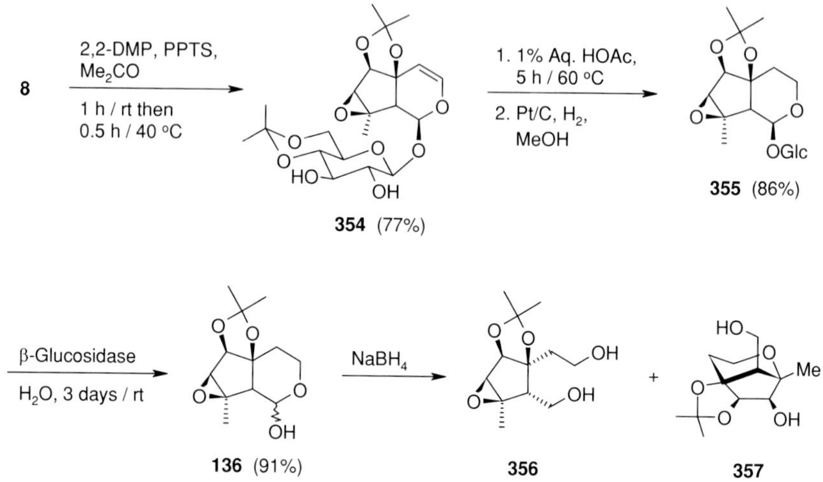

Scheme 60. Attempted pathway to bicyclic piperidines (20)

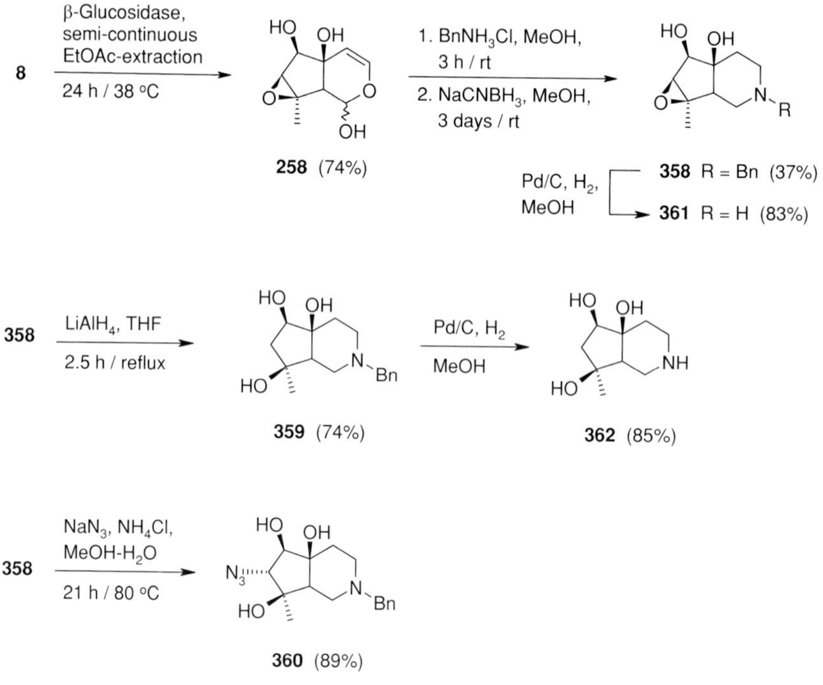

Scheme 61. Semi-synthesis of 3-azabicyclo[4.3.0]nonane alkaloids from antirrhinoside (**8**) (20)

5.7. Semi-Synthesis of Glucosidic Secoiridoid Alkaloids

Different methods for the conversion of secologanin (**13**) into bakankosin (**43**) have been investigated (*32*) (Scheme 62). Although it was possible to perform a reductive amination of **13** with benzylamine hydrochloride in the presence of sodium cyanoborohydride, the resulting *N*-benzylated lactam (**363**) proved to be inert towards debenzylation with

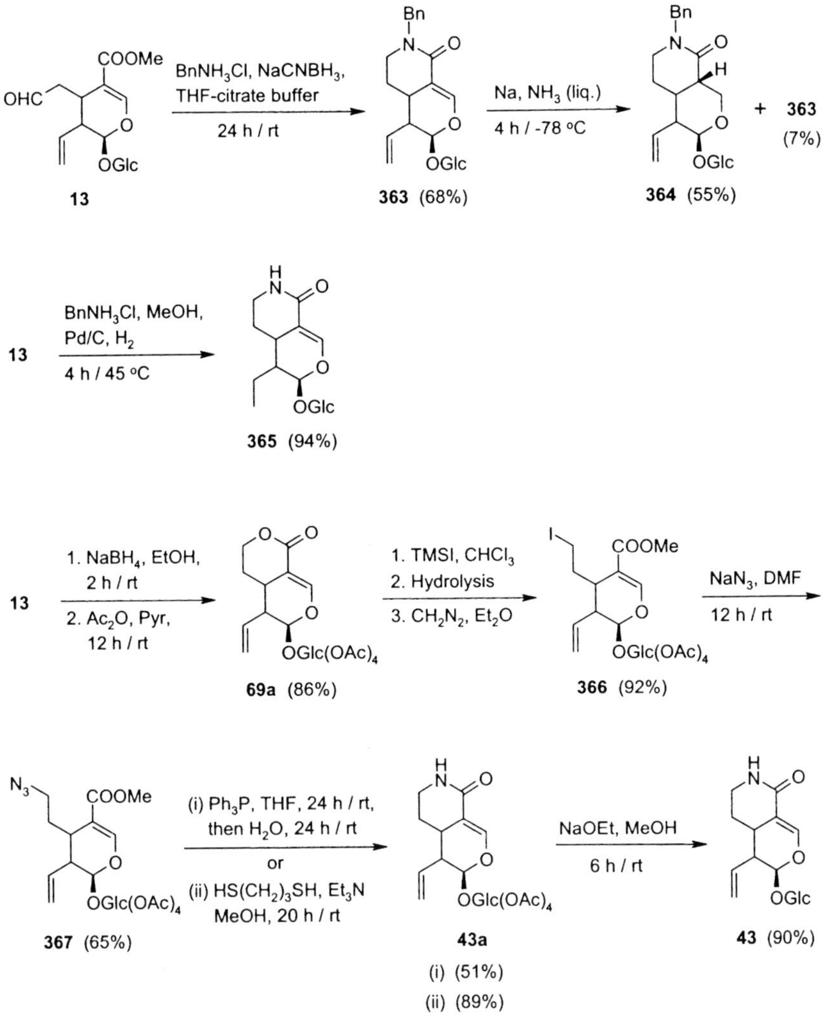

Scheme 62. Semi-synthetic conversion of secologanin (**13**) into bakankosin (**43**) (*32*)

sodium in liquid ammonia as only reduction of the conjugated enol ether (to **364**) was observed. As expected, an alternative reductive amination under hydrogenation conditions afforded only the 8,10-dihydro analogue (**365**). Instead, **13** was transformed into sweroside tetraacetate (**69a**) which in turn was treated with trimethylsilyl iodide. The intermediate 7-iodo 11-trimethylsilyl ester was hydrolyzed, and the resulting acid then methylated to yield the 7-iodo 11-methyl ester (**366**). At this point, an azido functionality was introduced at the 7-position (to give **367**), and its reduction to the amino stage (and concomitant cyclization to **43a**) was performed most efficiently with propanedithiol in the presence of triethylamine. Finally, deacetylation gave unprotected bakankosin (**43**) (*32*).

The monoterpenoid alkaloid xylostosidine (**368**) was prepared by a biomimetic condensation of secologanin (**13**) with cysteine that yielded tricyclic acid **369**, which upon heating in water underwent decarboxylation to **368** (*112*) (Scheme 63). The related sulfoxide loxylostosidine (**370**) was obtained from **368** by chemoselective oxidation with *m*-chloroperbenzoic acid. In a more direct approach, **368** was obtained from condensation of **13** and 2-aminoethanethiol. Notably, only the natural (7*S*)-isomer of **368** was produced by both procedures.

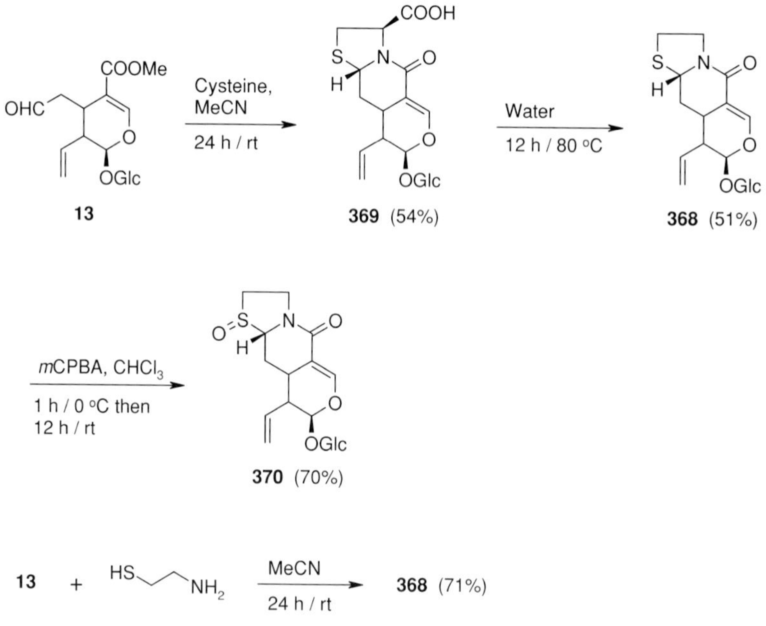

Scheme 63. Preparation of tricyclic alkaloids from secologanin (**13**) (*112*)

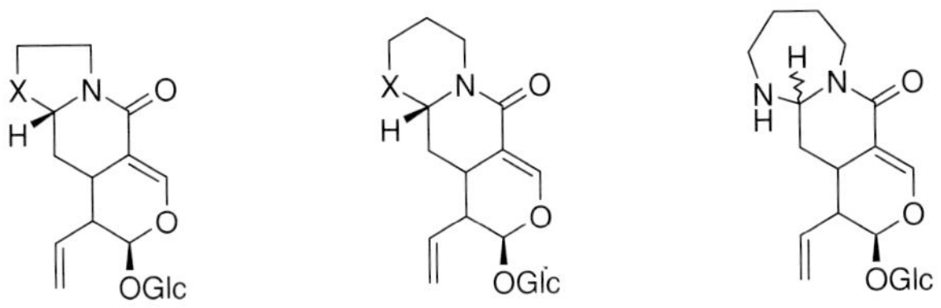

Synthetic Aspects of Iridoid Chemistry 69

371 X = NH	373 X = S	376
372 X = O	374 X = NH	
	375 X = O	

Chart 13. Unnatural tricyclic alkaloids prepared from secologanin (**13**) (*112*)

In a similar way, condensations between **13** and a range of C_2-C_4 amines allowed preparation of the corresponding unnatural tricyclic alkaloids **371–376** (in 42–76% yield) (*112*) (Chart 13).

6. Syntheses from Iridoids

A diverse range of chemical conversions of iridoids into many types of compounds will now be discussed. First, a rather heterogeneous array of short conversions into dyestuffs and novel secoiridoid-derived compounds will be reviewed. Recent transformations of iridoids into building blocks for the preparation of other classes of cyclopentanoid compounds of biological interest will also be treated thoroughly. In addition, a few miscellaneous modifications of the sugar moiety in iridoid glucosides will be summarized.

6.1. Formation of Colored Compounds

Iridoids undergo characteristic chromatic reactions in acidic media (*8, 113*) and this is exploited in their easy detection with acidic spray reagents (Section 3.3.). A number of red-colored fulvene dyestuffs associated with such color reactions have been prepared from the iridoid glucosides harpagide (**244**) and lamiol (**377**) (*114*) (Scheme 64). When treating these iridoid glucosides with acid in the presence of an aromatic aldehyde at low temperature and shielding the reaction from light, low but acceptable yields (10–15%) of several fulvoharpagides (**378–383**) and a fulvolamiol (**384**) were obtained. A possible route to such fulvoiridoids involves initial hydrolysis of the glucosidic bond; the resulting aglucone (*i.e.* **385** or **386**) then undergoes dehydration to a

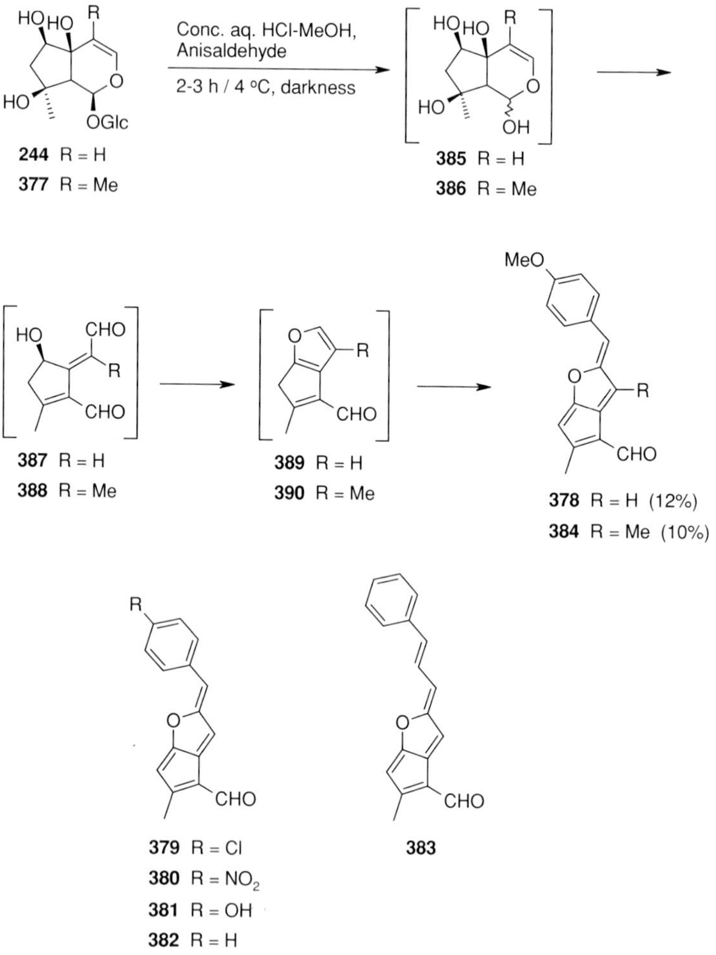

Scheme 64. Formation of red-colored compounds from harpagide (**238**) and lamiol (**377**) (*114*)

γ-hydroxy-α,β-unsaturated aldehyde (**387** or **388**) which cyclizes in the acidic medium to the corresponding furan derivative (**389** or **390**). Electrophilic attack at the α-position of the furan ring by the protonated anisaldehyde results in a substitution product that upon dehydration affords the corresponding fulvo-derivative (**378** or **384**). Five other aromatic aldehydes were examined in this reaction with **244**, but no significant differences with respect to reactivity or yield were observed.

Genipin (**29**) forms a blue pigment upon contact with peptides or amino acids and a report on its specific reaction with glycine has

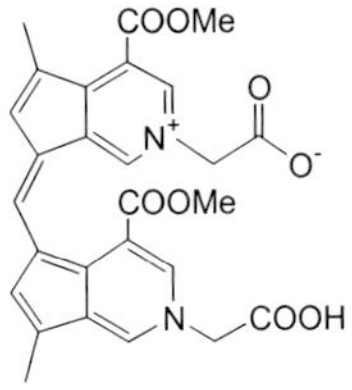

Synthetic Aspects of Iridoid Chemistry

391

Chart 14. Dimeric adduct between genipin (**29**) and glycine (*115*)

appeared (*115*). The structure of the dimeric adduct (**391**) was elucidated by NMR spectroscopy.

When genipin (**29**) was treated with methylamine hydrochloride under an inert atmosphere colored compounds were observed (*116*). First brownish-red compounds were formed in the reaction mixture, but upon contact with oxygen a polymeric blue pigment was readily obtained.

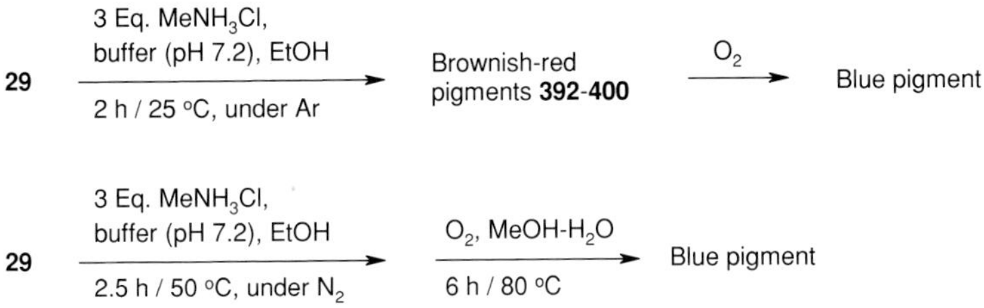

29 → [3 Eq. MeNH$_3$Cl, buffer (pH 7.2), EtOH, 2 h / 25 °C, under Ar] → Brownish-red pigments **392-400** → [O$_2$] → Blue pigment

29 → [3 Eq. MeNH$_3$Cl, buffer (pH 7.2), EtOH, 2.5 h / 50 °C, under N$_2$] → [O$_2$, MeOH-H$_2$O, 6 h / 80 °C] → Blue pigment

Scheme 65. Formation of coloured compounds from genipin (**29**) (*116,117*)

Nine such brownish-red pigments were obtained in sufficient amounts to allow their structure elucidation. Two monomeric 2-pyrindine compounds (**392** and **393**) were identified, but surprisingly **393** contained an extra methyl group at the 6-position. Four dimeric pigments (**394–397**) were obtained. Of these **394** was a symmetrical dimer of **392** while **395–397** consisted of a tetrahydropyridine unit coupled *via* C-1 to a 2-pyrindine unit. In addition two trimeric pigments (**398** and **399**) and a tetrameric pigment (**400**) were isolated, and they all seem related to **395–397**. During the formation of the brownish-red pigments a brief initial yellow colouring of the reaction mixture was observed (*117*) and a very labile yellow compound was isolated. This proved to be a dihydropyridine (**401**) which presumably arises from a Michael-type attack of methylamine at the 3-position in **29** followed by

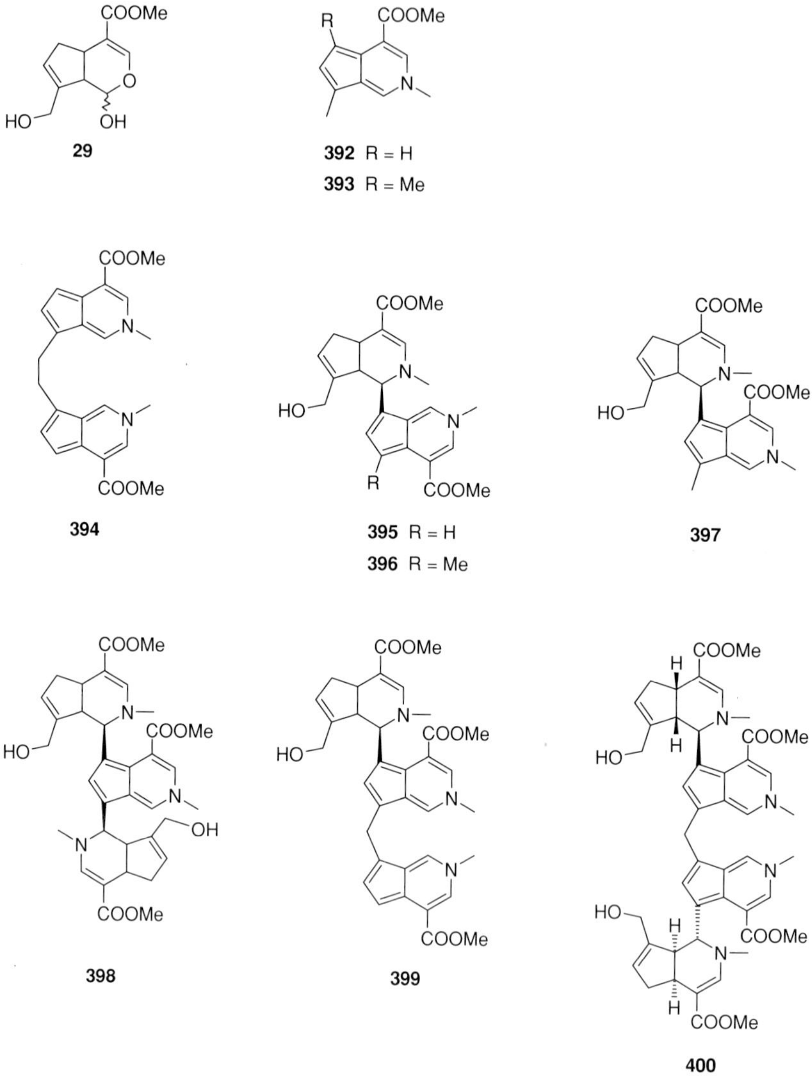

Chart 15. Nine brownish-red pigments (**392–400**) derived from genipin (**29**) (*116, 117*)

opening of the dihydropyran ring and subsequent cyclization of the secondary amino group with the free aldehyde at C-1. The role of intermediate (**401**) as a precursor of the brownish-red pigments was established by its conversion into pigments **392–398**. The source of the extra methyl group in **393** was shown to be the 10- hydroxymethyl group

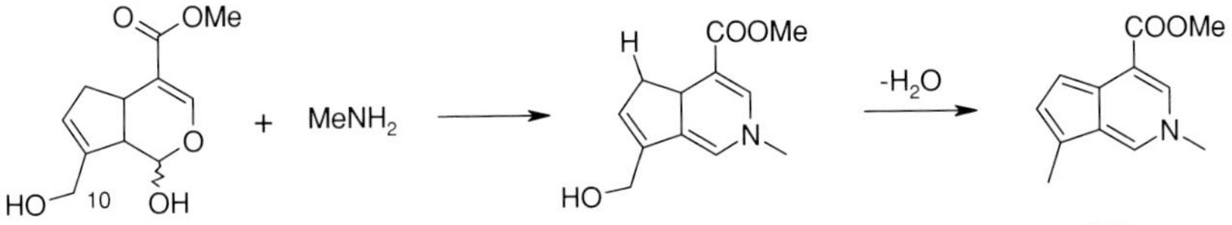

Scheme 66. Possible routes to intermediates **392** and **401** (*117*)

in **29**, which is in accordance with the presence of units apparently lacking C-10 in pigments **395** and **398–400**. Formation of **393** was rationalized as a heterolytic carbon-carbon bond cleavage of a dimeric intermediate, which also would give rise to a 10-*nor*-derivative; however, such a compound could not be isolated. A conceivable, but complex web of oxidation-reduction couplings between different forms of the simple pigments and several hypothetical intermediates has been outlined for the formation of these oligomeric brownish-red pigments (*117*). However, the exact scheme of formation is likely to be extremely complicated as it was shown that interconversion of the pigments occur under the reaction conditions (*117*).

6.2. Reactions of Secoiridoids

Among the readily available secoiridoids, secologanin (**13**) has been considered a suitable renewable raw material for synthetic purposes. This is exemplified by its conversion into bridged homoiridoids by a tandem-Knoevenagel-hetero-Diels-Alder reaction with *N,N*-dimethylbarbituric acid (**402**), Meldrum's acid (**403**), pyrazolones, isoxazolones or other 1,3-dicarbonyl compounds (*118*) (Scheme 67). First, Knoevenagel condensation of secologanin (**13**) with 1,3-dicarbonyl compounds **402–405** at room temperature afforded heterodiene products **406–409**, which without isolation were subjected to cyclization at elevated temperature to yield only the bridged homoiridoids **410–413**. However, when Meldrum's acid (**403**) and **13** were condensed the initially formed cycloadduct **411** could not be obtained in pure form since chromatography using a methanolic solvent mixture led to the formation of methyl ester **414** after loss of acetone. On the other hand, if water was employed as co-solvent decarboxylation to lactone **415** took place. The low yield of **413** could be ascribed to a lower reactivity of the oxabutadiene moiety in **409** which was partly recovered (28%).

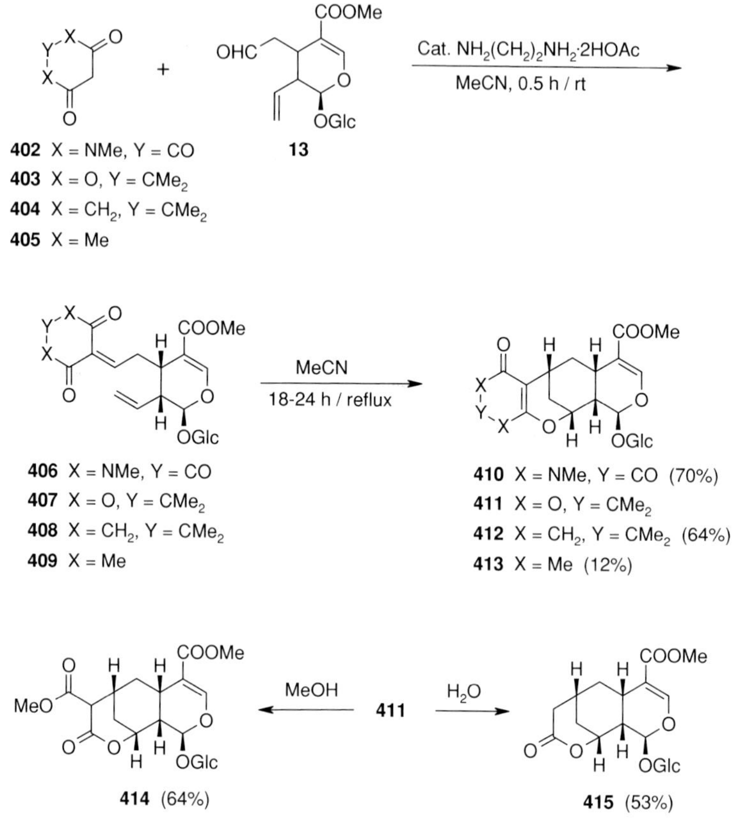
COOMe
OHC
Cat. $NH_2(CH_2)_2NH_2 \cdot 2HOAc$
MeCN, 0.5 h / rt
OGlc
402 X = NMe, Y = CO
403 X = O, Y = CMe_2
404 X = CH_2, Y = CMe_2
405 X = Me
13
MeCN
18-24 h / reflux
406 X = NMe, Y = CO
407 X = O, Y = CMe_2
408 X = CH_2, Y = CMe_2
409 X = Me
410 X = NMe, Y = CO (70%)
411 X = O, Y = CMe_2
412 X = CH_2, Y = CMe_2 (64%)
413 X = Me (12%)
MeO
MeOH
411
H_2O
414 (64%)
415 (53%)

Scheme 67. Tandem-Knoevenagel-hetero-Diels-Alder reactions of **13** with 1,3-dicarbonyl compounds (*118*)

In addition, isoxazolone **416** and pyrazolones **417–419** were tested as co-substrates (Scheme 68). However, due to the lower reactivity of the heterodiene moieties in the intermediary **420–423** only poor yields of **424** and **425** were obtained, whereas **422** and **423** proved so unreactive that cycloadducts were not formed at all. Generally, the above-mentioned cycloadditions exhibited high regio- and stereoselectivity with preferred attack of the oxadiene moiety at the *Si*-face of the methylene group in order to avoid steric interaction of the heterodiene with the glucosyl moiety (*118*).

Deglucosylation reactions of secoiridoids under basic conditions using a primary or a secondary amine were accompanied by incorporation of the amine into the products (*119*). Sweroside (**69**) and its 8,10-dihydro- and 8,10-dihydro aglycone derivatives (**426** and **427**)

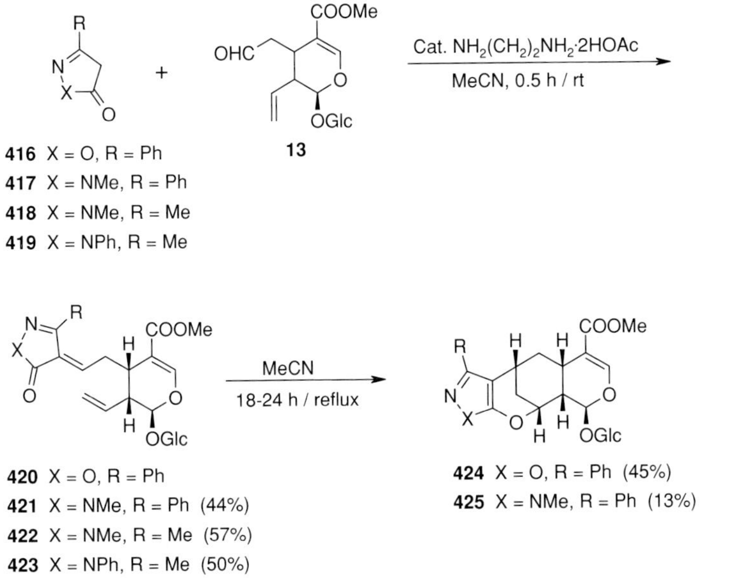

Scheme 68. Tandem-Knoevenagel-hetero-Diels-Alder reactions of **13** with isoxazolones and pyrazolones (*118*)

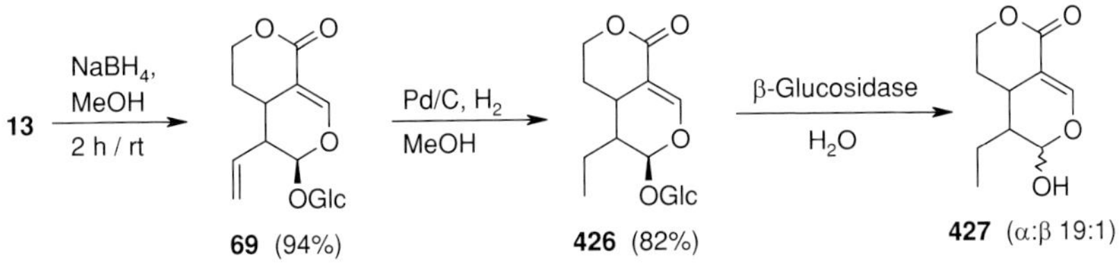

Scheme 69. Preparation of secoiridoid model compounds **69**, **426** and **427** (*119*)

were used as model compounds readily accessible from secologanin (**13**) (Scheme 69).

Piperidine and *n*-propylamine were selected as the nucleophiles. A higher reaction temperature proved necessary for the reactions with the less reactive piperidine. Sweroside (**69**) appeared to undergo an initial Michael-type attack at C-3 with both piperidine and *n*-propylamine (Scheme 70). The adduct (**428** or **429**) might then be stabilized by regeneration of the 3,4-double bond with simultaneous cleavage of the dihydropyran ring resulting in deglucosylation. This would afford a

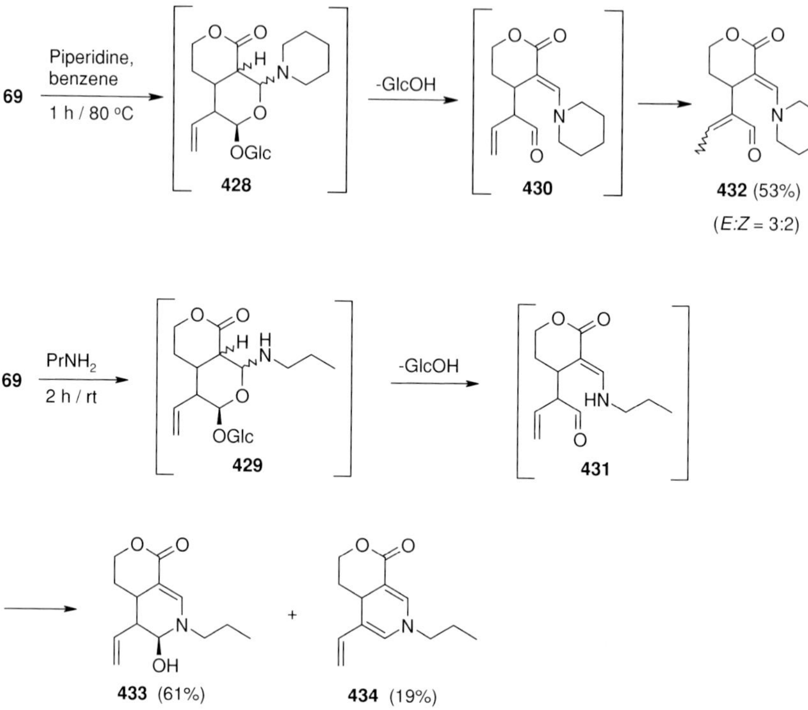

Scheme 70. Amine-promoted deglucosylations of sweroside (**69**) (*119*)

conjugated enamine (**430** or **431**). In the reaction with piperidine further cyclization was not possible, but the 8,9-double bond was shifted to the more stable α,β-position, and this resulted in E/Z-isomers **432**. In contrast, the putative **431** might undergo cyclization to hemiaminal **433** which was susceptible to dehydration to give the conjugated dihydropyridine **434**. However, due to the milder conditions used with *n*-propylamine, migration of the isolated double bond was not observed.

In the case of 8,10-dihydrosweroside (**426**), no pure isolates could be obtained after reaction with the secondary amine, but with the primary amine hemiaminal **435** (analogous to **433**) was obtained together with a lesser amount of a bicyclic aminal (**436**). However, the expected conversion of **435** into **436** under the applied reaction conditions could not be realized. Accordingly, an indepedent mechanism involving initial attack at C-1 was proposed for the formation of **436** (Scheme 71) (*119*).

When 8,10-dihydrosweroside aglucone (**427**) was allowed to react with piperidine, preferential attack at C-1 led conceivably to the

References, pp. 106–114

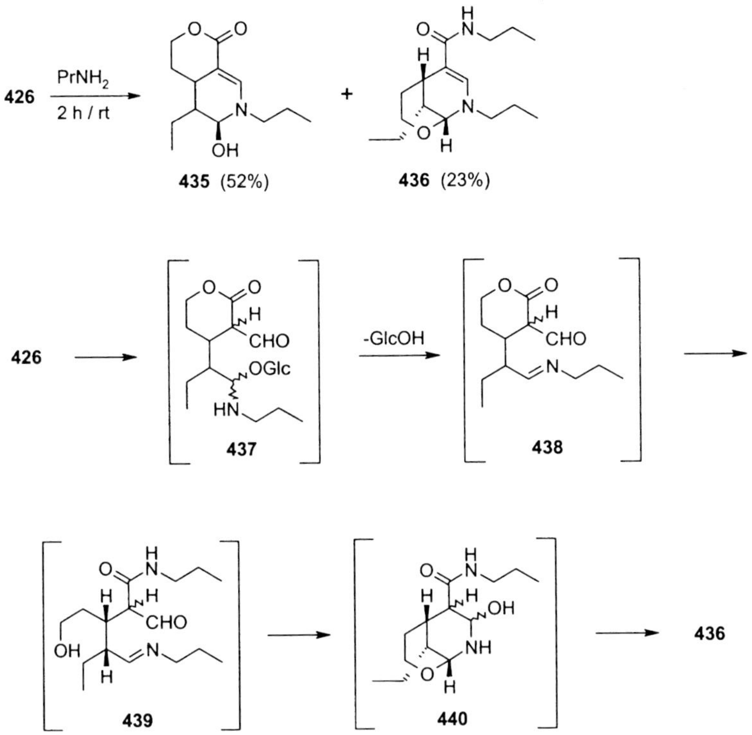

Scheme 71. Deglucosylation of 8,10-dihydrosweroside (**426**) induced by propylamine (*119*)

formation of cyclic aminal **443** (*via* **441** and **442**) (Scheme 72). Unexpectedly, the stereochemistry at C-9 and C-1 in the product (**443**) was inverted indicating that equilibration to the thermodynamically more stable compound had taken place. Similarly, **427** might well be converted into hemiaminal **444** by the action of *n*-propylamine, and *via* Schiff base **438**, formation of **436** might occur as for **426** (*119*).

Acid-catalyzed deglucosylations of the model compounds *N*-methylbakankosin (**445**) and its 8,10-dihydro derivative (**446**) have been investigated (*120*). Prognoses for the shortest and most probable mechanisms were based on a graph analysis of the complex reaction matrices consisting of 48 aglycones, 92 equilibria and 368 elementary steps for the deglucosylation of **445**, and 24 aglycones, 40 equilibria and 160 elementary steps for the deglucosylation of **446**. The model compounds were prepared sequentially from secologanin (**13**) by reductive amination with methylamine and subsequent hydrogenation (Scheme 73). Traditional deglucosylation of **445/446** by treatment with

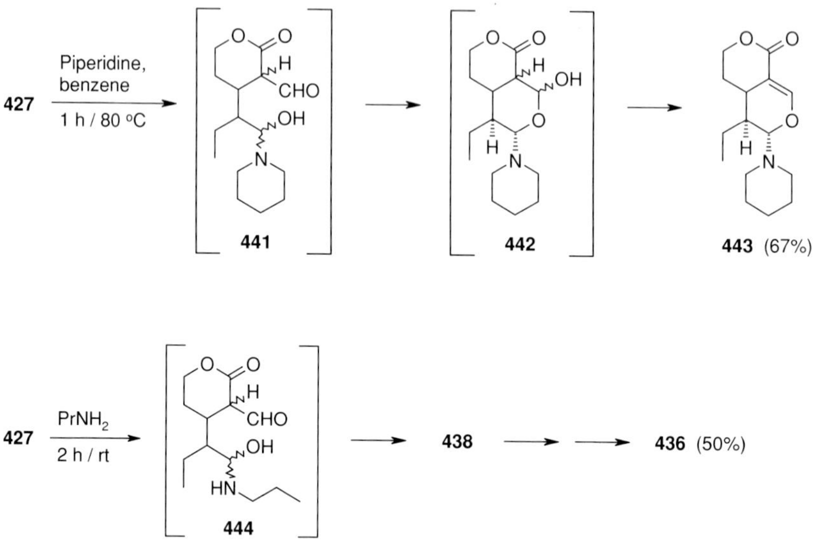

Scheme 72. Reaction of 8,10-dihydrosweroside aglucone (**427**) with amines (*119*)

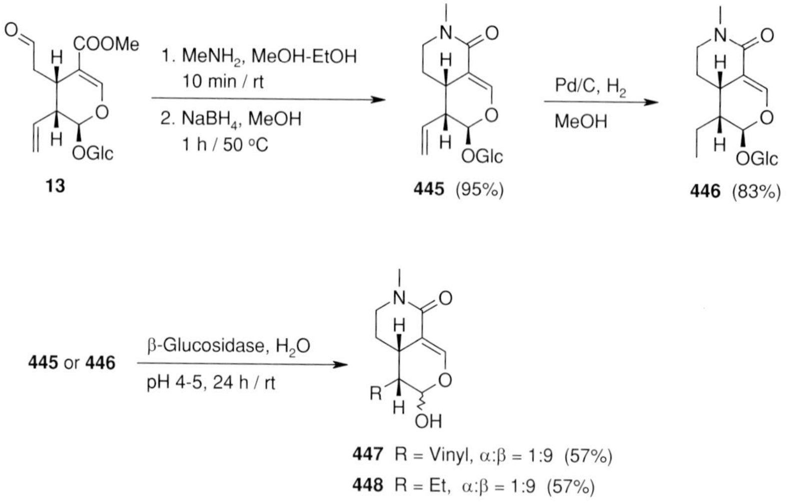

Scheme 73. Preparation of model compounds (**445** and **446**) and their reaction with β-glucosidase (*120*)

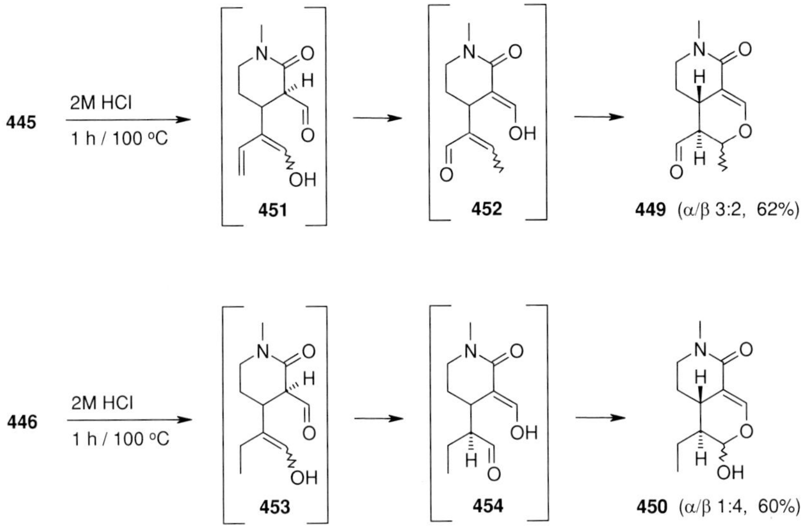

Scheme 74. Acid-catalyzed deglucosylation processes for secoiridoids **445** and **446** (*120*)

β-glucosidase afforded the expected simple aglycones **447/448** as anomer mixtures.

In contrast, deglucosylation of **445/446** in hot hydrochloric acid resulted in concurrent epimerization at C-9 for both compounds. The epimer mixture (**449**) obtained from **445** had a rearranged skeleton, whereas the aglycone product (**450**) from the dihydro derivative (**446**) was of the usual type (Scheme 74) (*120*).

Transformation of 3-substituted 3,4-dihydrosecologanin derivatives into trioxadamantane compounds under acidic conditions has recently been reported (*121*). From secologanin dimethyl acetal (**455**) and its 8,10-dihydro derivative (**456**) two types of substrates were prepared (Scheme 75). Treatment of **455/456** with methanol under alkaline conditions led to the *Michael*-type introduction of a methoxy group at C-3 and partial hydrolysis of the 11-methyl ester. After methylation, 3-methoxy derivatives **457/458** were obtained. Similarly, when **456** was exposed to methanolic bromine the corresponding 4-bromo-3-methoxy compound **459** was produced.

For the acid-catalyzed deglucosylation of these substrates, the water-toluene two-phase system containing a catalytic amount of hydrochloric acid proved to be most efficient (Scheme 76). In the case of **457/458** inseparable mixtures of trioxadamantane epimers (**462/463** and **464/465**, respectively) were obtained. The stereochemistry of these products was

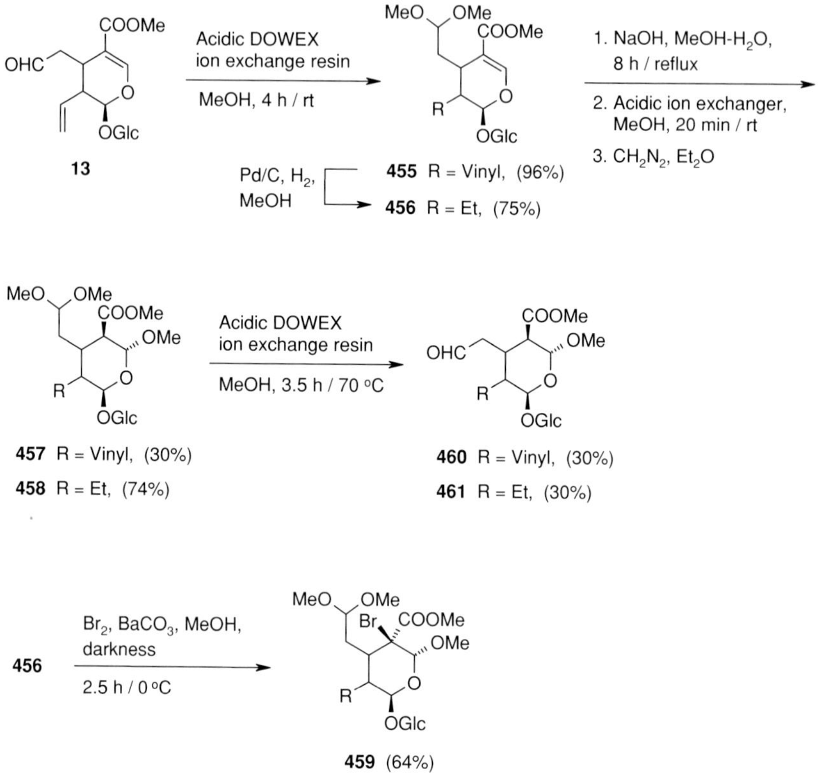

Scheme 75. Preparation of substrates for the formation of trioxadamantane compounds (*121*)

deduced from NMR data and further support for these structure assignments came from molecular mechanics calculations for the four possible 3,4-stereoisomers.

In the mechanism proposed (Scheme 77) for formation of *e.g.* **462/463** epimerization at C-4 occurs *via* an elimination-addition process. In contrast, only a single deglucosylation product (**466**) was obtained from 4-bromo-3-methoxy compound **459** in agreement with the exclusion of C-4 epimerization due to the presence of the bromo substituent. The reaction cascade most probably starts with hydrolysis of the dimethyl acetal at C-7, since this was shown to be removable under slightly milder conditions (*i.e.* to give **460**) (Scheme 75). Next, epimerization *via* **467** (to give **468**), subsequent formation of bicyclic hemiacetal **469**, and deglucosylation by transacetalization would lead to formation of tricyclic acetals **462/463**.

References, pp. 106–114

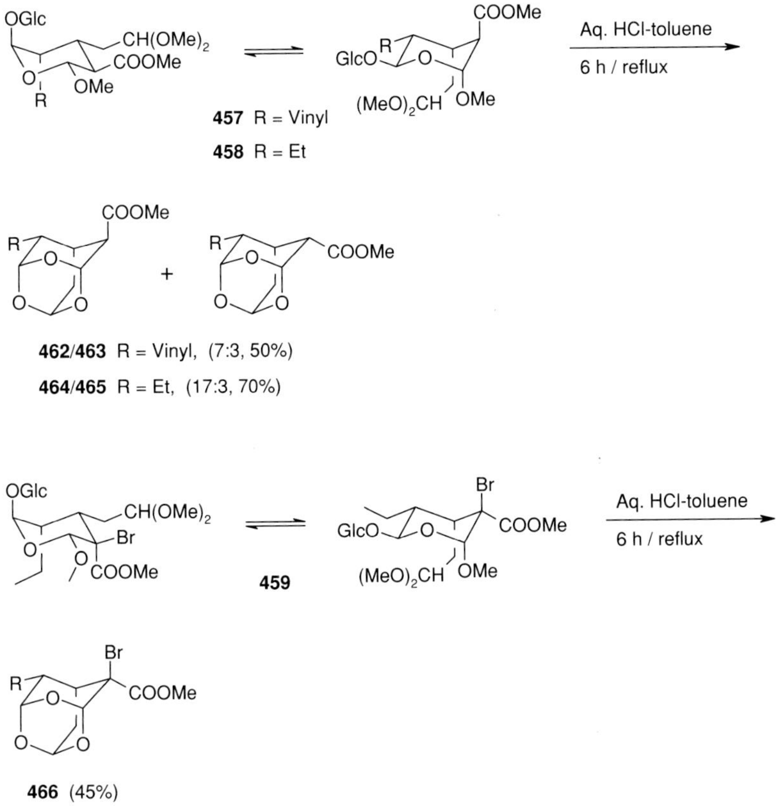

Synthetic Aspects of Iridoid Chemistry 81

457 R = Vinyl
458 R = Et

462/463 R = Vinyl, (7:3, 50%)
464/465 R = Et, (17:3, 70%)

459

466 (45%)

Scheme 76. Preparation of secoiridoid derived trioxadamantane compounds (*121*)

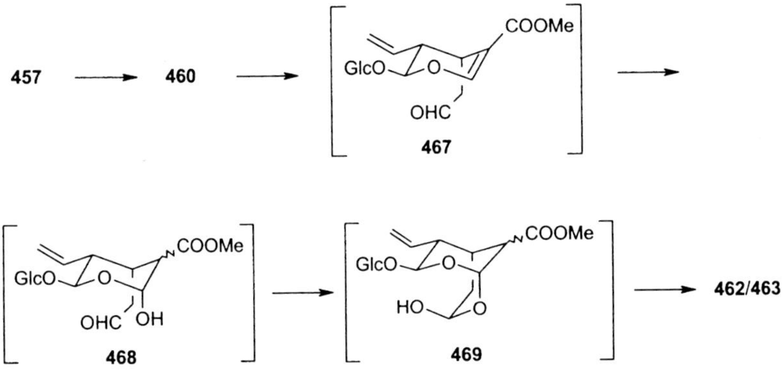

457 → 460 → [**467**] →

[**468**] → [**469**] → **462/463**

Scheme 77. Possible cascade leading to trioxadamantane compounds **462/463** (*121*)

6.3. Preparation of Marine Diterpenoids

A few representatives of a subgroup of biologically active marine diterpenoids isolated from algae have been the synthetic targets in the work of Isoe *et al.* (*9, 122–125*). These compounds all feature an iridoid-like framework linked to a geranyl side chain as seen in (−)-petiodial (**470**), udoteatrial (**471**), and halitunal (**472**). Genipin (**29**) was considered a suitable iridoid starting material for semi-synthetic preparations of such diterpenoids as it already possessed many of the essential functionalities.

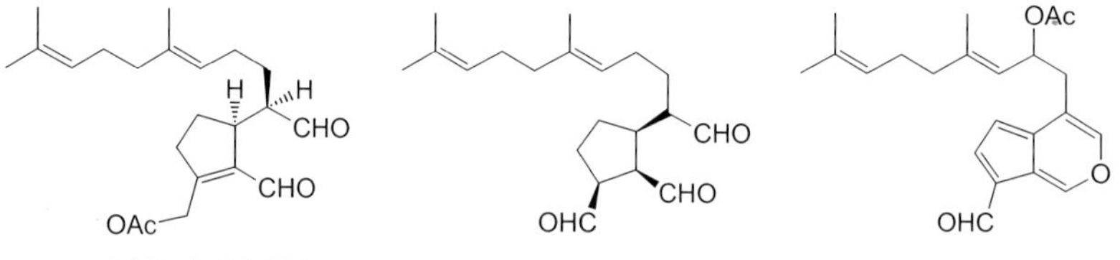

Chart 16. Marine diterpenoids synthesized from genipin (**29**) (*9, 122–125*)

The geranyl side chain had to be introduced at the 11-position of the iridoid skeleton, and **473** (obtained from **29** *via* **178**) proved an appropriate intermediate for this purpose in the partial synthesis of optically active petiodial (*9, 122*) (Scheme 78). The corresponding mesylate was generated *in situ* at low temperature where it was allowed to react with lithiated geranyl tolyl sulfone. The resulting diterpenoid **474** was selectively desilylated and the sulfone functionality was removed to give **475**. Acetylation and subsequent rapid treatment with TBAF interchanged the protection mode of the hydroxyl groups. Finally, the position of the cyclopentanoid double bond in hemiacetal **476** was isomerized to the requisite 8,9-position. The olefin transposition was accompanied by opening of the dihydropyran ring, and thus afforded separable diastereomers **477** and **478**. The natural (−)-petiodial (**470**) and the synthetic **477** were identical with respect to their NMR data, but they exhibited opposite optical rotations and thus constituted an enantiomeric pair (*122*).

In the synthesis of the hydrate of udoteatrial (**471**), the tricyclic *exo*-methylene lactone **479** was envisaged as the key intermediate onto which the geranyl side might be introduced by a *Michael*-type addition (*9, 123, 124*) (Scheme 79). Selective oxidation of the allylic hydroxyl group in genipin (**29**) afforded aldehyde **480**, which by hydrogenation with

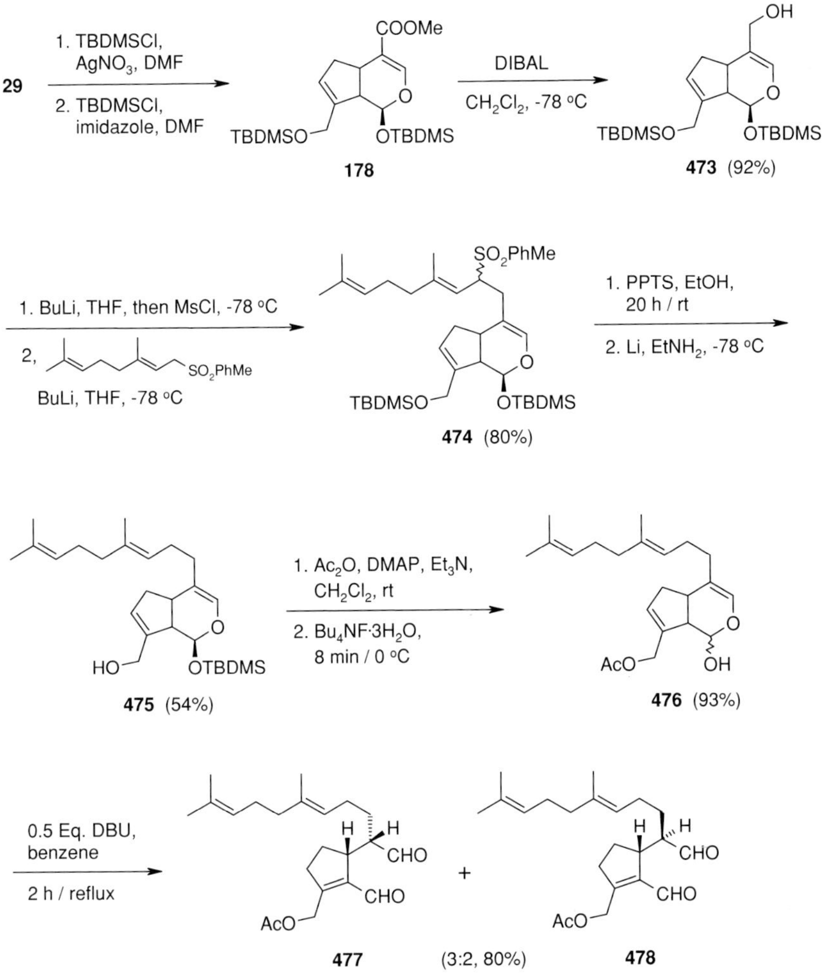

Scheme 78. Synthesis of (+)-petiodial (**477**) from genipin (**29**) (*9, 122*)

rhodium on alumina as catalyst gave tricyclic hemiacetal **481** as the major product. Conversion of **481** into methyl acetal **483** followed by reduction and acetylation gave acetate **484**. Bromohydrin formation with *N*-bromosuccinimide (NBS) in the presence of water and subsequent Swern oxidation of the resulting hemiacetal gave α-bromolactone **485**, which was subjected to reductive elimination with zinc in acetic acid.

The α,β-unsaturated lactone (**479**) obtained in this manner was treated with lithiated geranyl tolyl sulfone to give the 1,4-adduct (**486**)

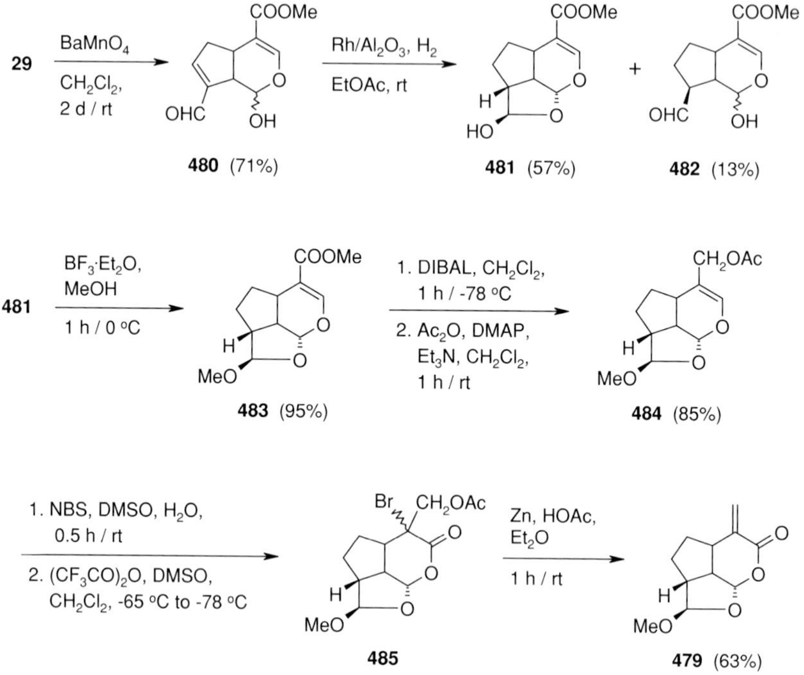

Scheme 79. Preparation of tricyclic intermediate **479** (*9*, *123*, *124*)

(Scheme 80). Since removal of the sulfone was unsuccessful at this stage, the lactone was temporarily reduced and protected as *O*-silyl acetal **487**. Birch reduction of the sulfone functionality followed by desilylation and oxidation to recover the lactone moiety afforded a separable diastereomeric mixture **488/489**. The major product (**488**) was isomerized under base-catalysis to give additional amounts of the desired product (**489**). Sequential reduction of the lactone and hydrolysis of the methyl acetal completed the synthesis of optically active udoteatrial hydrate (**490**). However, comparison of the physical data for the corresponding acetates with those of the acetylated natural compound showed the synthetic **490** to be the unnatural antipode.

The synthesis of the marine diterpenoid halitunal (**472**) has also been achieved with genipin (**29**) as starting material (*125*) (Schemes 81 and 82). Halitunal (**472**) consists of a geranyl moiety linked to an iridoid-like part, namely a 10π-aromatic cyclopentadieno[*c*]pyran ring system similar to that seen in *e.g.* cerbinal (**34**) and baldrinal (**149**) (Section 4.5.). Although the final target (**472**) only contains one stereogenic

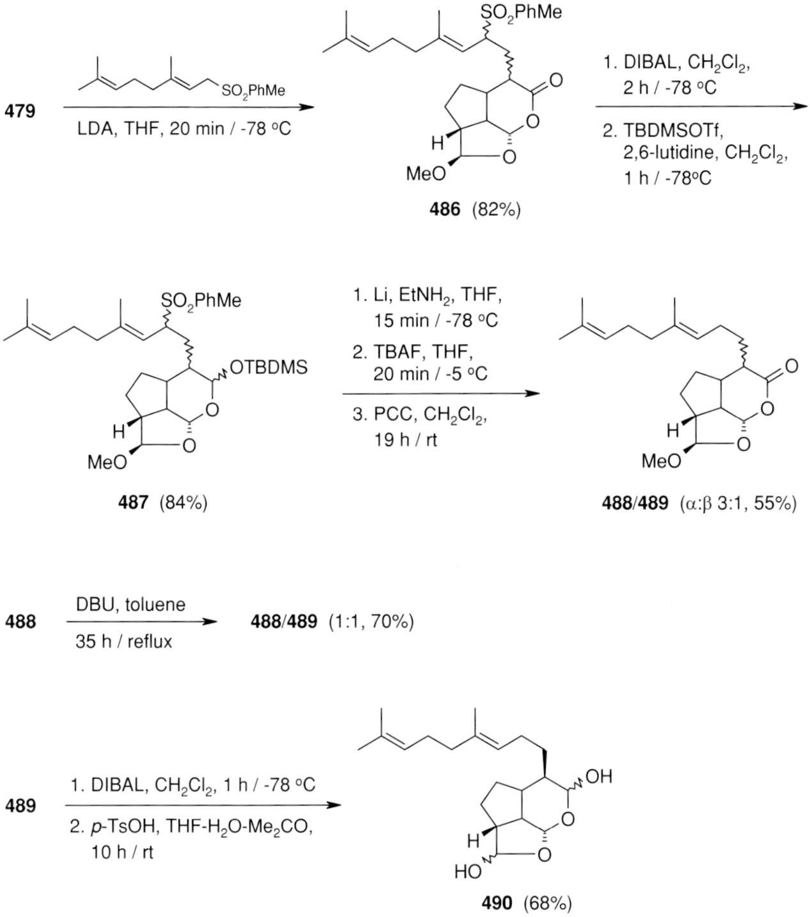

Scheme 80. Synthesis of *ent*-udoteatrial hydrate (**490**) (*9, 123, 124*)

carbon in the side chain (*i.e.* C-12), the chirality of genipin (**29**) was employed as a means of ensuring the diastereomeric purity of the intermediary 12-hydroxylated compounds. Again, the geranyl moiety was introduced on the intermediary disilyl ether **473** (Scheme 78) *via* its mesylate, which was treated with the carbanion of a geranial derived cyano ethoxyethyl ether (**491**) to give a 1:1-mixture of diastereomers (**492**). Acid-catalyzed dehydrocyanation with concomitant selective removal of the primary silyl group was followed by 1,2-reduction of the resulting ketone. This afforded the separable diastereomeric allylic alcohols **493** and **494** (Scheme 81). Each isomer of the pair **493/494** was

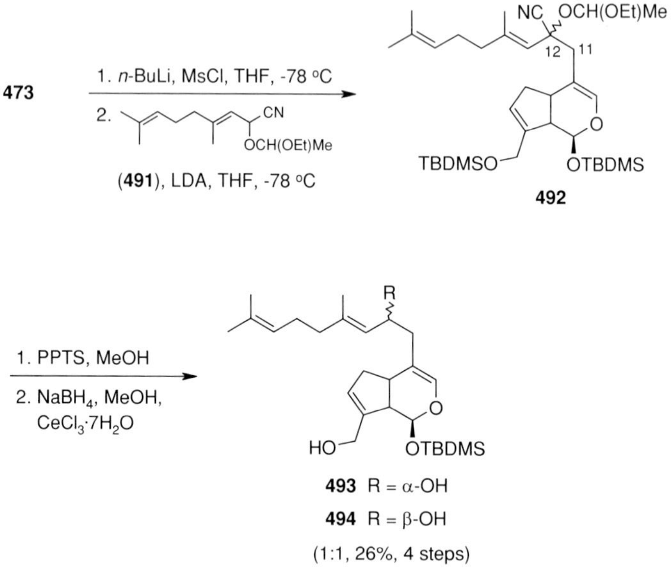

Scheme 81. Preparation of the separable diastereomeric allylic alcohols **493** and **494** (*125*)

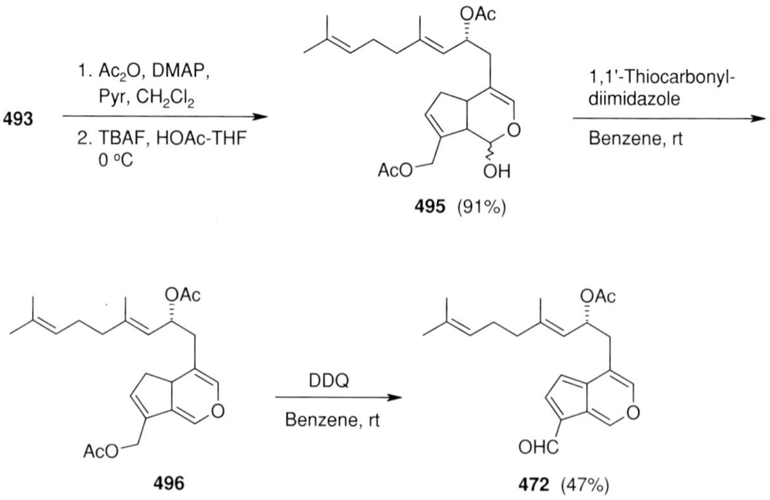

Scheme 82. Synthesis of optically active halitunal (**472**) from intermediate **493** (*125*)

converted into optically pure enantiomers of halitunal (**472**) by successive dehydration and oxidation (Section 4.5.) (Scheme 82). The natural and synthetic halitunal had identical NMR data, but the optical rotation was not available for the natural compound so the assignment of its absolute configuration still remains unsolved.

6.4. Building Blocks for Other Types of Cyclopentanoids

Utilization of iridoid glucosides as chiral synthons might lead to several classes of biologically interesting cyclopentanoid targets, but rather few iridoids have been examined with regard to this. The syntheses will be arranged primarily on the basis of the iridoid used as starting material, and subordinately on the basis of reaction type and aim. Generally it is the aglucone part of the initial glucosidic iridoid starting material that is employed as a building block in semi-synthetic work. Thus, the removal of the sugar moiety becomes a key step, for which enzymatic cleavage or acid-catalyzed hydrolysis are well-established methods as already outlined in the previous sections. A modified procedure for facilitated acid-catalyzed deglucosylation has been reported (*126*), and this involved successive periodate oxidation and borohydride reduction of the glucose moiety prior to hydrolysis. While the enzymatic method seems general but expensive, both the usual and modified acid-catalyzed hydrolyses suffer from the drawback that certain functionalities may undergo undesired reactions under these conditions. As will appear from the following examples, several other protocols have been developed for the direct conversion of iridoid glucosides into modified aglucone derivatives suitable for further synthesis.

Several types of cyclopentanoids have been prepared from catalpol (**7**). The first targets that were pursued were prostaglandin analogues (*127, 128*), but also enantiopure (−)-methyl jasmonate (*129*) and its C-9 epimer, (+)-methyl jasmonate (**497**) (*130*), have been obtained by similar synthetic strategies.

For the synthesis of **497** and other cyclopentanoids a convenient method for deglucosylation was developed (*29*) (Scheme 83). Hydrogenation of catalpol (**7**) under high pressure gave 5,7-dideoxy-3,4-dihydrocynanchoside (**498**) the 8,10-diol moiety of which was subsequently cleaved with sodium periodate. The intermediary 8-keto iridoid underwent elimination of the sugar moiety upon treatment with sodium bicarbonate to give keto enol ether **499** in excellent yield (*29*). Addition of benzylmercaptan to the enol ether afforded only one diastereomeric thioacetal (**500**) which upon mesylation gave elimination product **501**. Reduction afforded diols **502/503** the major isomer of which (**502**) was further transformed into partially protected hemiacetal **504** by benzyla-

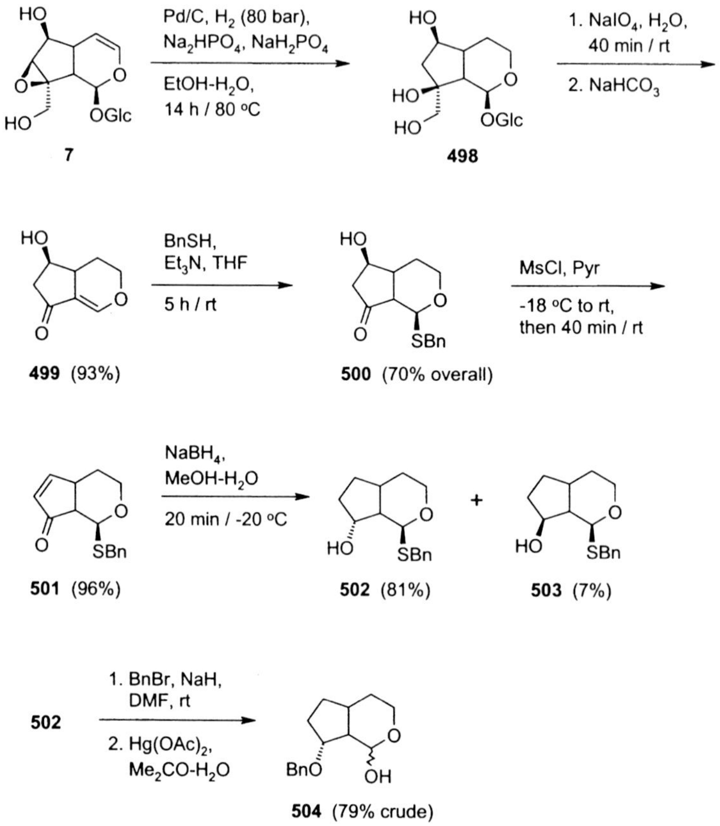

Scheme 83. Preparation of building blocks for synthesis of *e.g.* prostaglandins or methyl jasmonate (*29*, *129*)

tion and subsequent Hg(II)-catalyzed hydrolysis of the thioacetal (*29*, *129*) (Scheme 83).

A *Wittig* reaction between the ylide derived from methoxymethylene-triphenylphosphonium chloride and the latent aldehyde in hemiacetal **504** afforded *E/Z*-isomeric enol ether **505** (Scheme 84). The masked aldehyde in **505** was released by acidic hydrolysis, and then *via* another *Wittig* reaction with the ylide of propyltriphenylphosphonium chloride alcohol **506** was obtained. Debenzylation and oxidation with pyridinium dichromate (PDC) afforded an acid (**507**) which immediately was methylated to give (+)-methyl jasmonate (**497**) (*130*).

In addition, synthetic efforts have been made to realize the conversion of catalpol (**7**) into the structurally even more complex

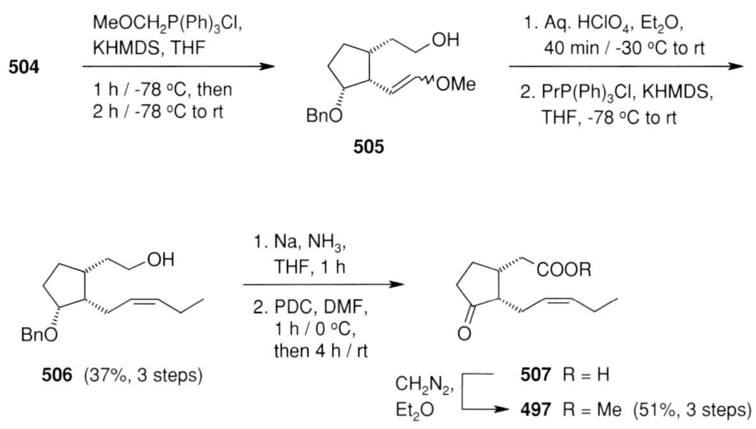

Scheme 84. Synthesis of (+)-methyl jasmonate (497) (130)

504 → (MeOCH$_2$P(Ph)$_3$Cl, KHMDS, THF; 1 h / -78 °C, then 2 h / -78 °C to rt) → **505** → (1. Aq. HClO$_4$, Et$_2$O, 40 min / -30 °C to rt; 2. PrP(Ph)$_3$Cl, KHMDS, THF, -78 °C to rt) → **506** (37%, 3 steps) → (1. Na, NH$_3$, THF, 1 h; 2. PDC, DMF, 1 h / 0 °C, then 4 h / rt) → **507** R = H → (CH$_2$N$_2$, Et$_2$O) → **497** R = Me (51%, 3 steps)

triquinane sesquiterpenes (*131–133*). In some preliminary work the Ramberg-Bäcklund ring-contraction was studied on bicyclic thioethers obtained from decarboxylated iridoid aglucone derivatives (*134*), and this reaction was employed as the crucial step in the semi-synthesis of (−)-hypnophilin (**508**) from the versatile keto enol ether **499** (*133*). Thus, silylation of **499** (to **509**) followed by 1,4-addition of lithium dimethylcuprate afforded a mixture of 9-epimers (**510**) that was subjected to elimination to give primarily *cis*-fused α,β-unsaturated ketone **511**. The hydroxy ether **512** obtained after 1,4-reduction of **511** was then protected and oxidized with *in situ* formed ruthenium(VIII) oxide to yield lactone **513**. The lactone moiety was reduced to give diol **514** which by mesylation and subsequent treatment with sodium sulfide was converted into cyclic sulfide **515**. Stepwise oxidation to the sulfoxide (**516**), α-chlorination, and oxidation to give a mixture of α-chlorosulfones was followed by Ramberg-Bäcklund ring contraction to yield bicyclo[3.3.0]octene **517**. The conversion of building block **517** into (−)-hypnophilin (**508**) required a further 16 steps (*133*).

Another approach towards aglucone building blocks was employed in the semi-synthesis of (+)-cyclosarkomycin (**518**), a stable precursor of the fungal metabolite (+)-sarkomycin (**519**) (*135*) (Scheme 86). Oxidation of the enol ether functionality in catalpol hexaacetate (**7a**) with *in situ* formed peroxyformic acid afforded an isomer mixture of 3-formyloxy-4-hydroxy derivatives (**520**). Reduction with lithium aluminum hydride was accompanied by ring-opening and deglucosylation to yield a protected polyol (**521**) after acetylation. Upon deacetyl-

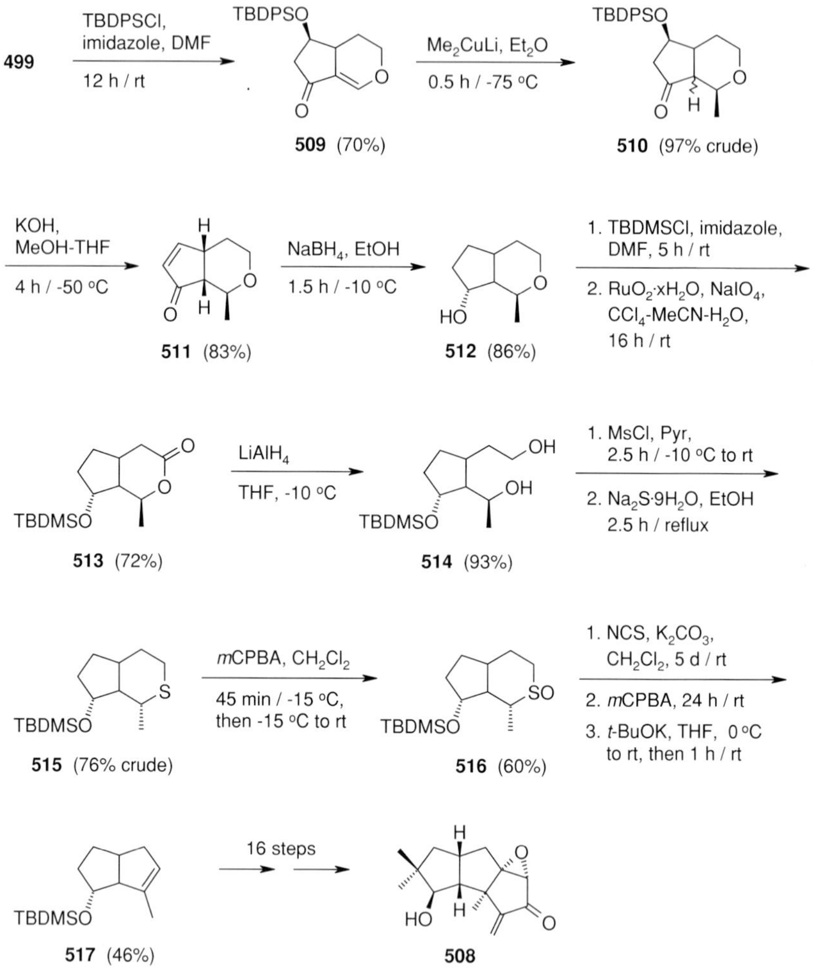

Scheme 85. Synthesis towards (−)-hypnophilin (**508**) (*133*)

ation, cleavage of the two vicinal diol systems accomplished the removal of both C-3 and C-10 to give the intermediary hemiacetal **522**. Conversion into the corresponding cyclic methyl acetal followed by dehydration *via* its tosylate led to α,β-unsaturated ketone **523**, which was subjected to successive hydrogenation and acetal hydrolysis. Finally, hemiacetal **524** was oxidized to yield (+)-cyclosarkomycin (**518**) (*135*).

Lastly, the naturally occurring 6-*O*-vanilloylcatalpol (**525**) has been transformed (*136*) into an analogue of (−)-specionin (**31**) which is a

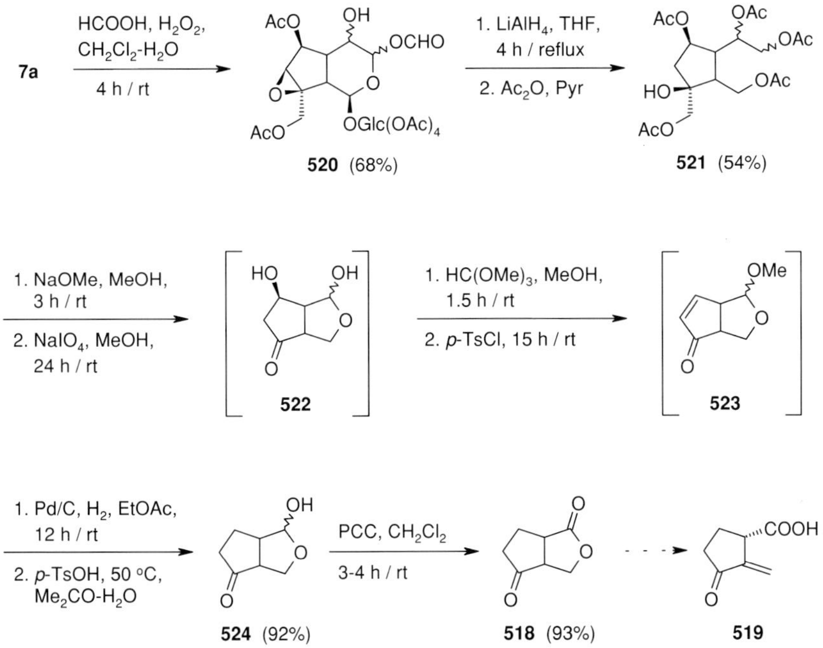

Scheme 86. Semi-synthesis of (+)-cyclosarkomycin (**518**) (*135*)

spruce budworm antifeedant substance already prepared from catalpol (**7**) (*137*) and from aucubin (**6**) (*60*) (Section 4.5.). The analogous, (−)-3′-methoxyspecionin (**526**) was prepared from hexaacetate **525a** in a two-step procedure (*136*) (Scheme 87). Hydroxymercuration of the enol ether functionality followed by reduction with zinc dust produced aglucone **527** which upon treatment with ethanolic triethyl orthoformate in the presence of trifluoroacetic acid and copper(II) bromide gave diacetate **528**. When the Cu(II)-catalyst was omitted addition of ethanol to the 3,4-enol ether did not take place, and only ethyl acetal **529** was formed. Furthermore, **529** could not be converted into **528** by subsequent treatment with ethanolic copper(II) bromide, but this transformation could be realized by successive ethoxy-mercuration and borohydride reduction. Finally, (−)-3′-methoxyspecionin (**526**) was obtained after a selective partial deacylation of **528**. The present protocol for the preparation of alkyl diacetalic aglucone derivatives appears to be generally applicable to decarboxylated iridoid peracetates as for example aucubin hexaacetate (**6a**) and catalpol hexaacetate (**7a**) gave similar results (*136*).

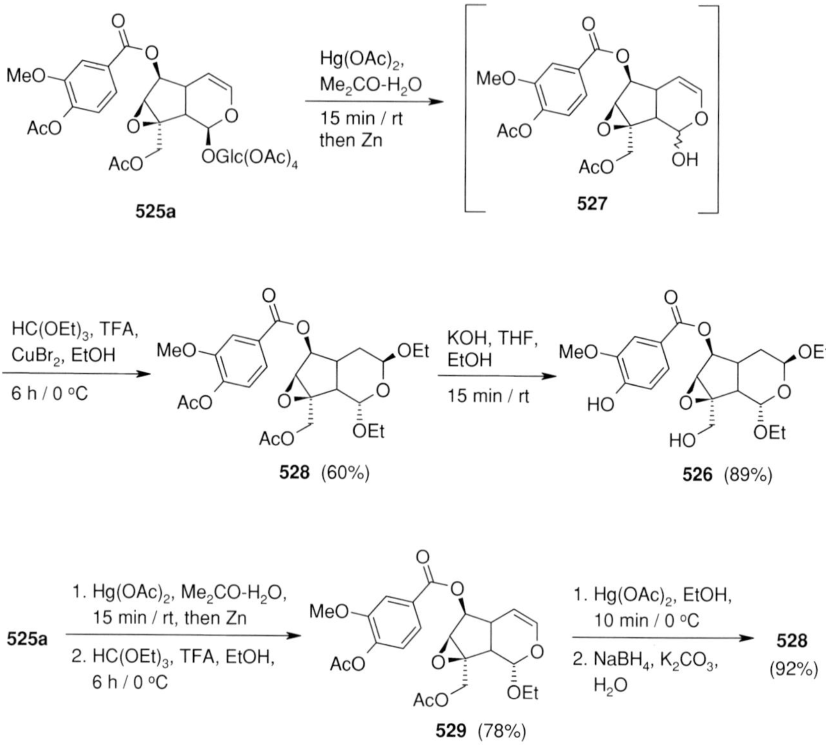

Scheme 87. Synthesis of (−)-3′-methoxyspecionin (**526**) from 6-*O*-vanilloylcatalpol hexaacetate (**525a**) (*136*)

Syntheses in which aucubin (**6**) has served as starting material have also dominated the literature on iridoids, especially in early iridoid chemistry as reviewed by BIANCO (*8*). Here only some major trends will be cited. A major topic has been the conversion of aucubin (**6**) into prostaglandins (*138–144*), and with this aim a number of intermediates analogous to Corey's lactone have been prepared from **6** (*27, 144–148*). Contemporary work includes palladium-catalyzed substitution on aucubin hexaacetate (**6a**) (*28*), and tranformation of **6** into a carbocyclic nucleoside analogue (*149*).

Although there are two allylic acetoxy groups in aucubin hexaacetate (**6a**), only the sterically less hindered 10-position was susceptible to replacement with nitrogen- and carbanion-nucleophiles under π-palladium-allyl complex formation conditions (Scheme 88) (*28*). Thus, both the piperidine and benzylamine derivatives (**530** and **531**) were obtained in good yield, while the reaction with the sodium salt of

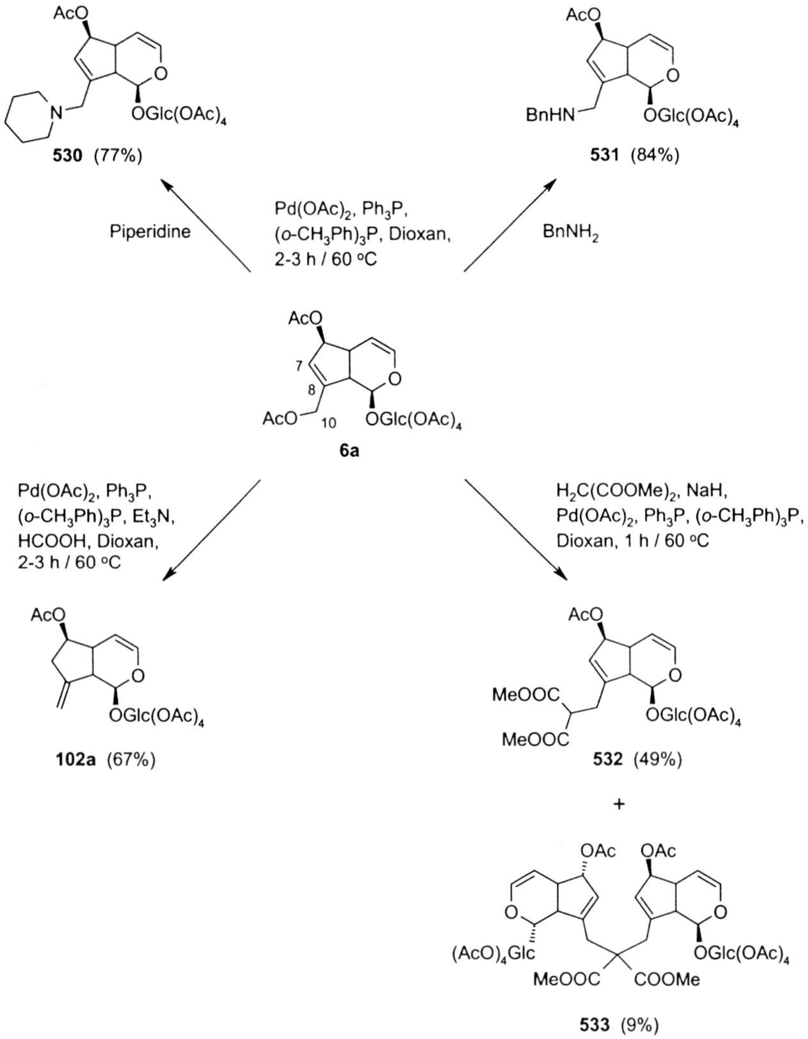

Scheme 88. Palladium-catalyzed substitutions on aucubin hexaacetate (**6a**) (28)

dimethyl malonate afforded a mixture of mono- and disubstituted malonates (*i.e.* **532** and **533**). Conversely, if ammonium formate was added as hydride donor nucleophilic attack occurred at the 7-position to give linarioloside pentaacetate (**102a**) as the main product. This is in contrast to usual transfer hydrogenolysis of **6a** with Pd/C-formic acid where **102a** is observed only as a minor product (Section 4.3.).

As a result of studies aiming at an inexpensive and efficient procedure for preparing cyclopentanoid building blocks from aucubin (**6**), a chemoselective hydroxymercuration-demercuration method has emerged (*150, 151*). First it was found that the hindered 7,8-double bond was inert towards treatment with one equivalent of mercury(II) acetate whereas the 3,4-enol ether was readily attacked to give an intermediate organomercurial (**534**) (Scheme 89). Reductive displacement of the mercury and successive reductions of the hemiacetal at C-3 and of the ensuing hemiacetal at C-1 (with loss of the glucose moiety) were performed with sodium borohydride (*150*). When these steps were performed in water-THF at room temperature polyols **535** and **536** were obtained as a 4:1 mixture. Acid-catalyzed cyclization afforded bicyclic ethers **537** and **538** in the same ratio. However, under weakly acidic

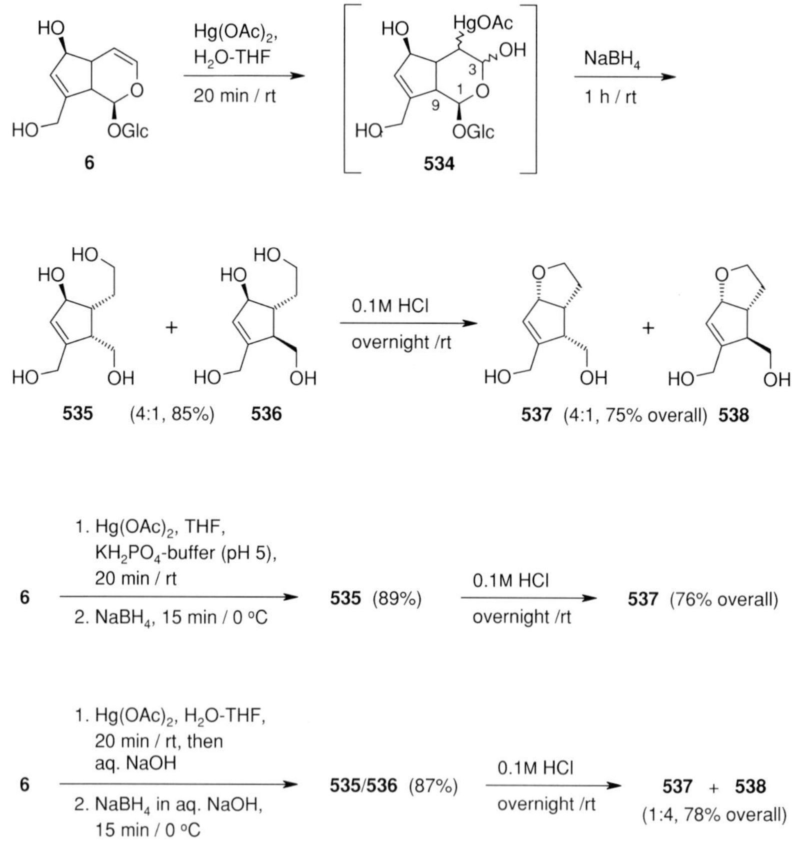

Scheme 89. Chemoselective hydroxymercuration-demercuration of aucubin (**6**) (*150*)

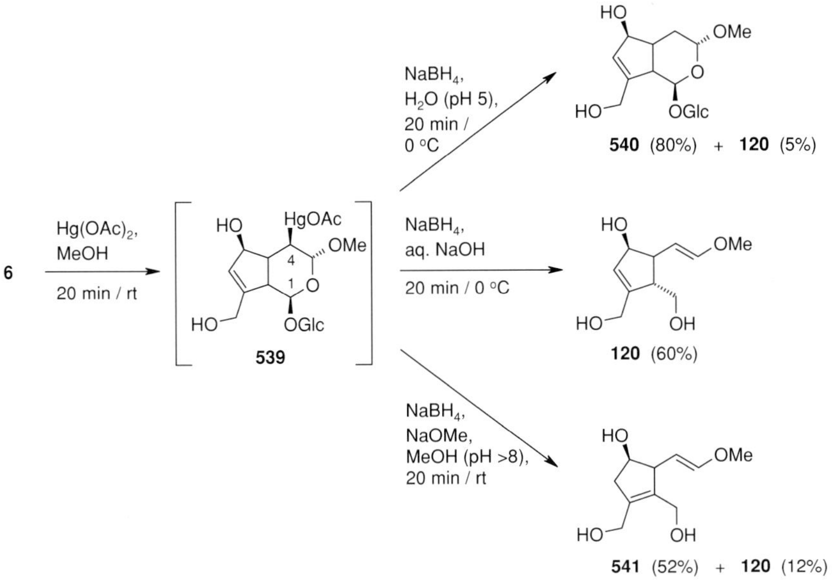

Scheme 90. Methoxymercuration-demercuration reactions of aucubin (**6**) (*152*)

conditions demercuration led exclusively to **535** with retained stereochemistry at C-9 due to a higher reduction rate. Conversely, the reductive demercuration proceeded so slowly in basic medium that extensive epimerization occurred at C-9 to give **536** as the predominant product (*150*).

When the solvent was changed to methanol this led to analogous methoxymercuration-demercuration reactions of aucubin (**6**) (*152*) (Scheme 90). However, in this case a mixture of three compounds was obtained under weakly alkaline conditions, but as discussed above the outcome of the reaction was shown to depend strongly on the pH and the composition of the solvent. Optimal conditions for selective formation of each product were found. Thus, under slightly acidic conditions reduction of the intermediate organomercurial **539** led primarily to the expected methyl acetal (**540**) possibly due to rapid protonation of an intermediate carbanion (at C-4). Alternatively, demercuration in alkaline medium resulted in ring opening and reduction of the intermediate aldehyde (at C-1) led to methyl vinyl ether **120**. By contrast, if the reaction was performed under anhydrous basic conditions, partial transpositioning of the 7,8-double bond occurred prior to reduction of the aldehyde at C-1. Hence, methyl vinyl ether **541** was obtained

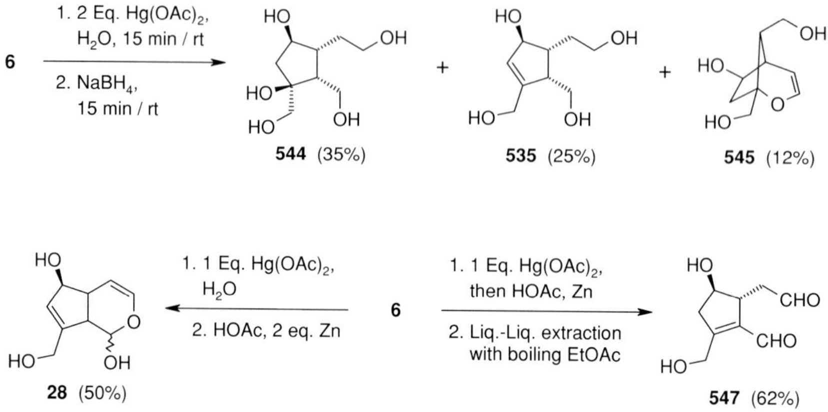

Scheme 91. Hydroxymercuration-demercuration reactions of **6** with one and two equivalents of Hg(OAc)$_2$ (*151*)

together with a smaller amount of **120** (*152*). A similar transformation was encountered in Section 4.5., where a thallium(III) mediated ring-opening of **6** and subsequent borohydride reduction allowed preparation of **120** (*56*).

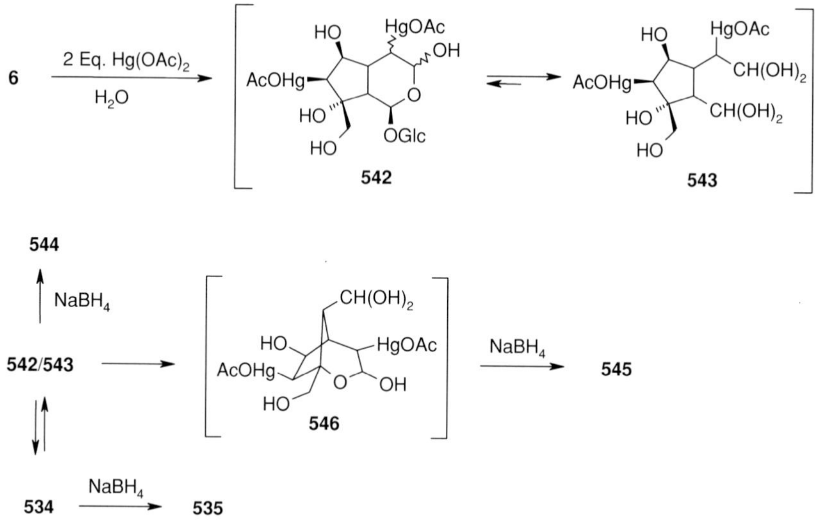

Scheme 92. Possible mechanism for hydroxymercuration-demercuration of **6** with 2 eq. Hg(OAc)$_2$ (*151*)

Synthetic Aspects of Iridoid Chemistry

Hydroxymercuration-demercuration reactions of **6** using both one and two equivalents of mercury(II) acetate have been investigated in more detail (*151*) (Schemes 91 and 92). It proved possible to detect both mono- and dimercurial intermediates (**534** and **542/543**) by ^{1}H NMR spectroscopy, and it was found that the open forms (*e.g.* **543**) of the mercurial intermediates predominated (Scheme 92).

Surprisingly, demercuration of **542/543** afforded three products: (i) the expected polyol **544** formed by straightforward reduction, (ii) isoeucommiol (**535**) corresponding to reduction of **534** which indicated a partial reversal of the hydroxymercuration at the 7,8-double bond, and (iii) a bicyclic enol ether (**545**) presumably formed by reduction of dimercurial **546** produced by internal acetalization in **543** (*151*). This reaction was also exploited for the non-enzymatic formation of

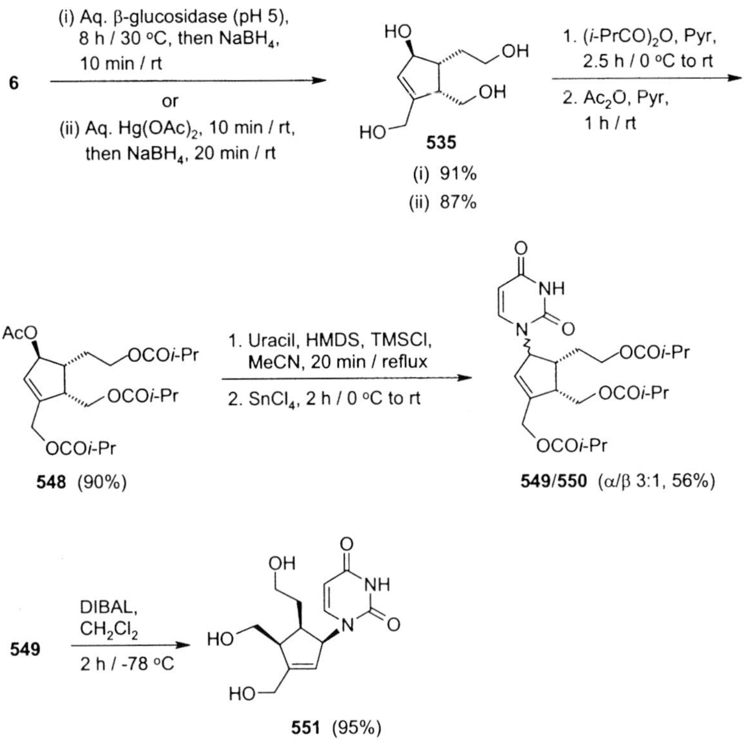

Scheme 93. Preparation of a carbocyclic nucleoside analogue (**551**) from aucubin (**6**) (*149*)

aucubigenin (**28**) and eucommial (**547**). Demercuration of monomercurial intermediate **534** carried out with zinc-acetic acid led to loss of the glucose moiety but with conservation of the masked 1,5-dialdehydic system. Attempted isolation of the resulting aucubigenin (**28**) by extraction with hot ethyl acetate led to eucommial (**547**) *via* transpositioning of the 7,8-double bond to the more favourable α,β-position.

Recently, a novel group of cyclopentanoid targets, namely the carbocyclic nucleoside analogues, has been added to the diverse range of biological interesting cyclopentanoids that are obtainable from iridoids (*149, 153*).

Polyol **535** was obtained from **6** either by the hydroxymercuration-demercuration method discussed above (*150, 151*) or by enzymatic hydrolysis and reduction. Selective protection of the three primary hydroxyl groups was achieved by acylation with isobutyric anhydride; subsequent acetylation of the remaining secondary alcohol afforded allylic acetate **548** in high yield (*149*). The nucleoside base, in the present case uracil, was introduced by use of the *Vorbrüggen* procedure. This afforded a mixture of epimers **549/550** from which the major compound (**549**) was separated. Deprotection of the isobutyric ester moieties yielded aucubovir (**551**). The somewhat unexpected stereochemical outcome of this reaction was rationalized by assuming that a cyclic acyloxonium ion (**552**) might be an intermediate in this reaction. Structurally, **551** resembles both the active compound carbovir (**553**) (*154*) and the inactive analogues **554/555** (*155*), but so far no reports regarding its possible antiviral activity have appeared.

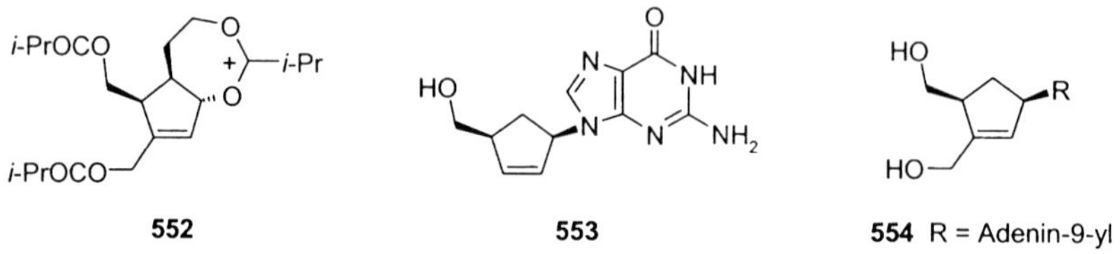

Chart 17. Possible intermediate cyclic acyloxonium ion (**552**). Nucleoside analogues related to **551**

Most methods for deglucosylation of iridoids encountered so far have been concerned with unprotected iridoid glucosides. The very polar aglucone products thus obtained pose both the problem of difficult

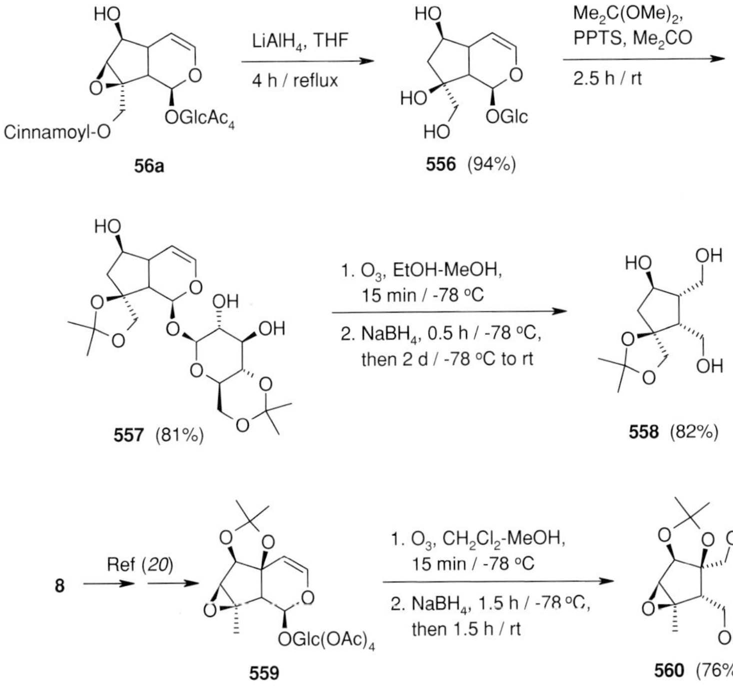

Scheme 94. Ozonolysis of partially and fully protected iridoid glucosides (21)

isolation/purification and the necessity of development of selective protection schemes to enable further synthesis. Moreover, the cyclopentanoid building blocks prepared from carbocyclic iridoids generally contain a C_2-chain as well as two C_1-substituents. The only exception to this trend has been the previously discussed peracid-periodate sequence (135). In order to be able to produce partially protected cyclopentanoid building blocks with only one-carbon side chains, a versatile methodology involving ozonolysis of partially and fully protected iridoid glucosides was developed (21) (Scheme 94).

The cinnamoyl ester scutellarioside I (**56**) was acetylated to **56a** and then reduced to 5,7-dideoxycynanchoside (**556**). Formation of diacetonide **557** proceeded smoothly; the subsequent ozonolysis carried out in a methanol-ethanol mixture at low temperature gave full conversion into a primary ozonolysis product. Immediate sodium borohydride reduction of this afforded a more polar product which proved to be triol **558**. Similarly, the fully protected antirrhinoside derivative **559** was subjected to ozonolysis in dichloromethane-methanol, which produced an

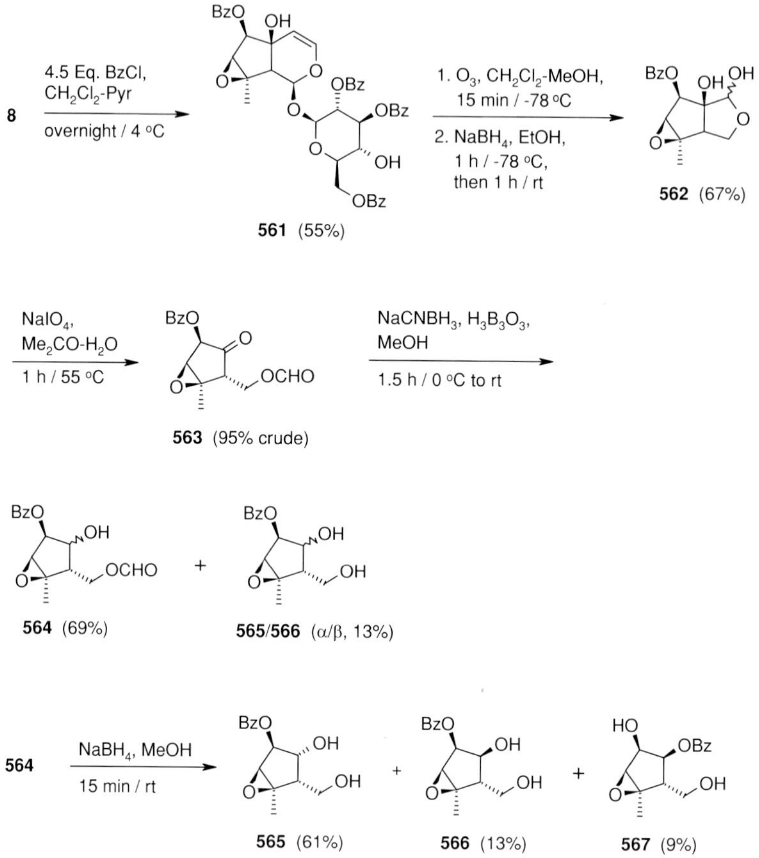

Scheme 95. Ozonolysis of antirrhinoside tetrabenzoate (**561**) (*21*)

analogous diol (**560**). Next a crystalline acylated derivative (**561**) of antirrhinoside (**8**) was obtained by partial benzoylation. In this case the reduction of the initial ozonolysis product was performed at a lower temperature and with a slightly shorter reaction time (Scheme 95). This resulted in selective formation of bicyclic hemiacetal **562** in reasonable yield. Periodate oxidation of **562** gave an unstable ketone (**563**) which by reduction with sodium cyanoborohydride was converted to a mixture of epimers (**564**). The formyl group in **564** was removed by a brief treatment with sodium borohydride to give diol **565** as the main product. When sodium borohydride was employed for the reduction of the keto group more complex mixtures were obtained (*21*).

References, pp. 106–114

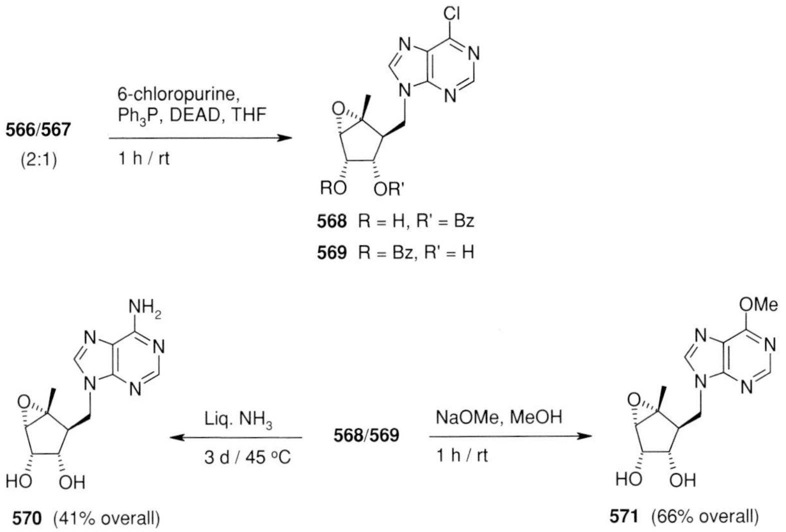

Scheme 96. Conversion of cyclopentane building blocks **566/567** into nucleoside analogues (*153*)

Further elaboration of cyclopentanes **566/567** into carbocyclic homo-*N*-nucleoside analogues was performed by coupling with 6-chloropurine under Mitsunobu conditions (*153*) (Scheme 96). Although two hydroxyl groups were left unprotected, reaction took place exclusively at the primary position to give a mixture of monobenzoates (**568/569**) that could not be separated completely from contaminant triphenylphosphine oxide. Ammonolysis and treatment with methanolic sodium methoxide afforded nucleoside analogues **570** and **571**, respectively.

In a different approach towards homo-*N*-nucleoside analogues, catalpol (**7**) was subjected to enzymatic deglucosylation, and the aglucone was subsequently reduced to afford tetrol **572** (*153*) (Scheme 97). Attempts at selective protection of **572** showed that the primary alcohol at C-1 was least reactive towards acylations. Partial benzoylation at low temperature allowed formation of tribenzoate **573**, which via the corresponding triflate was coupled with 6-iodopurine tetrabutylammonium salt (*153*). Protected nucleoside analogue **574** was obtained in good yield and subsequent ammonolysis afforded carbocyclic homo-*N*-nucleoside **575**.

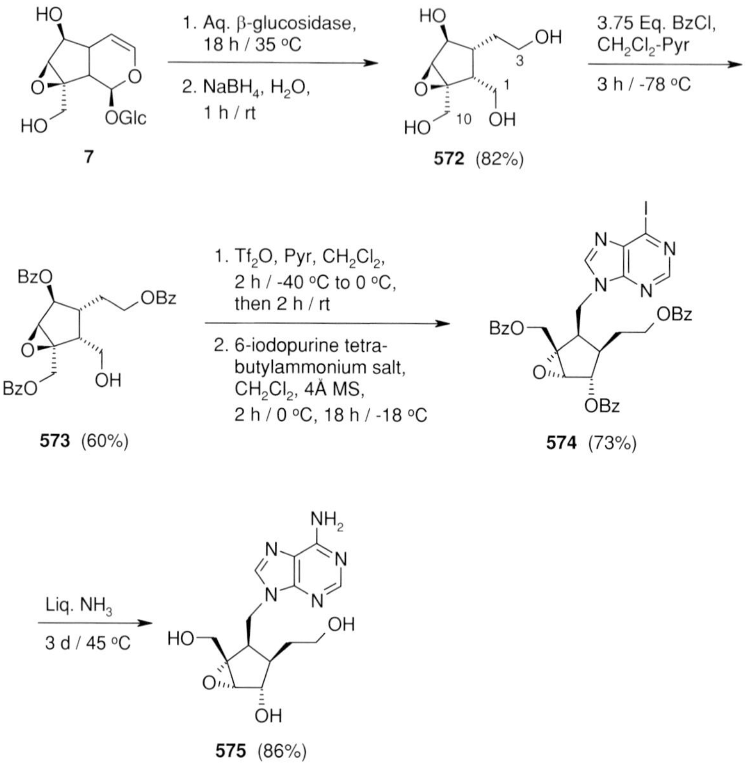

Scheme 97. Conversion of catalpol (**7**) into a homo-*N*-nucleoside analogue

The above nucleoside analogues **570** and **575** both have some structural features in common with the biologically active compounds neplanocin B and C (**576** and **577**) (*156*) and antiviral tests are in progress.

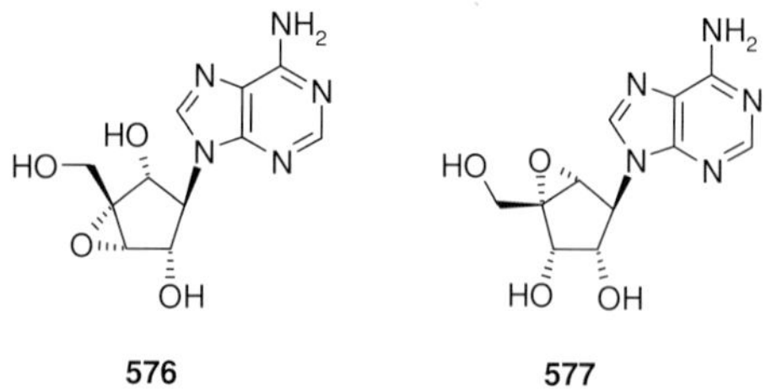

Chart 18. Neplanocin B and C (**576** and **577**)

6.5. Modifications of the Sugar Moiety in Iridoid Glucosides

In the course of a modification study of iridoid glucosides, the periodate oxidized analogues were investigated and were found to exhibit increased antitumor activity when compared with the corresponding aglucones (*157*). Ten different iridoids were subjected to oxidation with aqueous sodium periodate, and NMR spectroscopy indicated that the products had a dihemiacetalic sugar moiety (Scheme 98) (*157*).

In work concerning development of radioimmuno-assay of iridoid glucosides, three key intermediates in the biosynthesis of iridoids, namely deoxyloganin (**85**), 8-*epi*-deoxyloganin (**84**) and deoxygenipo-

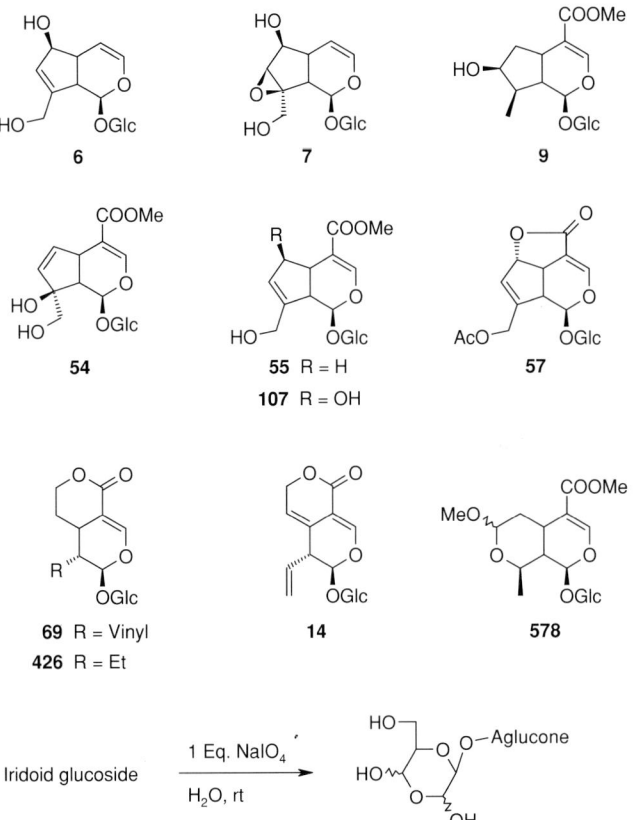

Scheme 98. Periodate oxidation of iridoid glycosides (*157*)

side (**74**) were prepared in optically pure form (Section 4.3.) (*45*). To permit their conversion into haptens and protein conjugates, these iridoids were modified in the glucose moiety (*45*). Oxidation of the 6′-position in **84/85** to give the corresponding acids **579/580** was carried out with oxygen by using platinum as catalyst (from prehydrogenated PtO_2). But this method failed in the case of **74**. Instead, **74** was selectively tritylated at the 6′-position using trityl pyridinium tetrafluoroborate, and after acetylation **581** was obtained in almost quantitative yield. Removal of the 6′-*O*-trityl group followed by Jones oxidation afforded the acetylated acid **582**. These acid intermediates (**579**, **580** and **582**) were condensed with methyl γ-aminobutyrate (Me-GABA), and

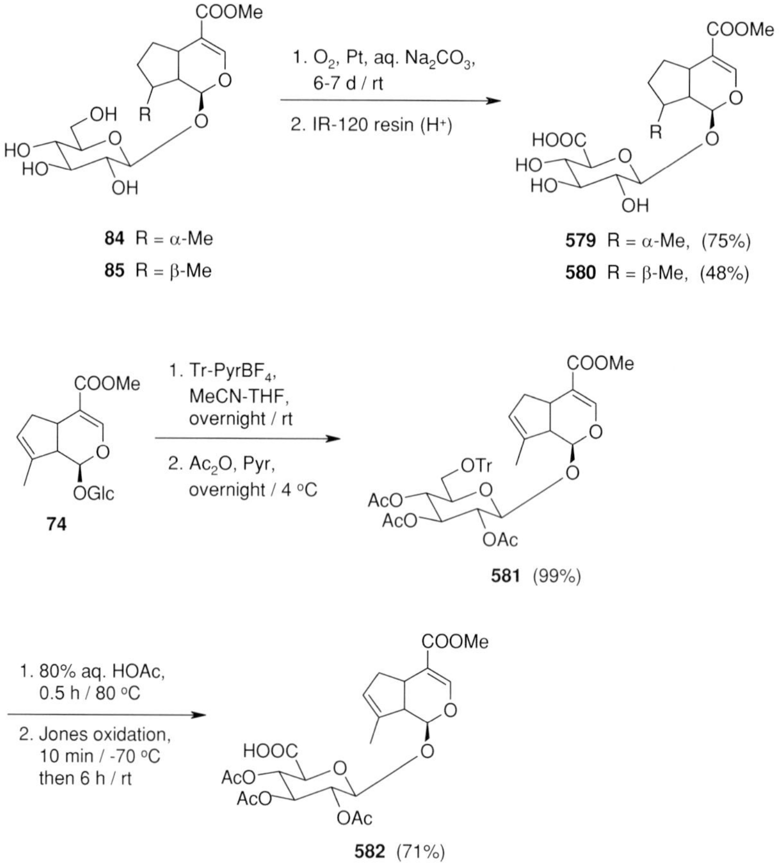

Scheme 99. Synthesis of intermediates for the preparation of haptens and protein conjugates (*45*)

after selective hydrolysis of the GABA methyl ester moiety they were attached to bovine serum albumin (BSA) (45).

A method for the preparation of 4′,5′-unsaturated iridoid glycosides has been reported (158, 159). Partially protected derivatives of loganin (**9**), ketologanin (**96**), and arbortristoside A (**583**) were obtained by selective tritylation of the 6′-position followed by acetylation and

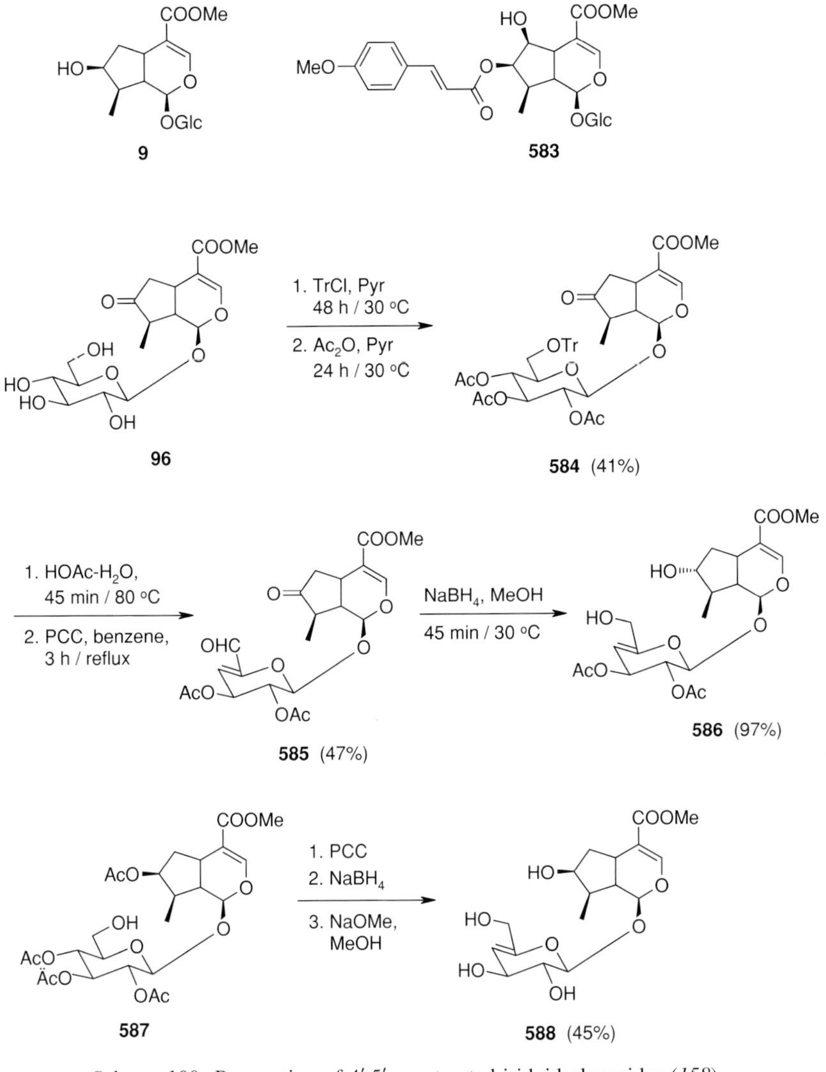

Scheme 100. Preparation of 4′,5′-unsaturated iridoid glycosides (158)

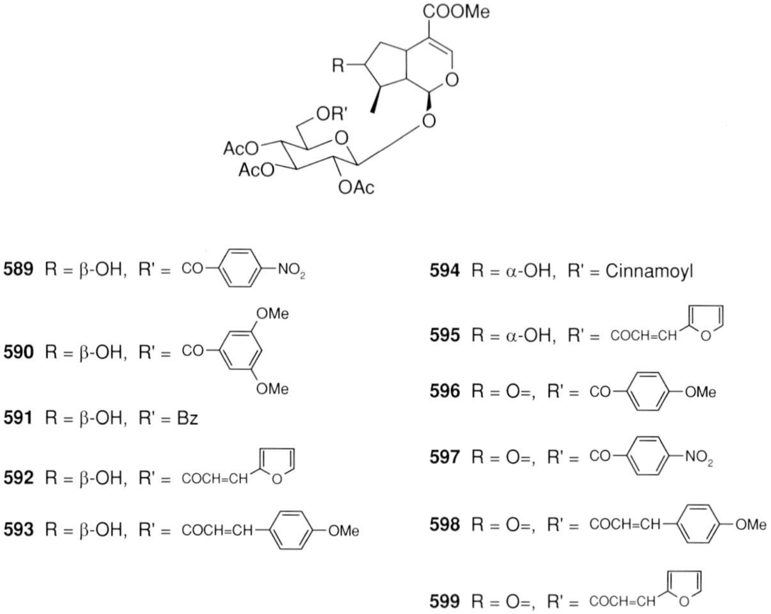

589 R = β-OH, R' = CO—⟨C₆H₄⟩—NO₂

590 R = β-OH, R' = CO—⟨C₆H₃(OMe)₂⟩ (OMe, OMe)

591 R = β-OH, R' = Bz

592 R = β-OH, R' = COCH=CH—⟨furyl⟩

593 R = β-OH, R' = COCH=CH—⟨C₆H₄⟩—OMe

594 R = α-OH, R' = Cinnamoyl

595 R = α-OH, R' = COCH=CH—⟨furyl⟩

596 R = O=, R' = CO—⟨C₆H₄⟩—OMe

597 R = O=, R' = CO—⟨C₆H₄⟩—NO₂

598 R = O=, R' = COCH=CH—⟨C₆H₄⟩—OMe

599 R = O=, R' = COCH=CH—⟨furyl⟩

Chart 19. Modification of iridoid glycosides by 6'-acylation (*159*)

hydrolysis. For example, ketologanin (**96**) was converted into **584**, which upon hydrolysis and subsequent oxidation with pyridinium chlorochromate (PCC) afforded 4',5'-unsaturated 6'-aldehyde **585**. Reduction with sodium borohydride gave 7-*epi*-loganin derivative **586** in high yield. Similarly, loganin derivative **587** was converted into the unprotected 4',5'-unsaturated analogue (**588**) in three steps (*158*). In addition, several 6'-*O*-acylated derivatives (**589–599**) were prepared from the 6'-*O*-tritylated iridoid acetates by successive hydrolysis and acylation with substituted benzoyl or cinnamoyl chlorides (*159*).

References

1. Jensen SR (1991) Plant Iridoids, their Biosynthesis and Distribution in Angiosperms. In: Harborne JB, Tomas-Barberan FA (eds) Ecological Chemistry and Biochemistry of Plant Terpenoids. Proceedings of the Phytochemical Society of Europe, vol 31. Clarendon Press, Oxford, p 133
2. Jensen SR (1992) Systematic Implications of the Distribution of Iridoids and other Chemical Compounds in the Loganiaceae and other Families of the Asteridae. Ann Missouri Bot Gard **79**: 284
3. Inouye H (1991) Iridoids. In: Dey PM, Harborne JB (eds) Methods in Plant Biochemistry, vol 7. Academic Press, New York, p 99

4. Grayer RJ, Chase MW, Simmonds MSJ (1999) A Comparison Between Chemical and Molecular Characters for the Determination of Phylogenetic Relationships among Plant Families: An Appreciation of Hegnauer's "Chemotaxonomie der Pflanzen". Biochem Syst Ecol **27**: 369
5. Ghisalberti EL (1998) Biological and Pharmacological Activity of Naturally Occurring Iridoids and Secoiridoids. Phytomedicine **5**: 147
6. Rimpler H (1991) Sequestration of Iridoids by Insects. In: Harborne JB, Tomas-Barberan FA (eds) Ecological Chemistry and Biochemistry of Plant Terpenoids. Proceedings of the Phytochemical Society of Europe, vol 31. Clarendon Press, Oxford, p 314
7. Bowers DM (1988) Chemistry and Coevolution: Iridoid Glycosides, Plants, and Herbivorous Insects. In: Spencer K (ed) Chemical Mediation of Coevolution. Academic Press, New York, p 133
8. Bianco A (1990) The Chemistry of Iridoids. In: Rahman AU (ed) Studies in Natural Products Chemistry, vol 7. Elsevier, Amsterdam, p 439
9. Isoe S (1995) Progress in the Synthesis of Iridoids and Related Natural Products. In: Rahman AU (ed) Studies in Natural Products Chemistry, vol 16. Elsevier, Amsterdam, p 289
10. Boros CA, Stermitz FR (1990) Iridoids. An Updated Review, Part I. J Nat Prod **53**: 1055
11. Boros CA, Stermitz FR (1991) Iridoids. An Updated Review, Part II. J Nat Prod **54**: 1173
12. Krull RE, Stermitz FR (1998) *Trans*-fused Iridoid Glycosides from *Penstemon mucrunatus*. Phytochemistry **49**: 2413
13. El-Naggar LJ, Beal JL (1980) Iridoids. A Review. J Nat Prod **43**: 649
14. Hegnauer R (1986) Chemotaxonomie der Pflanzen, vol. 7. Birkhäuser, Basel Boston Stuttgart, p 325
15. Junior P (1990) Recent Developments in the Isolation and Structure Elucidation of Naturally Occurring Iridoid Compounds. Planta Med **56**: 1
16. Cordell GA (1977) The Monoterpene Alkaloids. In: Manske RHF (ed) The Alkaloids. Academic Press, New York San Francisco London, p 432
17. Contin A, Van der Heijden R, Lefeber AWM, Verpoorte R (1998) The Iridoid Glucoside Secologanin is Derived from the Novel Triose Phosphate/pyruvate Pathway in a *Catharanthus roseus* Cell Culture. FEBS Letters **434**: 413
18. Eichinger D, Bacher A, Zenk MH, Eisenreich W (1999) Analysis of Metabolic Pathways *via* Quantitative Prediction of Isotope Labeling Patterns: A Retrobiosynthetic ^{13}C NMR Study on the Monoterpene Loganin. Phytochemistry **51**: 223
19. Bobbitt JM, Segebarth KP (1969) In: Battersby AR, Taylor WI (eds) Cyclopentanoid Terpene Derivatives. Marcel Dekker, New York, p 1
20. Franzyk H, Frederiksen SM, Jensen SR (1997) Synthesis of Monoterpene Piperidines from the Iridoid Glucoside Antirrhinoside. J Nat Prod **60**: 1012
21. Franzyk H, Jensen SR, Rasmussen JH (1998) Ozonolysis of Protected Iridoid Glucosides. Eur J Org Chem 365
22. Wysokinska H, Swiatek L (1990) Production of Iridoid Glucosides in Cell Suspension Cultures of *Penstemon serrulatus*. Planta Med **56**: 625
23. Wysokinska H, Skrzypek Z (1992) Studies on Iridoids of Tissue Cultures of *Penstemon serrulatus*: Isolation and their Antiproliferative Properties. J Nat Prod **55**: 58
24. Ueda S, Iwahashi Y (1991) Production of Anti-tumor-promoting Iridoid Glucosides in *Genipa americana* and its Cell Cultures. J Nat Prod **54**: 1677

25. Iyer RI, Mathuram V, Gopinath PM (1998) Establishment of Callus Cultures of *Nyctanthes arbor-tristis* from Juvenile Explants and Detection of Secondary Metabolites in the Callus. Current Sci **74**: 243
26. Koleva II (1997) Separation Methods for Iridoid Glycosides. Herba Polononica **43**: 322
27. Berkowitz WF, Sasson I, Sampathkumar PS, Hrabie J, Choudhry S, Pierce D (1979) Chiral Prostanoid Intermediates from Aucubin. Tetrahedron Lett 1641
28. Weinges K, Ziegler HJ (1991) Palladium-katalysierte Substitution an Hexaacetylaucubin. Liebigs Ann Chem 1109
29. Weinges K, Haremsa S, Huber-Patz U, Jahn R, Rodewald H, Irngartinger H, Jaggy H, Melzer E (1986) Ein einfaches Verfahren zur Herstellung von (1S,2S,6R,7R)-(–)-2-(Benzylthio)-7-hydroxy-3-oxabicyclo[4.3.0]nonan-9-on aus Catalpol – Strukturbeweis durch Röntgenbeugung. Liebigs Ann Chem 46
30. Weinges K, Schick H, Neuberger K, Ziegler HJ, Lichtenthäler J, Irngartinger H (1989) Isolierung und Strukturaufklärung neuer Inhaltsstoffe aus dem sodaalkalischen Extrakt von *Picrorhiza kurrooa*. Liebigs Ann Chem 1113
31. Nakatani K, Hiraishi A, Han Q, Isoe S (1992) Synthesis of Asperuloside Aglucon Silyl Ether and Garjasmine from (+)-Genipin *via* Gardenoside Aglucon Disilyl Ether as a Common Intermediate. Chem Lett 1851
32. Tietze LF, Bärtels C (1989) Synthesis of the Monoterpene Alkaloid Bakankosin from Secologanin. Tetrahedron **45**: 681
33. Briggs LH, Nicholls GA (1954) Chemistry of the *Coprosma* Genus. Part VIII. The Occurrence of Asperuloside. J Chem Soc 3940
34. Damtoft S, Franzyk H, Jensen SR (1994) Fontanesioside and 5-Hydroxysecologanol from *Fontanesia phillyreoides*. Phytochemistry **35**: 705
35. Damtoft S, Franzyk H, Jensen SR (1994) Biosynthesis of Iridoids in *Forsythia* spp. Phytochemistry **37**: 173
36. Damtoft S, Franzyk H, Jensen SR (1995) Biosynthesis of Iridoids in *Syringa* and *Fraxinus*: Secoriridoid Precursors. Phytochemistry **40**: 773
37. Damtoft S (1992) Iridoid Glucosides from *Lamium album*. Phytochemistry **31**: 175
38. Murai F, Tagawa M (1980) The Absolute Configuration of Boschnaloside and the Chemical Conversion of Genipin into Boschnaloside. Chem Pharm Bull **28**: 1730
39. Damtoft S, Jensen SR, Jessen CU, Knudsen TB (1993) Late Stages in the Biosynthesis of Aucubin in *Scrophularia*. Phytochemistry **33**: 1089
40. Damtoft S, Frederiksen LB, Jensen SR (1994) Biosynthesis of Iridoid Glucosides in *Thunbergia alata*. Phytochemistry **37**: 1599
41. Frederiksen LB, Damtoft S, Jensen SR (1999) Biosynthesis of Iridoids Lacking C-10 and the Chemotaxonomic Implications of their Distribution. Phytochemistry **52**: 1409
42. Jensen SR, Kirk O, Nielsen BJ (1989) Biosynthesis of the Iridoid Glucoside Cornin in *Verbena officinalis*. Phytochemistry **28**: 97
43. Damtoft S, Jensen SR, Nielsen BJ (1992) Biosynthesis of Iridoid Glucosides in *Lamium album*. Phytochemistry **31**: 135
44. Breinholt J, Damtoft S, Demuth H, Jensen SR, Nielsen BJ (1992) Biosynthesis of Antirrhinoside in *Antirrhinum majus*. Phytochemistry **31**: 795
45. Inoue K, Ono M, Nakajima H, Fujie I, Inouye H, Fujita T (1992) Radioimmunoassay of Iridoid Glucosides: Part I. General Methods for the Preparation of the Haptens and the Conjugates with a Protein of this Series of Glucosides. Heterocycles **33**: 673

46. Inouye H, Yoshida T, Tobita S, Okigawa M (1970) Studies on Monoterpene Glucosides IX. Chemical Correlation between Asperuloside and Loganin.Tetrahedron **26**: 3905
47. Damtoft S, Franzyk H, Jensen SR (1993) Biosynthesis of Secoiridoid Glucosides in Oleaceae. Phytochemistry **34**: 1291
48. Damtoft S, Franzyk H, Jensen SR (1995) Biosynthesis of Secoiridoids in *Fontanesia*. Phytochemistry **38**: 615
49. Damtoft S, Jensen SR, Schacht M (1995) Last Stages in the Biosynthesis of Antirrhinoside. Phytochemistry **39**: 549
50. Weinges K, Ziegler HJ (1990) Aucubin und Scandosid aus Catalpol. Liebigs Ann Chem 715
51. Jensen SR, Kirk O, Nielsen BJ (1987) Application of the Vilsmeier Formylation in the Synthesis of 11-^{13}C Labelled Iridoids. Tetrahedron **43**: 1949
52. Takeda Y, Morimoto Y, Matsumoto T, Honda G, Tabata M, Fujita T, Otsuka H, Sezik E, Yesilada E (1995) Nepetanudoside, an Iridoid Glucoside with an Unusual Stereostructure from *Nepeta nuda* ssp. *albiflora*. J Nat Prod **58**: 1217
53. Cachet X, Deguin B, Koch M, Makhlouf K, Tillequin F (1999) Efficient Conversion of Aucubin into 6-*epi*-Aucubin. J Nat Prod **62**: 400
54. Franzyk H, Jensen SR, Stermitz FR (1998) Iridoid Glucosides from *Penstemon secundiflorus* and *P. grandiflorus*: Revised Structure of 10-Hydroxy-8-*epi*-hastatoside. Phytochemistry **49**: 2025
55. Franzyk H, Jensen SR, Thale Z, Olsen CE (1999) Halohydrins and Polyols Derived from Antirrhinoside: Structural Revisions of Muralioside and Epimuralioside. J Nat Prod **62**: 275
56. Berdini R, Bianco A, Guiso M, Marini E, Nicoletti M, Passacantilli P, Righi G (1991) Isolation and Partial Synthesis of 7,8-Dehydro-6β,10-dihydroxy-11-*nor*-iridomyrmecin. J Nat Prod **54**: 1400
57. Kigawa M, Tanaka M, Mitsuhashi H, Wakamatsu T (1992) Synthesis of Iridolactones Isolated from *Silver Vine*. Heterocycles **33**: 117
58. Weinges K, Ziegler HJ, Maurer W, Schmidbauer SB (1993) Zwei einfache EPC-Synthesen mit chemischem Beweis der absoluten Konfiguration von (+)-Mitsugashiwa-Lacton aus (*S*)-(−)-Citronellol und Aucubin. Liebigs Ann Chem 1029
59. Franzyk H, Frederiksen SM, Jensen SR (1998) Synthesis of Antirrhinolide, a New Lactone from *Antirrhinum majus*. Eur J Org Chem 1665
60. Mohammad-Ali AK, Chan TH, Thomas AW, Jewett B, Strunz GM (1994) Spruce Budworm (*Choristoneura fumiferana*) Antifeedants 4. Synthesis of Specionin and Biological Studies. Canad J Chem **72**: 2137
61. Ge Y, Isoe S (1992) An Efficient Synthesis of Cerbinal, a 10π Aromatic Iridoid. Chem Letters 139
62. Nakatani K, Shimano K, Hiraishi A, Han Q, Isoe S (1993) Synthesis of Asperuloside Aglucon Silyl Ether and Garjasmine from (+)-Genipin *via* Gardenoside Aglucon Bis(silyl ether) as a Common Intermediate. Bull Chem Soc Japan **66**: 2646
63. Inouye H, Yoshida T, Nakamura Y, Tobita S (1970) Über die Monoterpenglucoside, XI. Chemische Korrelation des Asperulosids mit Swerosid. Chem Pharm Bull **18**: 1889
64. Inouye H, Yoshida T, Tobita S, Tanaka K, Nishioka T (1970) Absolute Struktur des Oleuropeins und einiger verwandter Glucoside. Tetrahedron Lett 2459
65. Inouye H, Yoshida T, Tobita S, Tanaka K, Nishioka T (1974) Über die Monoterpenglucoside und verwandte Naturstoffe, XXII. Absolutstrukturen des Oleuropeins, Kingisids und Morronisids. Tetrahedron **30**: 201

66. Bianco A, Naccarato G, Passacantilli P, Righi G, Scarpati ML (1992) Partial Synthesis of Oleuropein. J Nat Prod **55**: 760
67. Kuwajima H, Tanahashi T, Inoue K, Inouye H (1998) Synthesis of Four Possible Intermediates after Secologanin on the Biosynthesis of the Oleoside-, 10-Hydroxyoleoside and Ligustaloside-type Glucosides in Oleaceous Plants. Chem Pharm Bull **46**: 900
68. Shen Y, Chen C (1993) Enzymatic Transformation of 10-Hydroxyoleoside Type Secoiridoid Glucosides to Jasmolactones. Tetrahedron Lett **34**: 1949
69. Yamamoto H, Katano N, Ooi A, Inoue K (1999) Transformation of Loganin and 7-Deoxyloganin into Secologanin by *Lonicera japonica* Cell Suspension Cultures. Phytochemistry **50**: 417
70. Wildman WC, Le Men J, Wiesner K (1969) In: Battersby AR, Taylor WI (eds) Cyclopentanoid Terpene Derivatives. Marcel Dekker, New York, p 239
71. Gournelis D, Skaltsounis AL, Tillequin F, Koch M, Pusset J, Labarre S (1989) Plantes de Nouvelle-Caledonie, CXXI. Iridoïdes et Alcaloïdes de *Plectronia odorata*. J Nat Prod **52**: 306
72. Boros CA, Stermitz FR, Harris GH (1990) Iridoid Glycosides and a Pyridine Monoterpene Alkaloid from *Orthocarpus*. New Artifactual Iridoid Dienals. J Nat Prod **53**: 72
73. Benkrief R, Skaltsounis AL, Tillequin F, Koch M, Pusset J (1991) Iridoids and an Alkaloid from *Oxera morieri*. Planta Med **57**: 79
74. Benkrief R, Ranarivelo Y, Skaltsounis AL, Tillequin F, Koch M, Pusset J, Sévenet T (1998) Monoterpene Alkaloids, Iridoids and Phenylpropanoid Glycosides from *Osmanthus austrocaledonica*. Phytochemistry **47**: 825
75. Skaltsounis AL, Michel S, Tillequin F, Koch M, Pusset J, Chauvière G (1985) Plantes de Nouvelle-Calédonie. Helv Chim Acta **68**: 1679
76. Ranarivelo Y, Hotellier F, Skaltsounis AL, Tillequin F (1990) Biomimetic Synthesis of (−)-Deoxyrhexifoline, (−)-Tecostidine, and (−)-Actinidine. Heterocycles **31**: 1727
77. Frederiksen SM, Stermitz FR (1996) Pyridine Monoterpene Alkaloid Formation from Iridoid Glycosides. A Novel PMTA Dimer from Geniposide. J Nat Prod **59**: 41
78. Baghdikian B, Ollivier E, Faure R, Debrauwer L, Rathelot P, Balansard G (1999) Two New Pyridine Monoterpene Alkaloids by Chemical Conversion of a Commercial Extract of *Harpagophytum procumbens*. J Nat Prod **62**: 211
79. El-Sedawy AI, Shu YZ, Hattori M, Kobashi K, Namba T (1989) Metabolism of Swertiamarin from *Swertia japonica* by Human Intestinal Bacteria. Planta Med **55**: 147
80. Hattori M, Kawata Y, Kobashi K, Namba T (1990) Transformation of Iridoid and Secoiridoid Glucosides to Monoterpene Alkaloids by Human Intestinal Bacteria. Planta Med **56**: 625
81. Kawata Y, Hattori M, Akao T, Kobashi K, Namba T (1991) Formation of Nitrogen-Containing Metabolites from Geniposide and Gardenoside by Human Intestinal Bacteria. Planta Med **57**: 536
82. Baghdikian B, Guiraud-Dauriac H, Ollivier E, N'Guyen A, Dumenil G, Balansard G (1999) Formation of Nitrogen-Containing Metabolites from the Main Iridoids of *Harpagophytum procumbens* and *H. zeyheri* by Human Intestinal Bacteria. Planta Med **65**: 164
83. Cossy J, Belotti D, Leblanc C (1993) Total Synthesis of (±)-Actinidine and of (±)-Isooxyskytanthine. J Org Chem **58**: 2351
84. Shiao M, Chia W, Peng C, Shen C (1993) Facile Synthesis of two Pyridine Alkaloids *via* Functionalized 3,4-Dialkylpyridines. J Org Chem **58**: 3162

85. Aoyagi Y, Inariyama T, Arai Y, Tsuchida S, Matuda Y, Kobayashi H, Ohta A, Kurihara T, Fujihira S (1994) First Total Synthesis of (±)-Oxerine. Tetrahedron **50**: 13575
86. Jones K, Fiumana A (1996) Pyridine Radicals in Synthesis: a Formal Total Synthesis of (±)-Oxerine. Tetrahedron Lett **37**: 8049
87. Jones K, Escudero-Hernandez ML (1998) A Short Synthesis of (±)-Actinidine. Tetrahedron **54**: 2275
88. Stepanov AV, Lozanova AV, Veselovsky VV (1998) Chemistry of Natural Compounds and Bioorganic Chemistry. Stereocontrolled Synthesis of the Alkaloid (−)-Actinidine. Russ Chem Bull **47**: 2286
89. Lins AP, Felicio JD (1993) Monoterpene Alkaloids from *Tecoma stans*. Phytochemistry **34**: 876
90. Chi Y, Yan W, Chen D, Nogushi H, Iitaka Y, Sankawa U (1992) A Monoterpene Alkaloid from *Incarvillea sinensis*. Phytochemistry **31**: 2930
91. Chi Y, Hashimoto F, Yan W, Nohara T, Yamashita M, Marubayashi N (1997) Monoterpene Alkaloids from *Incarvillea sinensis*. VI. Absolute Stereochemistry of Incarvilline and Structure of a New Alkaloid, Hydroxyincarvilline. Chem Pharm Bull **45**: 495
92. Chi Y, Hashimoto F, Yan W, Nohara T (1995) Two Alkaloids from *Incarvillea sinensis*. Phytochemistry **39**: 1485
93. Chi Y, Hashimoto F, Yan W, Nohara T (1995) Incarvine A, a Monoterpene Alkaloid from *Incarvillea sinensis*. Phytochemistry **40**: 353
94. Chi Y, Hashimoto F, Yan W, Nohara T (1997) Four Monoterpene Alkaloid Derivatives from *Incarvillea sinensis*. Phytochemistry **46**: 763
95. Chi Y, Yan W, Li J (1990) An Alkaloid from *Incarvillea sinensis*. Phytochemistry **29**: 2376
96. Kan-Fan C, Sevenet T, Hadi HA, Bonin M, Quirion J, Husson H (1995) Monoterpene Alkaloids from *Kopsia macrophylla*. Nat Prod Lett **7**: 283
97. Kam T, Yoganathan K, Wei C (1996) Kinabalurine A, a New Monoterpene Alkaloid from a North Borneo *Kopsia*. Nat Prod Lett **8**: 231
98. Kam T, Yoganathan K, Wei C (1997) Monoterpene Alkaloids from *Kopsia pauciflora*. J Nat Prod **60**: 673
99. Alazard JP, Leboff A, Thal C (1989) Synthèse Stéréospécifique de la (±)-épi-7,7a-Tecomanine. Tetrahedron Lett **30**: 5267
100. Brayer JL, Alazard JP, Thal C (1990) Alcaloïdes Monoterpéniques II: Synthèse Stéréospécifique de la (±)-Δ-7,7a-4a-βH-Isotécomanine. Tetrahedron **46**: 5187
101. Cossy J, Leblanc C (1991) First Efficient Synthesis of Iso-Oxy-Skytanthine. Tetrahedron Lett **32**: 3051
102. Alazard JP, Leboff A, Thal C (1991) Alcaloïdes Monoterpéniques III: Synthese Stéréospécifique de la (±)-Épi-7,7a-Tecomanine. Tetrahedron **47**: 9195
103. Oppolzer W, Stevenson T (1986) Asymmetric Additions of 1-Alkenylcopper Reagents to Chiral Enoates: Enantioselective Synthesis of California Red Scale Pheromone. Tetrahedron Lett **27**: 1139
104. Oppolzer W, Jacobsen EJ (1986) Enantioselective Syntheses of (+)-α-Skytanthine, (+)-δ-Skytanthine and (+)-Iridomyrmecin by an Intramolecular Magnesium-Ene Reaction. Tetrahedron Lett **27**: 1141
105. Kametani T, Suzuki Y, Ban C, Honda T (1987) A Facile Synthesis of (+)-Tecomanine Using a Chiral Cyclopentane Derivative. Heterocycles **26**: 1491
106. Cid MM, Eggnauer U, Weber HP, Pombo-Villar E (1991) Synthesis of (−)-δ-*N*-Normethylskytanthine. Tetrahedron Lett **32**: 7233

107. Cid MM, Pombo-Villar E (1993) Enantioselective Synthesis of 3-Azabicyclo[4.3.0]-nonane Alkaloids. Helv Chim Acta **76**: 1591
108. Tsunoda T, Tatsuki S, Kataoka K, Itô S (1994) A Stereoselective Synthesis of (−)-Isoiridomyrmecin. Application of the Asymmetric Aza-Claisen Rearrangement. Chem Lett 543
109. Tsunoda T, Nagino C, Oguri M, Itô S (1996) Mitsunobu-type Alkylation with Active Methine Compounds. Tetrahedron Lett **37**: 2459
110. Tsunoda T, Ozaki F, Shirakata N, Tamaoka Y, Yamamoto H, Itô S (1996) Formation of Heterocycles by the Mitsunobu Reaction. Stereoselective Synthesis of (+)-α-Skytanthine. Tetrahedron Lett **37**: 2463
111. Cordell GA (1999) The Monoterpene Alkaloids. In: Cordell GA (ed) The Alkaloids, vol. 52. Academic Press, San Diego London Boston New York Sydney Tokyo Toronto, p 261
112. Tietze LF, Bärtels C, Fennen J (1989) Biomimetic Synthesis of the Monoterpene Alkaloids Xylostosidine and Loxylostosidine A and of Similar Unnatural Compounds by Transformations of the Monoterpene Glycoside Secologanin. Liebigs Ann Chem 1241
113. Bianco A (1994) Recent Developments in Iridoids Chemistry. Pure Appl Chem **66**: 2335
114. Bianco A, Cerichelli G, Guiso M, Lo Baido G, Mazzei RA (1993) Structure of Coloured Compounds Formed in a Chromatic Test of Iridoids. Gazz chim ital **123**: 437
115. Fujikawa S, Fukui Y, Koga K, Iwashita T, Komura H, Nomoto K (1987) Structure of Genipocyanin G_1, a Spontaneous Reaction Product Between Genipin and Glycine. Tetrahedron Lett **28**: 4699
116. Touyama R, Takeda Y, Inoue K, Kawamura I, Yatsuzuka M, Ikumoto T, Shingu T, Yokoi T, Inouye H (1994) Studies on the Blue Pigments Produced from Genipin and Methylamine. I. Structures of the Brownish-Red Pigments, Intermediates Leading to the Blue Pigments. Chem Pharm Bull **42**: 668
117. Touyama R, Inoue K, Takeda Y, Yatsuzuka M, Ikumoto T, Moritome N, Shingu T, Yokoi T, Inouye H (1994) Studies on the Blue Pigments Produced from Genipin and Methylamine II. On the Formation Mechanisms of Brownish-Red Intermediates Leading to the Blue Pigments. Chem Pharm Bull **42**: 1571
118. Tietze LF, Bärtels C (1991) Synthesis of Bridged Homoiridoids from Secologanin by Tandem-Knoevenagel-Hetero-Diels-Alder Reactions. Liebigs Ann Chem 155
119. Szabó-Pusztay K, Szabó LF, Podányi B (1994) Natural Products Chemistry: a Sort of Heterocyclic Chemistry. Reactions of Sweroside with Amines. ACH − Models in Chemistry **131**: 475
120. Schwartz A, Szabó LF, Podányi B (1997) Chemistry of Secologanin. Part 3. Graph Analysis of the Acidic Deglucosylation of Secologanin Derivatives. Tetrahedron **53**: 10489
121. Krajsovszky G, Kocsis A, Szabó LF, Podányi B (1997) Formation of Trioxadamantane Type Aglucones of 3-Methoxy Secologanin Derivatives. Tetrahedron **53**: 11659
122. Isoe S, Ge Y, Yamamoto K, Katsumura S (1988) Synthesis of Optically Active Petiodial and Determination of its Absolute Structure. Tetrahedron Lett **29**: 4591
123. Ge Y, Kondo S, Odagaki Y, Katsumura S, Nakatani K, Isoe S (1993) Synthesis of the Antipode of Udoteatrial Hydrate Using (+)-Genipin as a Chiral Building Block: Determination of the Absolute Configuration of Udoteatrial Hydrate. Tetrahedron Lett **34**: 2621

124. Ge Y, Kondo S, Katsumura S, Nakatani K, Isoe S (1993) Absolute Configuration of Novel Marine Diterpenoid Udoteatrial Hydrate. Synthesis and Cytotoxities of *ent*-Udoteatrial Hydrate and its Analogues. Tetrahedron 49: 10555
125. Shimano K, Ge Y, Sakaguchi K, Isoe S (1996) Synthesis of Both Enantiomers of Halitunal. Tetrahedron Lett 37: 2253
126. Tanaka M, Kigawa M, Mitsuhashi H, Wakamatsu T (1991) The Practical Method for the Preparation of Iridoid Aglycons from their Glycosides. Heterocycles 32: 1451
127. Weinges K, Braun G, Oster B (1983) Chemie und Stereochemie der Iridoide, III. Über die Synthese von 12-*epi*-Prostaglandinen. Liebigs Ann Chem 2197
128. Weinges K, Huber W, Huber-Patz U, Irngartinger H, Nixdorf M, Rodewald H (1984) Chemie und Stereochemie der Iridoide, IV. Synthese und Röntgenstrukturanalyse von 15-Methyl-12-*epi*-prostaglandin $F_{2\beta}$. Liebigs Ann Chem 761
129. Weinges K, Gethöffer H, Huber-Patz U, Rodewald H, Irngartinger H (1987) EPC-Synthese von (1R,2R,2″Z)-(−)-Methyljasmonat aus Catalpol − Kristall- und Molekularstruktur von Methyl dehydrojasmonat-semicarbazon. Liebigs Ann Chem 361
130. Weinges K, Lernhardt U (1990) Synthese von enantiomerenreinem (1R,2S,2″Z)-(+)-Methyljasmonat aus Catalpol. Liebigs Ann Chem 751
131. Weinges K, Iatridou H, Stammler H, Weiss J (1989) Chirale Bausteine zur Synthese von Triquinan-Sesquiterpenen: Derivate des 2-Methylbicyclo[3.3.0]octan-3-ols aus Catalpol. Angew Chem 101: 485
132. Weinges K, Dietz U, Oeser T, Irngartinger H (1990) Synthese von enantiomerenreinem (−)-Hypnophilin. Angew Chem 102: 685
133. Weinges K, Iatridou H, Dietz U (1991) EPC-Synthese von (−)-Hypnophilin. Liebigs Ann Chem 893
134. Weinges K, Haremsa S (1987) Chemistry and Stereochemistry of Iridoids X. Enantiomerically Pure Hexahydropentalene Derivatives − Building Units for the Synthesis of Cyclopentanoid Natural Products. Liebigs Ann Chem 679
135. Weinges K, Ziegler HJ, Schick H (1992) Enantiomerenreines (+)-Cyclosarkomycin aus Catalpol. Liebigs Ann Chem 1213
136. Weinges K, Neuberger K, Schick H, Reifenstahl U, Irngartinger H (1991) Eine effiziente Synthese enantiomerenreiner (−)-Specionin-Analoga aus peracetylierten Iridoidglucosiden. Kristall- und Molekülstruktur von (−)-3′-Methoxyspecionin. Liebigs Ann Chem 477
137. Van der Eycken E, Janssens A, Vandewalle M (1987) Iridoids: an Efficient Conversion of (−)-Catalpol into (−)-Specionin. Tetrahedron Lett 28: 3519
138. Naruto M, Ohno K, Naruse N (1978) The Synthesis of Useful Chiral Prostanoid Intermediates and Naturally Occurring Prostaglandins from Aucubin. Chem Lett 1419
139. Naruto M, Ohno K, Naruse N, Takeuchi H (1978) (+)-11-Deoxy-13,14-dihydro-13β,11α-epoxymethano-12-isoprostaglandin $F_{2\alpha}$ from Aucubin. Chem Lett 1423
140. Naruto M, Ohno K, Naruse N, Takeuchi H (1979) Synthesis of Prostaglandins and their Congeners I. (+)-11-Deoxy-11α-hydroxymethyl Prostaglandin $F_{2\alpha}$ from Aucubin. Tetrahedron Lett 251
141. Ohno K, Naruto M (1979) A Simple Synthesis of (+)-11-Deoxy-11α-hydroxymethyl $PGF_{1\alpha}$ and its 12-Isomer from Aucubin. Chem Lett 1015
142. Ohno K, Naruto M (1980) Synthesis of Novel Prostanoids Having a Cyclopenta[*c*]-furan Structure with a Hemithioacetal Side Chain from Aucubin. Chem Lett 175
143. Tixidre A, Rolland Y, Garnier J, Poisson J (1982) Aucubin, a Source of Prostanoid Synthons − New Hemisynthetic Pathways. Heterocycles 19: 253

144. Bernini R, Davini E, Iavarone C, Trogolo C (1986) Synthesis of a Corey Lactone Analogue from Iridoid Glucoside Aucubin and Its Utilization in the Synthesis of a New 12-*epi*-PGF$_{2\alpha}$ Modified at C-11. J Org Chem **51**: 4600

145. Bonini C, Di Fabio R (1982) Synthesis of a Corey's Lactone Analogue from the Iridoid Aucubin. Tetrahedron Lett **23**: 5199

146. Bonadies F, Gubbiotti A, Bonini C (1985) New Routes for the Conversion of Aucubin to 11-Deoxy-11-methylprostaglandin Intermediates. Gazz chim ital **115**: 45

147. Bonini C, Iavarone C, Trogolo C, Di Fabio R (1985) One-pot Conversion of 6-Hydroxy-Δ^7-iridoid Glucosides into *cis*-2-Oxabicyclo[3.3.0]oct-7-enes and Transformation into Corey's Lactone Analogue. J Org Chem **50**: 958

148. Davini E, Iavarone C, Mataloni F, Trogolo C (1988) Conversion of Aucubin to a Useful Corey Lactone Analogue for the Synthesis of 11-Methyl PGA$_2$. J Org Chem **53**: 2089

149. Bianco A, Mazzei RA (1997) Synthesis of a New Carbocyclic Nucleoside Analog. Tetrahedron Lett **38**: 6433

150. Carnevale G, Davini E, Iavarone C, Trogolo C (1988) Organomercury Chemistry of Iridoid Glucosides. 1. Chemoselective Hydroxymercuration-Demercuration of Aucubin: A Cheaper and Efficient Approach to Epimeric Isoeucommiols and 6,7-Bis(hydroxymethyl)-*cis*-2-oxabicyclo[3.3.0]oct-7-enes. J Org Chem **53**: 5343

151. D'Annibale A, Iavarone C, Trogolo C (1993) Organomercury Chemistry of Iridoid Glucosides, Part 3. Heterocycles **36**: 701

152. Carnevale G, Davini E, Iavarone C, Trogolo C (1990) Organomercury Chemistry of Iridoid glucosides, Part 2. Chemoselective Methoxymercuriation-Demercuriation of Aucubin: A New Approach to Optically Active Cyclopentanols. J Chem Soc Perkin Trans 1, 989

153. Franzyk H, Rasmussen JH, Mazzei RA, Jensen SR (1998) Synthesis of Carbocyclic Homo-*N*-nucleosides from Iridoids. Eur J Org Chem 2931

154. Vince R, Hua M, Brownell J, Daluge S, Lee F, Shannon WM, Lavelle GC, Qualls J, Weislow OS Potent and Selective Activity of a New Carbocyclic Nucleoside Analog (Carbovir-NSC-614846) against Human Immunodeficiency Virus In Vitro (1988) Biochem Biophys Res Comm **156**: 1046

155. Wachtmeister J, Classon B, Samuelsson B, Kvarnström I (1995) Synthesis of 2',3'-Dideoxycyclo-2'-pentenyl-3'-*C*-hydroxymethyl Carbocyclic Nucleoside Analogues as Potential Anti-viral Agents. Tetrahedron **51**: 2029

156. Hayashi M, Yaginuma S, Muto N, Tsujino M (1980) Structures of Neplanocins, New Antitumor Antibiotics. Nucleic Acids Symp Ser **8**: 65

157. Ishiguro K, Yamaki M, Tagaki S, Ikeda Y, Kawakami K, Ito K, Nose T (1988) Studies on Iridoid-related Compounds. V. Antitumor Activity of Iridoid Dervs. Periodate Oxidation Products. J Pharmacobio-Dyn **11**: 131

158. Raj K, Mathad VT, Bhaduri AP (1995) Modified Iridoid Glycosides (Part 1). Syntheses of 4',5'-Unsaturated Iridoid Glycosides from Loganin and Arbortristoside A. Nat Prod Lett **7**: 51

159. Raj K, Mathad VT, Bhaduri AP, Pandey CP, Patnaik GK (1996) Modified Iridoid Glycosides, Part 3. Syntheses and Hepatoprotective Evaluation of Modified Iridoid Glycosides. Indian J Chem **35B**: 1056

(*Received August 25, 1999*)

The Defensive Chemistry of Ants

S. Leclercq[1], J. C. Braekman[1], D. Daloze[1], and J. M. Pasteels[2]

[1] Laboratory of Bio-Organic Chemistry, Department of Organic Chemistry, Free University of Brussels, Brussels, Belgium
[2] Laboratory of Cellular and Animal Biology, Free University of Brussels, Brussels, Belgium

Contents

1. Introduction 116

2. Alkaloids 118

    2.1. Structures, Occurrence, and Function 118
        2.1.1. Piperidines and Pyridines 118
        2.1.2. Pyrrolidines and Pyrrolines 120
        2.1.3. Pyrrolizidines 122
        2.1.4. Indolizidines 124
        2.1.5. Tetraponerines 126
        2.1.6. Other Alkaloids 127

    2.2. Synthesis 128

        2.2.1. Piperidines 128
        2.2.2. Pyrrolidines 131
            A. Synthesis of Racemic Pyrrolidines 131
            B. Syntheses of Nonracemic Pyrrolidines 139
        2.2.3. Pyrrolines 149
        2.2.4. Pyrrolizidines 152
            2.2.4.1. 3,5-Dialkylpyrrolizidines 152
                A. Xenovenine 152
                B. (5E,8E)-3-Butyl-5-hexylpyrrolizidine 159
            2.2.4.2. 3-Methyl-5-alkenylpyrrolizidines and
                      3,5-Dialkenylpyrrolizidines 160
        2.2.5. Indolizidines 163
            2.2.5.1. Monomorine I 163
                A. Syntheses of Racemic Monomorine I 163
                B. Syntheses of Nonracemic Monomorine I 173
            2.2.5.2. 3,5-Dialkylindolizidines 190
                A. 3-Butyl-5(4-penten-1-yl)indolizidine 190

         B. 3-Ethyl- and 3-Hexyl-5-methylindolizidines 193
         C. Myrmicarin 237A and 237B 194
         D. Myrmicarin 217 198
    2.2.6. Tetraponerines 200
         A. Syntheses of Racemic Tetraponerines 200
         B. Syntheses of Nonracemic Tetraponerines 205

3. Nonalkaloidal Compounds 211

4. Biosynthesis 217

References 221

1. Introduction

Nearly 10,000 living ant species are known today and many are renowned for their fierce aggressiveness. In these highly social insects, defense (offense) is a collective behaviour coordinated by recruiting or dispersing signals, *e.g.* trail and alarm pheromones. Excellent accounts of ants' defensive behaviour are provided by BUSCHINGER and MASCHWITZ (*1*) and HÖLLDOBLER and WILSON (*2*). Mechanical and chemical weapons are used by ants during fighting, and both can be combined in a single defensive device, *e.g.* secretions from glands associated with the mandibles or the sting. Chemical defense is prominent and no less than six different exocrine glands have been reported as sources of defensive compounds (*1*). All these glands are not necessarily present in all species and when present, they can fulfil other functions than defense, *e.g.* in communication.

Stinging is the most notorious, but by no means the only or even the most frequent defensive reaction in ants. Two major glands are associated with the sting, the poison and the Dufour glands. These glands are present in all Aculeata Hymenoptera (stinging Hymenoptera) whether they are social or not. The poison gland secretes the venom which accumulates in the poison sac before being injected into the prey or enemy. Many injected venoms are aqueous solutions or proteins, polypeptides and biogenic amines. These proteinaceous venoms are considered to be the ancestral state. However many ants do not sting (more than 50 % of all living ant species according to MASCHWITZ (*3*)), the sting being either reduced, sometimes completely absent as in the Formicinae, and/or modified into a device which facilitates

References, pp. 221–229

smearing or spilling the venom on the enemy, *e.g.* spatula or pen-like structures. Venoms are either projected or applied on the enemy. This probably evolved in ants because most ants are rather opportunistic in their feeding behaviour and stinging behaviour cannot be optimally adapted for hunting specific prey (*3*). In many instances, the ants' main enemies are other competing ants which are difficult to fight by stinging, but more easily by spraying deterrents, by applying contact poisons or by emitting sticky secretions. In these ants either the poison glands secrete non-proteinous venoms, *e.g.* formic acid in the formicines or alkaloids in some myrmicines, or chemical defense is taken over by the Dufour gland (see below), or by still other exocrine glands not associated with the sting, *e.g.* the anal gland of dolichoderines secreting iridoids, mandibular, sternal or metapleural glands in other species (*1*). In leafcutter ants and in army ants it is the mandibles which are the main fighting weapons.

The second gland associated with the sting, the Dufour gland, usually produces lipophilic compounds, *i.e.* hydrocarbons, linear esters or ketones, and sesquiterpene derivatives. It has been suggested that lubrification of the sting apparatus could be its original function, but this remains highly speculative. In ants, the Dufour's gland secretions play various biological roles (*e.g.* alarm pheromone, trail pheromone, home range marking pheromone, wetting agents to enhance the toxicity of the poison gland products etc.). In some non-stinging ants, however, *e.g. Crematogaster* ants, it is the enlarged Dufour gland instead of the poison gland which secretes the contact venom.

In this review, we will limit ourselves to non-proteinous poisons produced by either the venom or the Dufour glands. It is in the knowledge of the chemistry of those secretions that the most spectacular progress has been made recently. We will review the diversity of compounds identified so far, their synthesis and their biosynthesis. Focus will be on toxins, but compounds chemically related to toxins, with either unknown functions or acting as pheromones will be briefly mentioned. Information on proteinaceous venoms can be found in SCHMIDT (*4, 5*) and reviews on other ants' exocrine secretions, including mandibular and anal glands secretions can be found in ATTYGALLE and MORGAN (*6*) and in BLUM (*7*).

As the topic of ant alkaloids has been exhaustively reviewed in 1987 by NUMATA and IBUKA (*8*), and in 1989 by two of us (*9*), we will report only the literature that has been published since. The literature was covered up to the end of June, 1998.

2. Alkaloids

2.1. Structures, Occurrence, and Function

2.1.1. Piperidines and Pyridines

In formicids, piperidine and pyridine alkaloids have been found in the venom of the sub-familiy Myrmicinae. A complete list of these alkaloids covering the literature until 1989 can be found in the reviews of NUMATA and IBUKA (8) and of BRAEKMAN and DALOZE (9). Since then, only a few further alkaloids of this type have been discovered in ants (Fig. 1).

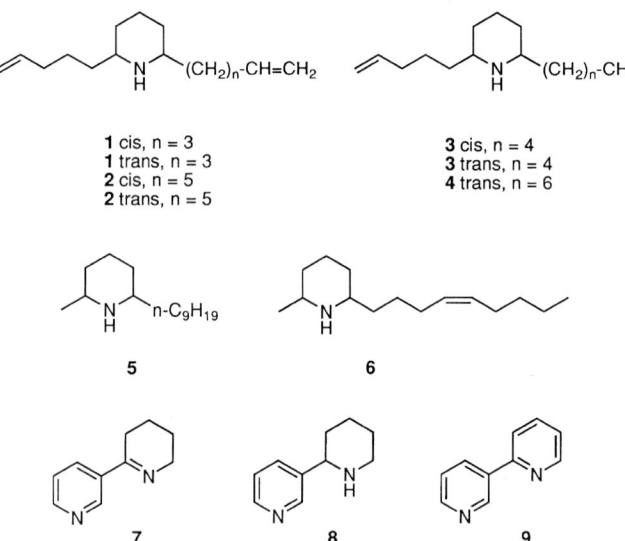

Fig. 1. Piperidine and pyridine alkaloids, isolated from ants

In the Myrmicinae, 2-alkyl-6-methylpiperidines are characteristic of the venom of *Solenopsis* species and 2,5-dialkylpyrrolidines predominate in the venom of *Monomorium* species. In contrast to this prevalent distribution, the South African species *Monomorium delagoensis* was found by JONES et al. (10) to contain the novel 2,6-dialkylpiperidines **1** to **4**. The structures of these compounds, first established on the basis of spectral properties, were confirmed by synthesis. These alkaloids possess insecticidal and repellent properties against foraging ant workers in the genera *Pheidole* and *Iridomyrmex*. This is the first identification of 2,6-dialkylpiperidines in ants outside the genus *Solenopsis*.

References, pp. 221–229

Piperidine [*cis*- and *trans*-2-methyl-6-nonylpiperidine (**5**) and *cis*-*trans*-2-methyl-6-((Z)-4-nonenyl) piperidine (**6**)] together with indolizidine alkaloids have been detected in the venom of four undetermined species of *Solenopsis* (*Diplorhoptrum* subgenus) from Puerto Rico (*11*). Conspicuous difference between alkaloids found in workers and in queens of the different collections were detected. *Cis*- and *trans*-**5** had already been isolated from *Solenopsis* ants (*8, 9*) while the corresponding dehydro derivatives **6** are new compounds. The structures of the latter were ascertained by synthesis. Compound *cis*-**5** has also been recognized in the venom of several populations of *Solenopsis* (*Diplorhoptrum*) from California (*12*).

Both *cis*- and *trans*-2,6-dialkylpiperidines occur as constituents of the venom of *Solenopsis* and *Monomorium* ants and several methods were already known to assign the *cis* or *trans* configuration of these compounds (*e.g. 13–15*). GARRAFFO *et al.* (*16*) have reported a further general method for distinguishing these two stereoisomers. The method is based on the appearance in the FTIR spectra of weak Bohlmann bands at $2800-2600\,cm^{-1}$ in *cis* but not in *trans* stereoisomers. Moreover, a procedure has been developed by LECLERCQ *et al.* (*17*) to allow the assignment of absolute configuration to 2-alkyl-6-methylpiperidines. The method is applicable to micro quantities of material and is based on the transformation of the alkaloids into diastereoisomeric amides by reaction with (*R*)-MTPA, followed by comparison of the chromatographic behaviour of the natural amides thus formed with those of standards with established absolute configuration. The method has been applied to alkaloids isolated from three samples of ants: *Solenopsis geminata* workers, *S. invicta* workers and *S. invicta* alates. The absolute configuration of the *trans* alkaloids is always (2*R*,6*R*) while that of the *cis* alkaloids is (2*R*,6*S*).

The presence of anabaseine (**7**) in the venom gland of *Aphaenogaster fulva* and *A. tennesseensis* had been reported in 1981 by WHEELER *et al.* (*18*). Since then, this alkaloid together with anabasine (**8**) has been found in the poison gland of a few other Myrmicinae (Table 1). In the species *A. rubis* compounds **7** and **8** are accompanied by small amounts of 2,3'-bipyridyl (**9**) and *N*-isopentyl-2-phenylethylamine. This mixture of alkaloids acts as a trail pheromone (*19*).

The main alkaloid of the queen poison gland of *Solenopsis invicta* is *cis*-2-methyl-6-undecylpiperidine. It has been demonstrated that this alkaloid is deposited on the eggs as they are laid. It acts as an inhibitor of entomopathogenic fungi, thus giving the eggs protection in their subterranean habitat (*23*). It has also been shown that the venom alkaloid profiles of *Solenopsis* species, which contain primarily *cis*- or

Table 1. *Distribution of Anabaseine* (**7**) *and Anabasine* (**8**) *in Myrmicinae*

	Anabasine %	Anabaseine %	Reference
Aphaenogaster fulva	–	90	*(18)*
A. rudis	+	+++	*(19)*
A. tennesseensis	–	90	*(18)*
Messor bouvieri	10	90	*(20)*
M. capensis	5	92	*(21)*
M. ebeninus	90	–	*(22)*

trans-2-methyl-6-alkyl or alkenylpiperidines, do have taxonomic value and can be used as chemical characters in defining populations of fire ants especially when utilized in conjunction with classical taxonomy (*24* and references therein).

2.1.2. Pyrrolidines and Pyrrolines

While 2,5-dialkylpyrrolidines and the corresponding 1-pyrrolines have been found in the venom of some *Solenopsis* ants, these compounds are particularly characteristic of the venom of ants of the genus *Monomorium*. A complete list of these alkaloids covering the literature until 1989, can be found in the reviews of NUMATA and IBUKA (*8*) and of BRAEKMAN and DALOZE (*9*). Since then, a few further pyrrolidine and pyrroline alkaloids have been discovered in ants (Fig. 2). Thus, the major alkaloids present in the venom of an unidentified Australian species of *Monomorium* (*Monomorium* #1066) are *trans*-2-ethyl-5-undecylpyrrolidine (**10**) and *trans*-2-ethyl-5-(12-tridecen-1-yl)pyrrolidine (**11**) (*25*). The four corresponding pyrrolines (**21** to **24**) were also detected in varying amounts in the venom. These compounds possess insecticide activity when evaluated against termite workers.

GC/MS analysis of venom of several unspecified European *Monomorium* species led to the detection of two 1-pyrrolines (**25** and **26**) and three pyrrolidines (**12** to **14**) (*26*). Compound **26** is the only new natural compound; the others had already been isolated from myrmicine ants (*8, 9*).

A comparison of the venom produced by twenty eight New Zealand populations of ants belonging to the *Monomorium* (=*Chelaner*) *antarcticum* complex has been undertaken by JONES *et al.* (*27*). Four pyrrolidines (**12, 15** to **17**) were detected together with several pyrrolizidine alkaloids. Compounds **16** and **17** were new compounds while **12** and **15** were available from previous studies. The alkaloids **12**

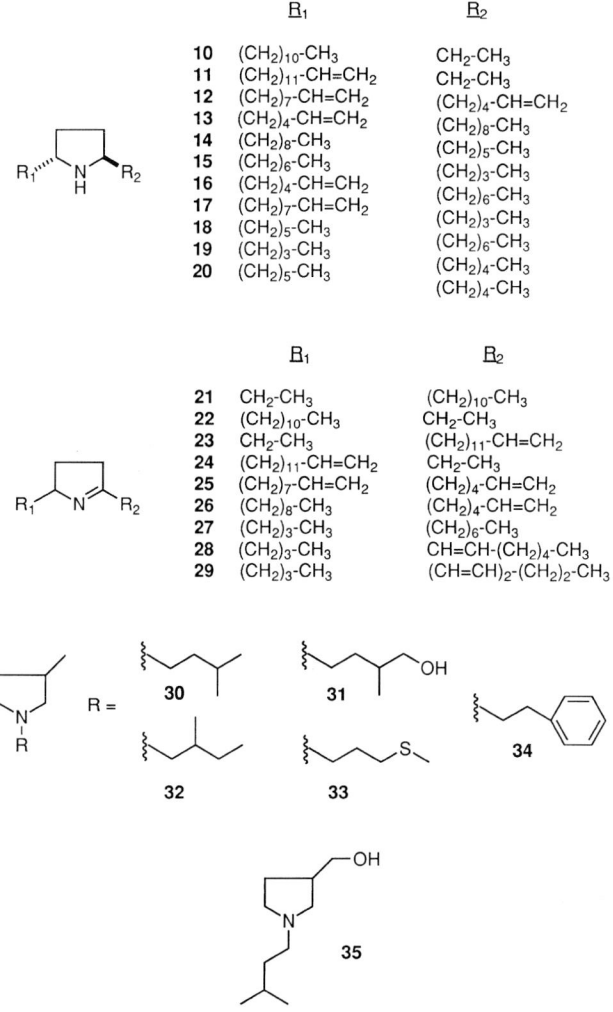

Fig. 2. Pyrrolidine alkaloids, isolated from ants

and **17** were also recognized in the venom of another *Monomorium* endemic to New Zealand, *M. smithii* (*28*).

Chemical analyses of three species in the neotropical ant genus *Megalomyrmex* have identified this taxon as the third Myrmicinae genus besides the genera *Solenopsis* and *Monomorium* to produce alkaloids as major venom products (*29*, *30*). In particular, workers of *Megalomyrmex leoninus* and workers and ergatoids of *M. goeldii* produce one or more of

Table 2. *Distribution of the different N-alkyl-3-methylpyrrolidines in some species of the subtribe Leptothoracini (31)*

	Harpagoxenus sublaevis (ng)	Leptothorax acervorum (ng)	Leptothorax muscorum (ng)	Doronomyrmex goesswaldi (ng)
30	15	10	5	0.5
31	0.5	0.2	0.1	0.05
32	<0.01	<0.01	<0.01	–
33	0.2	0.03	0.01	–
34	0.1	0.01	0.005	1
35	0.5	0.05	0.01	–

four *trans*-2,5-dialkylpyrrolidine (**15, 18** to **20**) previously identified in other myrmicine genera (*8, 9, 27*). Moreover the venom of *M. foreli* from Costa Rica (*30*) revealed the presence of four further five-membered ring alkaloids: two already known from myrmicine ants (**15** and **27**) and two new derivatives [2-butyl-5-((*E*)-1-heptenyl)-5-pyrroline (**28**) and 2-butyl-5-(1*E*,3*E*)-heptadienyl)-5-pyrroline (**29**)] whose structures were assigned from their spectral and chemical behaviour and unambiguous syntheses. The genera *Megalomyrmex* and *Monomorium* are considered as closely related so that the presence of pyrroline and pyrrolidine alkaloids as major compounds in the venom of both ants may not be unexpected.

We have already mentioned that GC/FTIR spectroscopy provides a convenient and rapid method to distinguish *cis*- and *trans*-α,α'-disubstituted piperidines (*16*). The method is also applicable to pyrrolidine alkaloids. But in this case *N*-methylation is required first.

Several myrmicine ants of the subtribe Leptothoracini were found to contain unique *N*-alkyl-3-methylpyrrolidines (*31*). Six alkaloids differing from each other in the substituent of the nitrogen atom were isolated and identified (**30** to **35**, see Table 2 and Fig. 2). The absolute configuration at C-3 was determined as *R* (*32*). The basic components of the poison gland secretion of female display a male attracting behaviour (*33*).

2.1.3. Pyrrolizidines

Occurrence of pyrrolizidine alkaloids in the venom of Myrmicinae is rare and until 1989 only (5*Z*,8*E*)-3-heptyl-5-methylpyrrolizidine (**36**) from the North American thief ant *Solenopsis xenovenenum* (*34*) and (5*E*,8*Z*)-3-(1-non-8-enyl)-5-((*E*)-1-prop-1-enyl) pyrrolizidine (**40**) from the New Zealand ant *Monomorium* (=*Chelaner*) *antarcticum* (*35*) had been isolated. Since then, the 3,5-dialkylpyrrolizidines **37** to **41**, **43** and

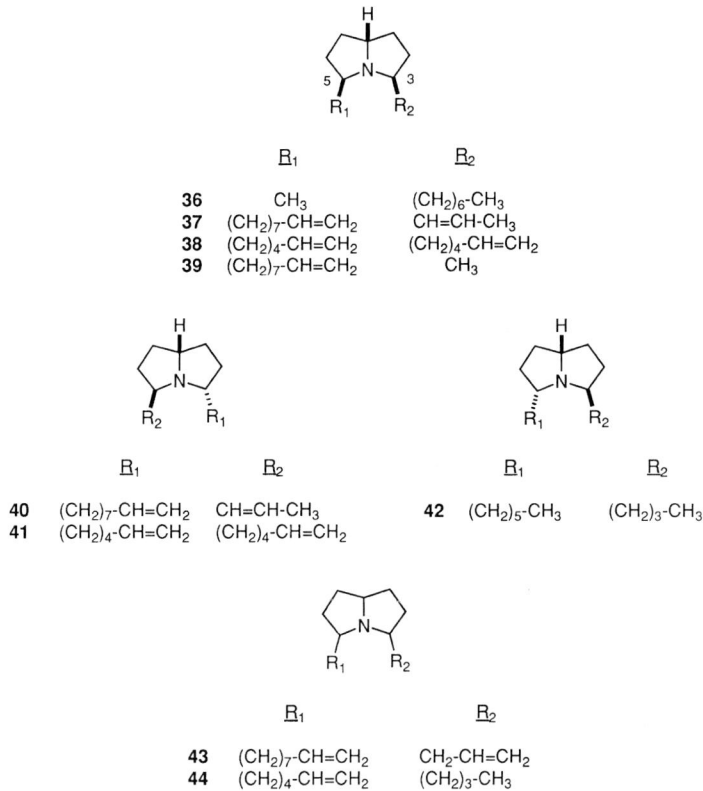

Fig. 3. Pyrrolizidine alkaloids, isolated from ants

44, together with some *trans*-2,5-dialkylpyrrolidines, were identified in the alkaloidal extracts of twenty eight single-nest collections of New Zealand ants of the *Monomorium antarcticum* complex (Fig. 3) (*27*). While structures of compounds **37** to **41** could be established by comparison with synthetic samples, those of compounds **43** and **44** were only suggested by their mass spectra. Moreover, it has been shown that there are major differences in the alkaloids produced by different populations. It appears from the observed distribution that elucidation of the *Monomorium antarcticum* complex in New Zealand needs much more precise morphological studies in conjunction with further chemical data on the alkaloidic content of the venom. The alkaloid **41** was also detected in the venom of *Monomorium smithii* (*28*).

As already mentioned, two species of the genus *Megalomyrmex* produce venoms containing 3,5-dialkylpyrrolidines (*29*). Pyrrolizidine **42** is the only alkaloid produced by the species *M. modestus* (*29*).

2.1.4. Indolizidines

Since the discovery of monomorine I (**45**) in the secretion of the poison gland of *Monomorium pharaonis* (*36*), of 3-ethyl-5-methylindolizidine (**46**) from *Solenopsis conjurata* (*37*), and of (5Z,9Z)-3-hexyl-5-methylindolizidine (**47**) in the secretion of queens of *Solenopsis* sp. AA (*37*), several new indolizidine alkaloids have been reported from Myrmicinae ants (Fig. 4).

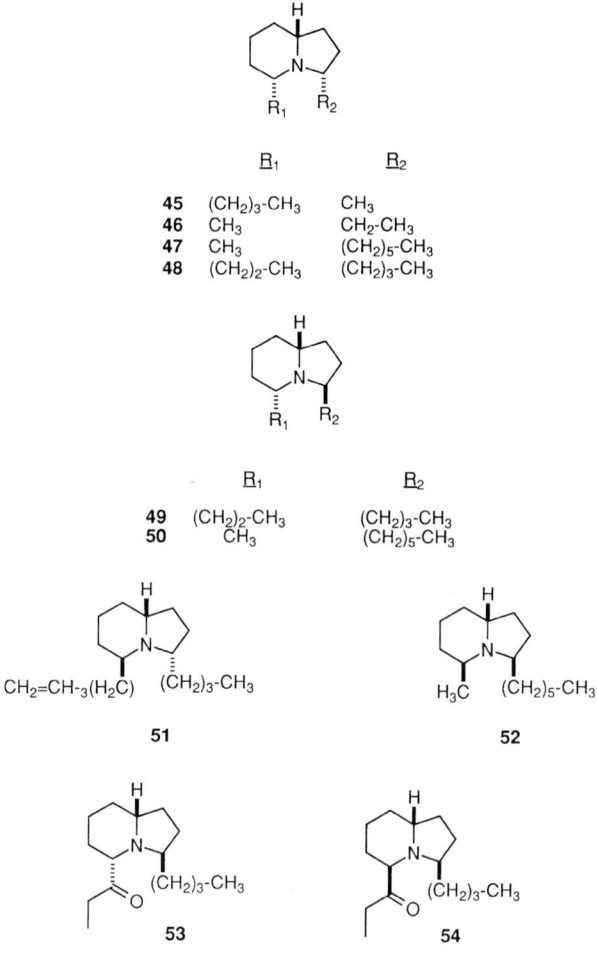

Fig. 4. Indolizidine alkaloids, isolated from ants

In 1990, JONES et al. (28) reported the isolation from the secretion of *Monomorium smithii* of compound **51** along with other alkaloid types. Moreover, three indolizidines (**47, 48,** and **49**) were identified in venom extracts of four Puerto Rican species of ants of the genus *Solenopsis* (*Diplorhoptrum*) (*11*), while the venoms of workers of nine collections of *Solenopsis* (*Diplorhoptrun*) from California contain either the indolizidines **50** or **52** along with *cis*-2-methyl-6-nonylpiperidine (**5**) (*12*).

The venom of ants of the genus *Myrmicaria* (Myrmicinae) is also made up of alkaloids. The secretion of the remarkably large poison gland of the widespread African ant *Myrmicaria eumenoides* contains a mixture of monoterpene hydrocarbons dominated by (+)-limonene and two major alkaloids that were identified as the stereoisomers **53** (myrmicarin 237A) and **54** (myrmicarin 237B), whose structures and absolute configurations were established by asymmetric synthesis (Fig. 5) (*38*). The poison gland secretion of another African *Myrmicaria* (*M. opaciventris*) contains three families of alkaloids containing 15, 30 and

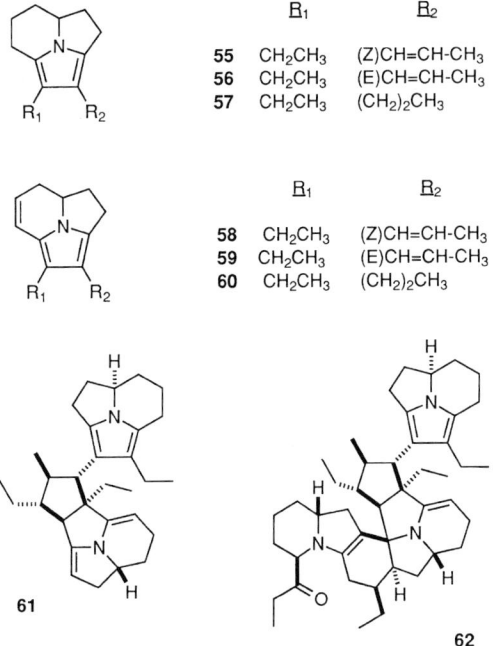

	R₁	R₂
55	CH₂CH₃	(Z)CH=CH-CH₃
56	CH₂CH₃	(E)CH=CH-CH₃
57	CH₂CH₃	(CH₂)₂CH₃

	R₁	R₂
58	CH₂CH₃	(Z)CH=CH-CH₃
59	CH₂CH₃	(E)CH=CH-CH₃
60	CH₂CH₃	(CH₂)₂CH₃

Fig. 5. Myrmicarins, isolated from *Myrmicaria* ants

45 carbon atoms respectively (*39*). The $C_{15}N$ alkaloids are pyrrolo[2,1,5-cd]indolizines while the two other families can be considered as dimers ($C_{30}N_2$) and trimers ($C_{45}N_3$) of the $C_{15}N$ basic skeleton. Myrmicarin 215A (**55**), 215B (**56**) and 217 (**57**) are the main alkaloids of the venom of colonies from Kenya. The determination of their structures was based mainly on their spectral properties. The structure of myrmicarin 217 has been confirmed by synthesis (*40*). Myrmicarin 213A (**58**), 213B (**59**) and 215C (**60**) are less abundant but a secretion that had been exposed to air showed higher amounts of these compounds. This suggests that they may simply be products of non-enzymatic oxidation during storage and / or isolation (*39*). The main component of the $C_{30}N_2$ family is myrmicarin 430A (**61**) whose structure was established by extensive two-dimensional NMR experiments (*39, 41*). The compound is very sensitive to air, showing more than 90% decomposition after only 1 h at ambient temperature. Myrmicarin 663 (**62**) is the major alkaloid in colonies of *Myrmicaria opaciventris* from Cameroun and of *M. striata* (*39, 42*). It was shown by two-dimensional NMR that it is a decacyclic compound representing a new class of alkaloids. It is by far the most complex alkaloid isolated until now from insects. Like myrmicarin 430A, myrmicarin 663 is also very sensitive to air. The carbon skeleton of these two alkaloids can be seen to consist of two and three unbranched C_{15} chains, respectively, forming hexahydropyrrolo[2,1,5-cd]indolizine systems as in myrmicarin 215A (Scheme 1).

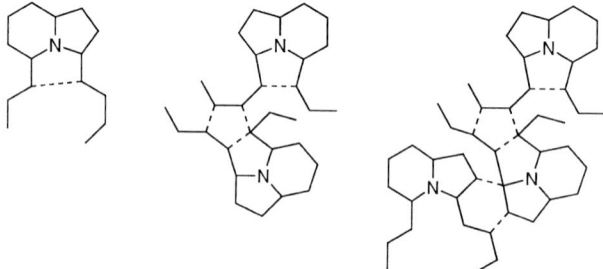

Scheme 1. Structural relationships between the myrmicarins

2.1.5. Tetraponerines

The New Guinean pseudomyrmecine ant *Tetraponera* sp. utilizes its modified sting to smear upon enemies a contact poison with strong deterrent and toxic properties (*43*). The venom which originates from the poison gland contains a mixture of eight structurally related alkaloids, tetraponerines-1 to -8 (T1 to T8) (Fig. 6) (*44*). The structure of T8, one

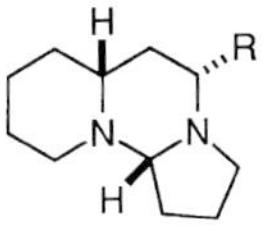

63	R = C$_5$H$_{11}$ (T8)	65	R = C$_5$H$_{11}$ (T7)
64	R = C$_3$H$_7$ (T4)	66	R = C$_3$H$_7$ (T3)

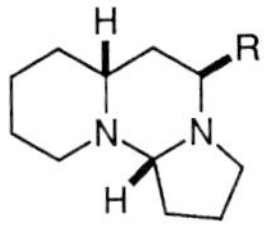

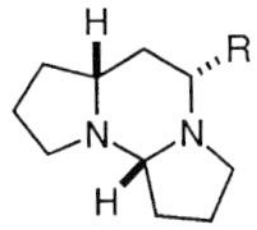

67	R = C$_5$H$_{11}$ (T6)	69	R = C$_5$H$_{11}$ (T5)
68	R = C$_3$H$_7$ (T2)	70	R = C$_3$H$_7$ (T1)

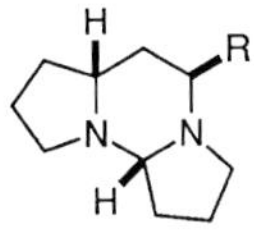

Fig. 6. Tetraponerines, isolated from *Tetraponera* ants

of the major constituents of the venom has been determined as **63** by X-ray diffraction (*43*). The absolute configuration of T8 was determined as (5R,9S,11R) by chemical degradation to (+)-pipecolic acid (*45*) and by enantioselective synthesis (*46*). Upon topical application on *Myrmica* ants, the LD$_{50}$ of T8 was 2.0×10^{-9} mol/ant mg. In this test, T8 was ten times more toxic than nicotine (*43*).

Several syntheses of T8 and its homologue T4 (**64**) confirmed the proposed structures (*47* to *51*). Structures of four other members of the series (T7, T6, T5 and T3) were subsequently proposed by comparing their spectral properties with those of T8 and of model compounds (*44*). But recently, extensive NMR studies at high field, total syntheses and CD measurements indicated that these structures were not correct (*52, 53*). It follows from these studies that the structures of T7 and T3 had to be corrected to **65** and **66** respectively. Moreover, the absolute configuration of T4 was determined to be (5R,9S,11R) and that of T7 and T3 to be (5R,9R,11R) (*52*). In addition, the structures of T5 and T6 were corrected to **69** and **67** respectively (*53*). At the same time, the structures of T1 and T2 which are minor constituents of the secretion, were determined to be **70** and **68** respectively (*53*).

2.1.6. Other Alkaloids

Besides the alkaloids mentioned above, a few other nitrogen-containing compounds have been isolated from ants. But, in most cases

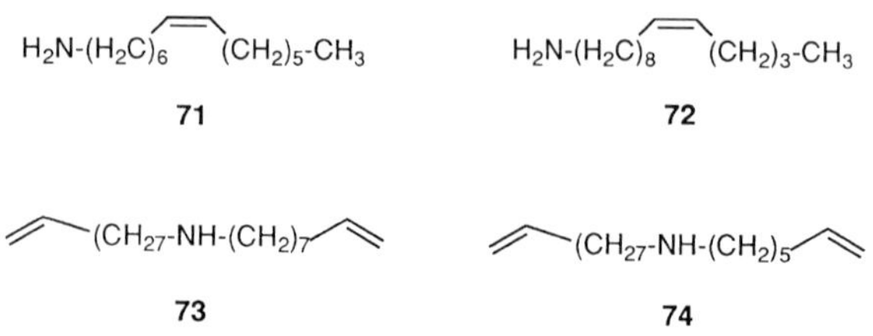

H₂N-(H₂C)₆ /=\ (CH₂)₅-CH₃   H₂N-(H₂C)₈ /=\ (CH₂)₃-CH₃
         71   **72**

/=\ (CH₂)₇-NH-(CH₂)₇ \=\   /=\ (CH₂)₇-NH-(CH₂)₅ \=\
         73   **74**

Fig. 7. Long-chain primary and secondary amines, isolated from ants

these compounds are not utilized as means of defense, but rather as trail [*e.g.* alkylpyrazines (*54*)] or alarm pheromones [*e.g.* actinidine (*55*)]. Furthermore, the defensive roles of some other compounds have not yet been determined clearly, for example, (Z)-7-tetradecenylamine (**71**) and (Z)-9-tetradecenylamine) (**72**) isolated from queens and workers of four different collections of *Monomorium floricola* (*56*) and two unsaturated secondary amines (**73** and **74**) isolated from the Dolichoderine ant *Technomyrmex albipes* (Fig. 7) (*57*).

2.2. Synthesis

2.2.1. Piperidines

A number of substituted piperidine are chemical weapons of several species of ants. The most important of these are the 2-methyl-6-alkylpiperidines or solenopsins which have been the subject of numerous syntheses. The syntheses reported prior to 1996 have been recently reviewed by us (*58*). In this chapter, we bring this topic up-to-date by the latest synthesis of (+)-solenopsin A (*59*), and by the two most recent syntheses of racemic 2,6-dialkylpiperidines containing unsaturation in the side chains (*10*, *11*).

In 1997 CHAKALAMANNIL and WANG reported a chiral synthesis of *trans*-(+)-solenopsin A, using nitrone **82** prepared from *N*-Cbz-L-alanine methyl ester (**75**) (Scheme 2) (*59*) in 34% yield. Amino acid derivative **75** was converted to the corresponding aldehyde **76** by treatment with diisobutylaluminium hydride. Wittig reaction of **76** with the commercially available phosphonium salt **77** mediated by *t*-BuOK yielded *cis*-alkene **78**. Palladium catalyzed reduction and treatment of the resulting primary amine **79** with an excess of benzoyl peroxide gave hydroxylamine benzoate **80** in 69% overall yield. Nitrone **82** was then obtained by saponification of **80** followed by cyclization of the resulting

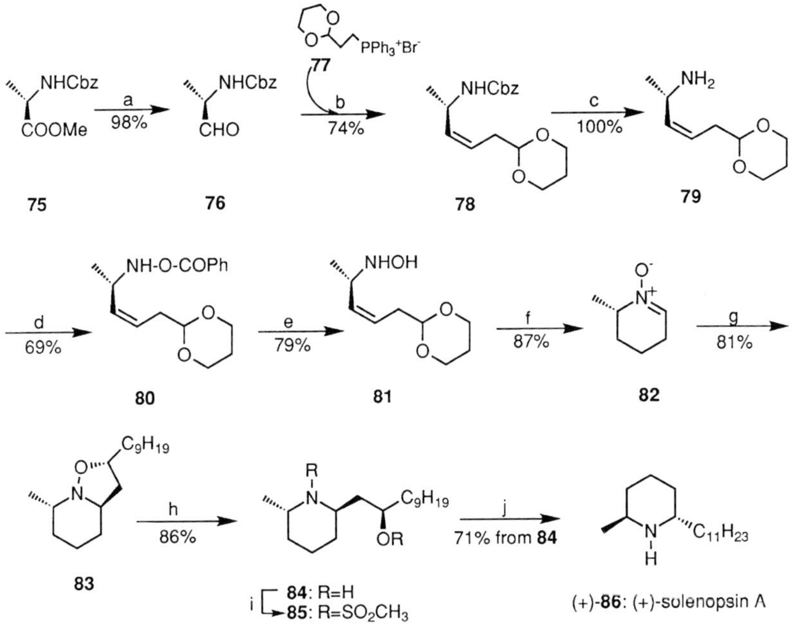

(a) DIBALH, CH$_2$Cl$_2$; (b) 77, ^{t}BuOK, THF; (c) H$_2$, Pd/C, EtOH, AcOEt; (d) (PhCOO)$_2$, CHCl$_3$, K$_2$CO$_3$; (e) LiOH, THF/MeOH; (f) dioxane, HCl; (g) 1-undecene; (h) Zn/AcOH; (i) MsCl, NEt$_3$, CH$_2$Cl$_2$; (j) LiAlH$_4$, THF.

Scheme 2. Asymmetric synthesis of (+)-*trans*-solenopsin A from L-alanine (59)

hydroxylamine derivative **81** in acidic medium. Nitrone **82** was transformed into (+)-solenopsin A using CARRUTHERS' protocol (60). Thus, reaction of nitrone **82** with an excess of 1-undecene at 145°C gave isoxazolidine **83** in 81% yield which underwent a reductive cleavage with zinc and acetic acid to afford amino alcohol **84**. The latter was converted into the bis-mesyl derivative **85** and subsequent lithium aluminium hydride reduction gave (+)-solenopsin A (**86**) in 71% yield from **84**. The overall yield of this ten-step synthesis is 17%.

In 1990, JONES *et al.* reported structure elucidation of a novel group of 2,6-dialkylpiperidines produced by the workers of *Monomorium delagoense* Forel which contain one or two double bonds in the side chains, such as *cis*- and *trans*-**1**, -**2**, -**3** (*10*). In order to assign their relative configuration, synthetic samples of (±)-**1**, (±)-**2** and (±)-**3** were prepared from the appropriate pyridines, **87** and **89** (Scheme 3) (*10*). The 2,6-dialkylpyridines were obtained in moderate yield by alkylation of the

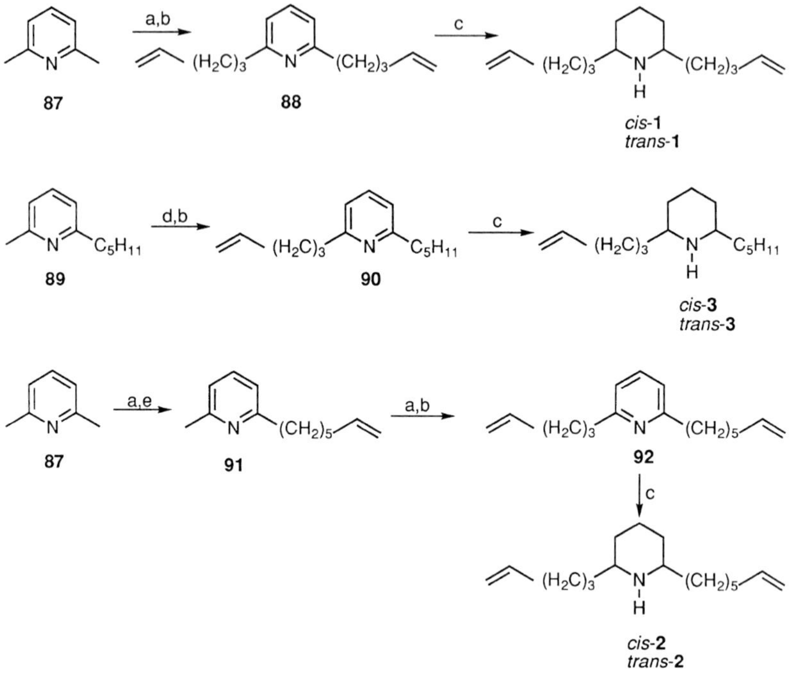

Scheme 3. Synthesis of (±)-cis- and (±)trans–1, -2, and -3 (10)

(a) 2 BuLi; (b) $CH_2=CH-CH_2-CH_2$-Br; (c) Na/EtOH; (d) BuLi; (e) $CH_2=CH-(CH_2)_4$-Br.

lithium salt of 2,6-lutidine anion, or in the case of **88**, by dialkylation of the dilithium salt of 2,6-lutidine. Reduction of the pyridine ring with sodium in ethanol provided a mixture of cis- and trans-piperidines without disturbing the side chain terminal double bonds. No yield was mentioned in the paper.

Recently, JONES and co-workers (11) isolated a mixture of cis- and trans-2-methyl-6-((4Z)-nonenyl)piperidine **6** from the poison gland of some Solenopsis species from Puerto Rico. In order to locate the side chain double bond and to determine its stereochemistry, the authors undertook the synthesis outlined in Scheme 4. The acetal derivative **93** was synthesized from lutidine anion and 2-(2-bromoethyl)-1,3-dioxolane. **93** Was deprotected to the aldehyde, which smoothly produced 2-methyl-6-((Z)-4-nonenyl)pyridine **94** through a Wittig reaction with the phosphonium ylide from 5-bromopentane. The pyridine ring of **94** was reduced, first with sodium in ethanol and then, after acidification to pH 6, by treatment with sodium cyanoborohydride to afford a mixture

References, pp. 221–229

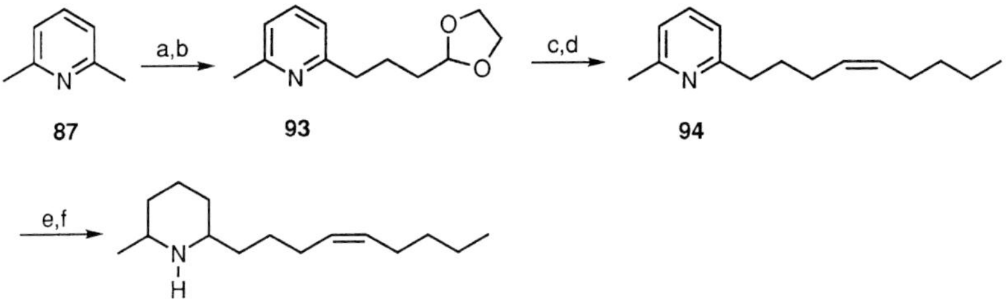

(a) BuLi, ether; (b) 2-(2-bromoethyl)-1,3-dioxolane; (c) HCl, HClO$_4$; (d) BuLi, Ph$_3$PC$_5$H$_{11}$; (e) Na/EtOH; (f) NaBH$_3$CN, pH=6.

Scheme 4. Synthesis of ($\pm$)-*cis*-**6** (*11*)

containing *cis*- and *trans*-**6** in a 3:1 ratio, respectively. No yield was mentioned in the original paper.

2.2.2. Pyrrolidines

A. Synthesis of Racemic Pyrrolidines

Many methods have been reported for the synthesis of racemic pyrrolidine alkaloids, including procedures based on the reductive amination of 1,4-diketones, the nitrone-alkene cycloaddition reaction, the 1,4-addition of an oxygen function to a diene, the intramolecular Michael type *N*-heterocyclization, the α-alkylation through the use of an enamidine, an alkyllactam or an acyl nitronate.

The first approach to synthesis of racemic pyrrolidines was based on the reductive amination of 1,4-diketones. This approach has been widely used by JONES *et al.* to confirm the structure of naturally occurring alkaloids isolated from ants (Scheme 5) (*61, 62*). Using this methodology, the authors synthesized ($\pm$)-*trans*-**10**, ($\pm$)-*trans*-**11**, ($\pm$)-*trans*-**16**, and ($\pm$)-*trans*-**17** as a mixture of *cis* and *trans* isomers. The 1,4-diketone **97** was prepared by condensation of 1-tetradecen-3-one (**95**) with propanal (**96**) in the presence of a thiazolium salt catalyst and triethylamine (*63*). Treatment of **97** with ammonium acetate, NaBH$_3$CN, KOH and subsequent reduction with NaBH$_4$ produced the *cis*- and *trans*-2,5-disubstituted pyrrolidines ($\pm$)-*trans*-**10** and ($\pm$)-*cis*-**10** in a 1:1 ratio, in 80% GC-yield (*61*). Pyrrolidines ($\pm$)-*trans*-**16**, ($\pm$)-*trans*-**17** and ($\pm$)-*trans*-**11** were synthesized in a similar manner. The 1,4-diketones **104**, **105** and **106** containing a terminal double bond were obtained from 1-alken-3-ones **98–100** and alkenals **101–103**, respectively (Scheme 6) (*61, 62*).

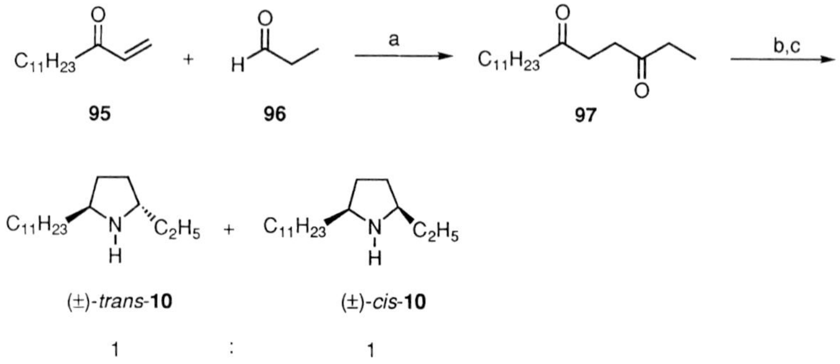

$C_{11}H_{23}$ 95 + H 96 →a→ $C_{11}H_{23}$ 97 →b,c→

(±)-trans-**10**   (±)-cis-**10**
  1   :   1

(a) Et$_3$N, 5-(2'-hydroxyethyl)-4-methyl-3-benzylthiazolium chloride; (b) NaBH$_3$CN, NH$_4$OAc, KOH, MeOH; (c) NaBH$_4$.

Scheme 5. Synthesis of (±)-cis- and trans-**10** (61, 62)

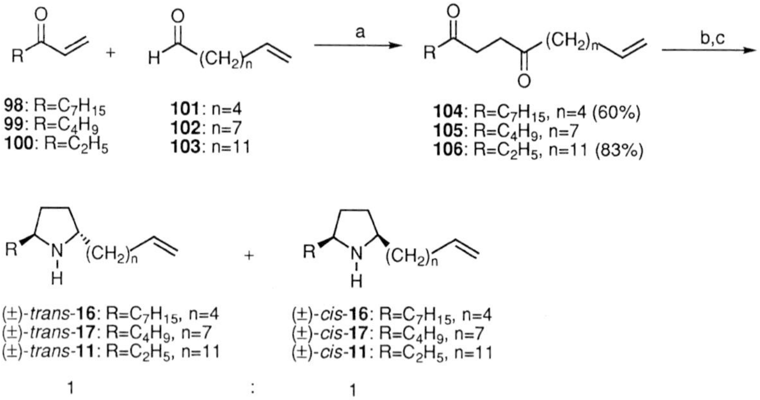

98: R=C$_7$H$_{15}$
99: R=C$_4$H$_9$
100: R=C$_2$H$_5$

101: n=4
102: n=7
103: n=11

104: R=C$_7$H$_{15}$, n=4 (60%)
105: R=C$_4$H$_9$, n=7
106: R=C$_2$H$_5$, n=11 (83%)

(±)-trans-**16**: R=C$_7$H$_{15}$, n=4
(±)-trans-**17**: R=C$_4$H$_9$, n=7
(±)-trans-**11**: R=C$_2$H$_5$, n=11

(±)-cis-**16**: R=C$_7$H$_{15}$, n=4
(±)-cis-**17**: R=C$_4$H$_9$, n=7
(±)-cis-**11**: R=C$_2$H$_5$, n=11

  1   :   1

(a) Et$_3$N, 5-(2'-hydroxyethyl)-4-methyl-3-benzylthiazolium chloride; (b) NaBH$_3$CN, NH$_4$OAc, KOH, MeOH; (c) NaBH$_4$.

Scheme 6. Synthesis of (±)-trans-**11**, (±)-trans-**16**, and (±)-trans-**17** (61, 62)

In the approach of TUFFARIELLO and PUGLIS (64), the alkaloids trans-2-butyl-5-heptylpyrrolidine [(±)-trans-**15**] and trans-2-hexyl-5-pentylpyrrolidine [(±)-trans-**20**] were synthesized by the nitrone-alkene addition reaction which consists of α,α'-dialkylation of cyclic amines through the use of nitrone methodology (Scheme 7). Cycloaddition of 1-pyrroline 1-oxide (**107**) and 1-heptene afforded adduct **108** in 87% yield. Oxidative cleavage (by m-chloroperoxybenzoic acid) of **108** yielded nitrone **110** regiospecifically (66%). A second cycloaddition was then effected

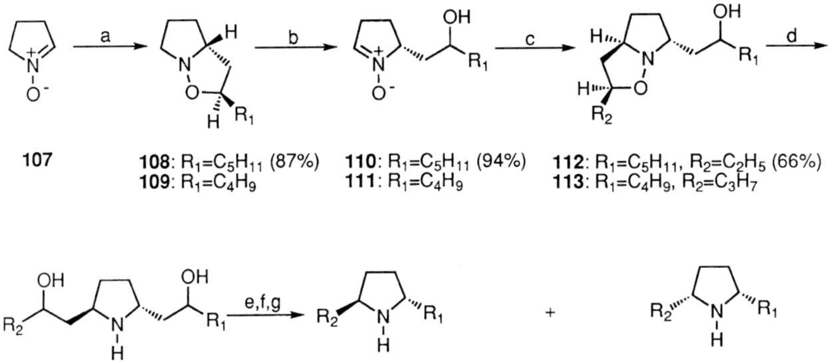

107

108: $R_1=C_5H_{11}$ (87%)   **110**: $R_1=C_5H_{11}$ (94%)   **112**: $R_1=C_5H_{11}$, $R_2=C_2H_5$ (66%)
109: $R_1=C_4H_9$ **111**: $R_1=C_4H_9$ **113**: $R_1=C_4H_9$, $R_2=C_3H_7$

114: $R_1=C_5H_{11}$, $R_2=C_2H_5$ (±)-*trans*-**15**: $R_1=C_7H_{15}$, $R_2=C_4H_9$ (±)-*cis*-**15**: $R_1=C_7H_{15}$, $R_2=C_4H_9$
115: $R_1=C_4H_9$, $R_2=C_3H_7$ (±)-*trans*-**20**: $R_1=C_6H_{13}$, $R_2=C_5H_{11}$ (±)-*cis*-**20**: $R_1=C_6H_{13}$, $R_2=C_5H_{11}$
                                           87   :   13

(a) for **108**: 1-heptene; for **109**: 1-hexene; (b) MCPBA, CH_2Cl_2; (c) for **112**: 1-butene; for **113**: 1-pentene; (d) $LiAlH_4$, THF; (e) MsCl, NEt_3, CH_2Cl_2; (f) $LiEt_3BH$; (g) sodium bis-(2-methoxyethoxy)aluminium hydride.

Scheme 7. Synthesis of (±)-*trans*-**15** and (±)-*trans*-**20** (*64*)

between **110** and 1-butene to provide **112**, which was reduced by $LiAlH_4$ to aminodiol **114**. The latter was exhaustively mesylated, after which both the mesyloxy (by $LiEt_3BH$) and the sulfonamide groups (by sodium *bis*-2-methoxyethoxyaluminium hydride) were subsequently removed by hydrogenolysis. The resulting 2-butyl-5-heptylpyrrolidine [(±)-*trans*-**15**] was 87% *trans* and was identical with the venom constituent of *Solenopsis fugax*. The second venom component, *trans*-2-hexyl-5-pentylpyrrolidine [(±)-*trans*-**20**], from *Solenopsis molesta* and *Solenopsis texana*, was synthesized in an analogous manner (**107**→ **109** → **111**→**113**→**115**→(±)-*trans*-**20**+(±)-*cis*-**20**).

Another approach to synthesis of pyrrolidine alkaloids consists in the stereocontrolled *anti*-1,4-addition of an amino and an oxygen function to a diene by palladium catalysis. By this methodology, BÄCKVALL *et al.* synthesized (±)-*trans*-**15** in 9 steps (Scheme 8) (*65*). (*E,E*)-Diene **120** was prepared in 65% yield from (*E*)-1-iodo-1-hexene (**117**) and the 1-alkenylaluminium derivative **119**, using the procedure described by NEGISHI (*66*). Palladium-catalysed chloroacetoxylation of **120** was highly 1,4-syn selective (>96%) producing the (R^*,R^*) isomer **121** as a mixture of two regioisomers in which the chlorine atom is close to either the C_4H_9 or the C_7H_{15} group. The chlorine was then substituted by NaNHTs with inversion of configuration without affecting the acetate

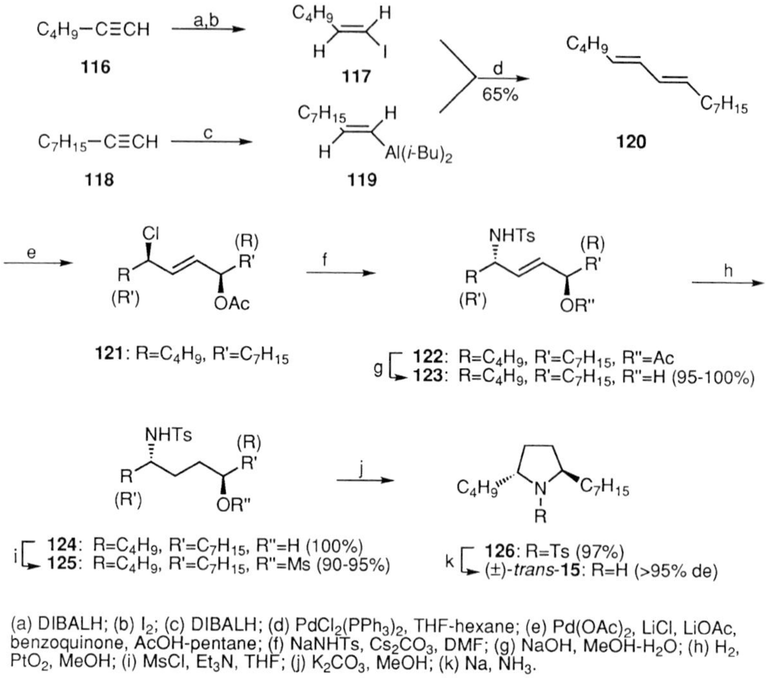

Scheme 8. Synthesis of (±)-*trans*-**15** (*65*)

(a) DIBALH; (b) I₂; (c) DIBALH; (d) PdCl₂(PPh₃)₂, THF-hexane; (e) Pd(OAc)₂, LiCl, LiOAc, benzoquinone, AcOH-pentane; (f) NaNHTs, Cs₂CO₃, DMF; (g) NaOH, MeOH-H₂O; (h) H₂, PtO₂, MeOH; (i) MsCl, Et₃N, THF; (j) K₂CO₃, MeOH; (k) Na, NH₃.

or the geometry of the double bond. Hydrolysis of **122** followed by reduction of the double bond afforded sulfonamido alcohol (S*,R*)-**124** almost quantitatively. Hydrogenation required a hydrogen pressure of at least 5 atm in order to avoid isomerization at C–O or C–N. Alcohol **124** was transformed to mesylate **125**, which was cyclized to *trans*-pyrrolidine **126** in 97% yield (>95% *trans*). Removal of the tosyl group by treatment with sodium in liquid ammonia provided (±)-*trans*-**15**.

Intramolecular Michael-type *N*-heterocyclization has been exploited by D'ANGELO and DUMAS to synthesize (±)-*trans*-2-heptyl-5-ethylpyrrolidine protected as its sulfonamide derivative **135** (Scheme 9) (*67*). (*S*)-1-phenylethylamine was first added to ketoenoate **127**, in 75% yield to give adduct **128** as a mixture of four diastereoisomers in the ratio 44:44:6:6. The mixture **128** was then quantitatively reduced by LiAlH₄ to a 7:3:1 mixture of *trans*- (**129**/**130**) and *cis*-alcohols (**131**/**132**), which were easily separated by flash chromatography on silica gel and shown to be a nearly equimolar mixture of two diastereoisomers. This is due to the absence of stereocontrol between the residual

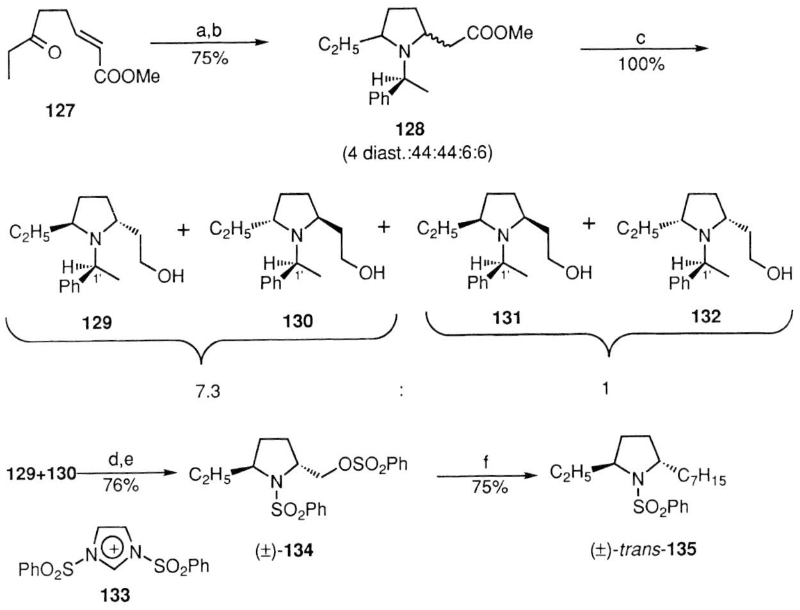

Scheme 9. Synthesis of (±)-*trans*-**135** (67)

(a) (1)-Phenylmethylamine; (b) NaBH$_3$CN; (c) LiAlH$_4$; (d) H$_2$, Pd/C; (e) **133**; (f) *n*-C$_5$H$_{11}$MgBr, Li$_2$CuCl$_4$.

stereogenic center at C-1' on the benzylic appendage and the C-2/C-5 centers on the ring. Cleavage of the benzylic N-appendage of the *trans* mixture **129/130** by catalytic hydrogenation led to a racemic aminoalcohol which was bis-sulfonated using Vilka's reagent (**133**) to give **134** in 76% overall yield. The latter derivative was then coupled with *n*-pentylmagnesium bromide, according to SCHLOSSER'S procedure (68), leading to (±)-**135** in 75% yield.

MEYERS *et al.* (69) have reported that saturated heterocycles, as their *tert*-butyl formamidines, may be transformed into enamidines by metalation-selenation-elimination. These enamidines are valuable precursors to 2-substituted, 2,4-disubstituted, 2,4,6-trisubstituted piperidines, 2-substituted and 2,5-disubstitued pyrrolidines prepared in a regiospecific manner. In particular, (±)-*trans*-2-heptyl-5-ethylpyrrolidines [(±)-*trans*-**142**] has been synthesized from *tert*-butyl formamidine (**137**) in six steps and 41% overall yield (Scheme 10). Metalation of **137** with *tert*-butyllithium, followed by addition of diphenyl diselenide generated the α-selenophenyl derivative **138** which readily underwent

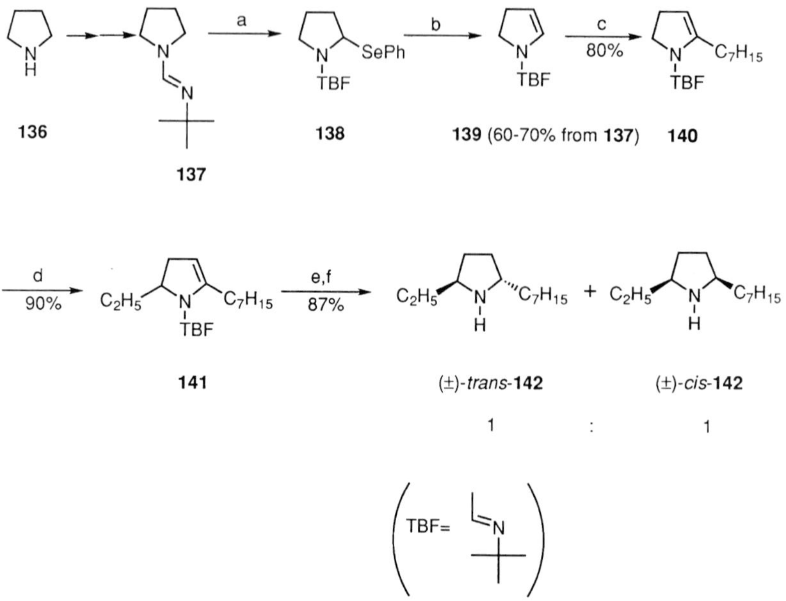

Scheme 10. Synthesis of (±)-trans-**142** (69)

(a) tert-BuLi, (PhSe)$_2$; (b) HCO$_3^-$; (c) n- or tert-BuLi, n-C$_7$H$_{15}$X; (d) tert-BuLi, C$_2$H$_5$I, TMEDA; (e) N$_2$H$_4$; (f) LiAlH$_4$.

elimination at room temperature by treatment with bicarbonate to give the cyclic enamidine **139** in 60–70% overall yield from **137**. The facile elimination of phenylselenol is obviously due to its acetal-like nature. Metalation of enamidine **139** was accomplished by using n-butyllithium or tert-butyllithium. Addition of heptyl halide gave excellent yields of 2-substituted enamidine **140** which was alkylated at the α-methylene carbon by addition of t-butyllithium and ethyl iodide in the presence of TMEDA to afford the disubstituted dihydropyrrole **141**. Removal of the formamidine with hydrazine resulted in the cyclic imine which was reduced directly with LiAlH$_4$ to the red fire ant venom component, (±)-trans-**142**, accompanied by its cis-isomer in a 1:1 ratio.

In the approach of LHOMMET and co-workers (70) (Scheme 11) ω-alkyllactam **143** can be transformed into ω-alkyl cyclic β-enaminoesters **147–149** which are good precursors of ant venom alkaloids. In this manner, the authors stereoselectively synthesized three natural 2,5-dialkylpyrrolidines (±)-trans-**153**, (±)-trans-**154** and (±)-trans-**155**. Lactim ether **144**, prepared by reaction of lactam **143** with dimethylsulfate, was condensed with Meldrum's acid in chloroform with a

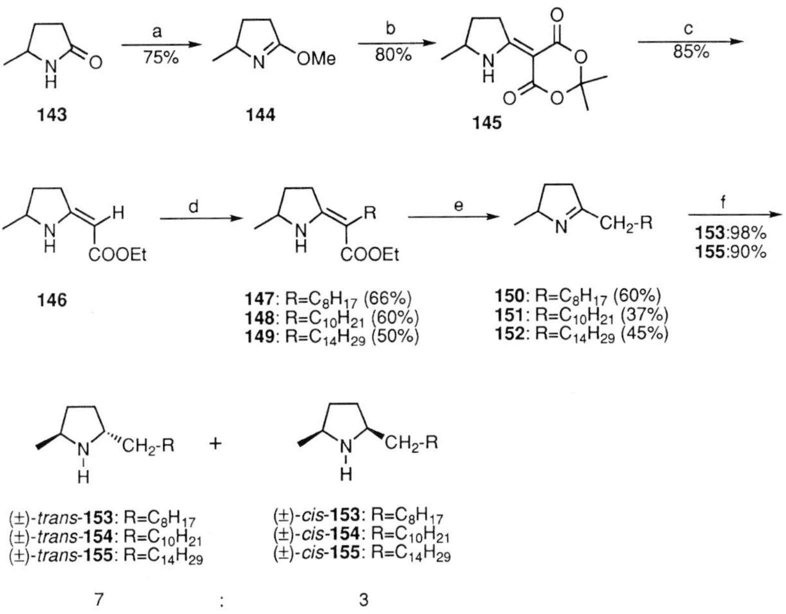

Scheme 11. Synthesis of (±)-*trans*-**153**, -**154** and -**155** (*70*)

(a) Me$_2$SO$_4$; (b) Meldrum's acid, Ni(acac)$_2$, CHCl$_3$; (c) EtOH, Δ; (d) RBr, NaH, toluene; (e) B$_3$BO$_3$, Δ; (f) NaBH$_3$, AcOH.

catalytic amount of Ni(acac)$_2$ to give compound **145** in 60% yield. A monodecarboxylating transesterification of β-enamino diester **145** led to acetate **146** in 85% yield. Afterwards, treatment of **146** with sodium hydride and addition of alkyl bromides yielded the corresponding C-alkylated β-enamino esters **147–149**, which were readily decarboxylated using boric acid to give imines **150–152**, respectively. Reduction of **150** with sodium borohydride in acetic acid led to a mixture of *trans*- and *cis*-pyrrolidines (±)-**153** in a 7:3 ratio. Mixtures of *trans*- and *cis*-pyrrolidines (±)-**154**, and of *trans*- and *cis*-analogues (±)-**155**, were obtained in a similar manner from imines **151** and **152**, respectively.

A short-step synthesis of unsymmetrical 2,5-dialkylpyrrolidines has been briefly reported by YOSHIKOSHI (*71*). It uses an acyl nitronate as starting point and has been used to synthesize several ant venom components (Scheme 12). Interestingly, hydrogenation of acetyl nitronates **160–161**, readily prepared from ketone enolates **156** and **157** and nitroalkenes **158** and **159**, was found to give different results depending on the catalyst employed. In particular, hydrogenation of **160**

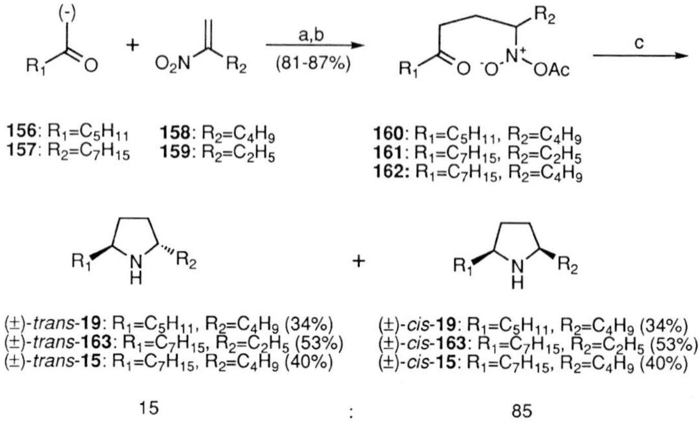

156: $R_1=C_5H_{11}$
157: $R_2=C_7H_{15}$

158: $R_2=C_4H_9$
159: $R_2=C_2H_5$

160: $R_1=C_5H_{11}$, $R_2=C_4H_9$
161: $R_1=C_7H_{15}$, $R_2=C_2H_5$
162: $R_1=C_7H_{15}$, $R_2=C_4H_9$

(±)-*trans*-**19**: $R_1=C_5H_{11}$, $R_2=C_4H_9$ (34%)
(±)-*trans*-**163**: $R_1=C_7H_{15}$, $R_2=C_2H_5$ (53%)
(±)-*trans*-**15**: $R_1=C_7H_{15}$, $R_2=C_4H_9$ (40%)

(±)-*cis*-**19**: $R_1=C_5H_{11}$, $R_2=C_4H_9$ (34%)
(±)-*cis*-**163**: $R_1=C_7H_{15}$, $R_2=C_2H_5$ (53%)
(±)-*cis*-**15**: $R_1=C_7H_{15}$, $R_2=C_4H_9$ (40%)

15 : 85

(a) LDA, THF; (b) Ac$_2$O; (c) H$_2$, 5% Rh/Al$_2$O$_3$, MeOH.

Scheme 12. Synthesis of (±)-*trans*-**15**, -**19** and -**163** *(71)*

over 5% Rh on Al$_2$O$_3$ in MeOH directly produced a mixture of *cis*- and *trans*-pyrrolidine **19** in which the former predominated (ratio 85:15). Alkaloids (±)-*trans*-**15** and (±)-*trans*-**163** were obtained in a similar manner. Taking into account that various 2,5-dialkylpyrrolidines are accessible from ketones **156** and **157** and nitroalkenes **158** and **159** in two steps, the yields of alkaloids are acceptable.

The pyrrolidines **30**–**35** have been isolated from the poison glands of female *Leptothoracini* ants (Fig. 2) *(31)*. These compounds are variously *N*-alkylated and possess a C-3 methyl (or a hydroxymethyl group for **35**). The structures of all compounds were established by synthesis of reference compounds (Scheme 13) *(31)*. Lactam **164** was reduced with LiAlH$_4$ to afford 3-methylpyrrolidine (**165**) which was then alkylated with the appropriate alkyl bromide, yielding the isoprenoid alkaloids (±)-**30**, (±)-**32** or (±)-**34**. Two other *N*-alkylated-3-methylpyrrolidines, (±)-**33** and (±)-**31**, were prepared from 3-methylpyrrolidine (**165**) and 3-methyltetrahydrofuran-2-one (**167**) respectively, as depicted in Scheme 13 *(31)*. Finally, *N*-isopentyl-3-hydroxymethylpyrrolidine [(±)-**35**] was prepared by reduction of **172** which was almost quantitatively obtained from 3-methylbutylamine (**171**) and diester **170**.

Recently, Veith *et al.* have reported a simple synthesis of (*S*)- and (*R*)-**30**, starting from (*S*)-4,5-dihydro-3-methyl-2(3H)furanone and dimethyl (*R*)-2-methylsuccinate, respectively *(72)*.

References, pp. 221–229

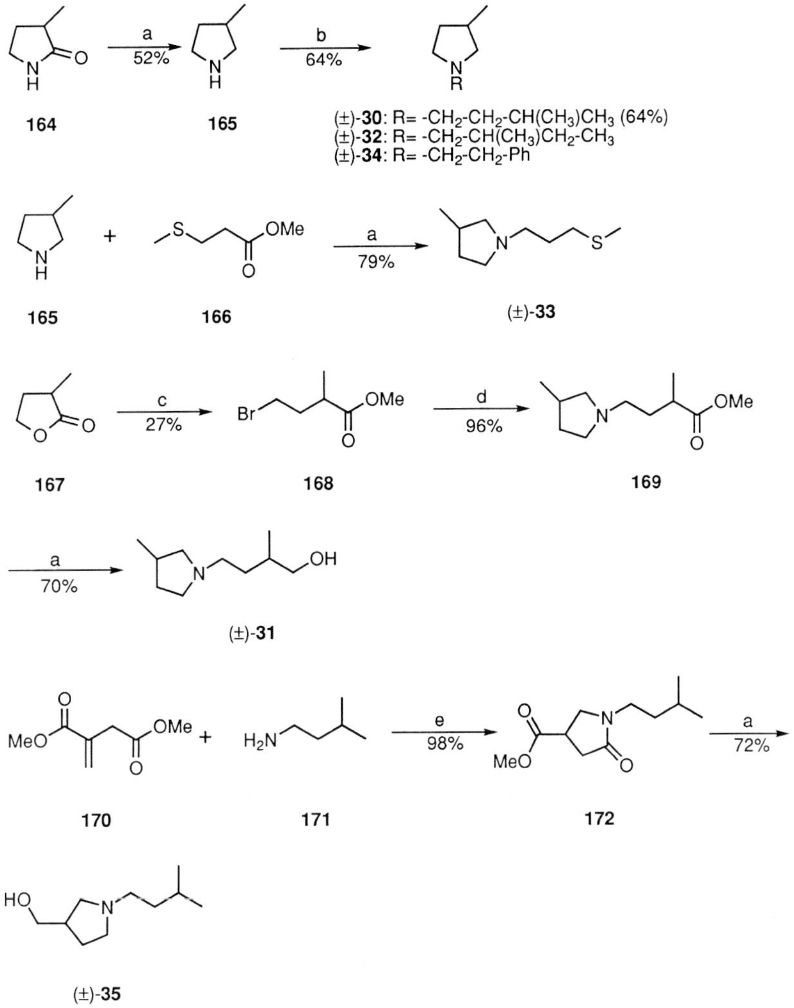

Scheme 13. Synthesis of (±)-**30**–**35** (*31*)

(a) LiAlH₄, ether; (b) RBr, K₂CO₃; (c) HBr, MeOH; (d) **165**, MeOH; (e) MeOH.

B. Syntheses of Nonracemic Pyrrolidines

Six of the seven syntheses of nonracemic ant pyrrolidine exploit the innate chirality of readily available α-aminoacids. In each case, the strategy requires transformation of the chosen α-aminoacid to a chiron that can undergo cyclization and/or a chain elongation. The last enantioselective synthesis of pyrrolidines we will describe in this section relies upon the chirality of D-mannitol.

a) From L-Norleucine

In 1990, MOMOSE *et al.* (*73*) reported the synthesis of (2*S*,5*S*)-*trans*-5-butyl-2-alkylpyrrolidines **15** and **19** in 9 steps, starting from L-norleucine **173** (Scheme 14). The α-butylated 4-pentenylcarbamate (*S*)-**177** was prepared from (*S*)-**173** by a procedure reported by SCHLESSINGER and IWANOWICZ (*74*) in 40% overall yield. **177** Underwent a cyclization mediated by mercuric acetate to provide the organomercurial bromide **178**, which was reduced by NaBH$_4$ in the presence of oxygen to give a 25:1 mixture of diastereoisomeric *cis*- and *trans*-2,5-disubstituted pyrrolidines (2*S*,5*S*)- and (2*S*,5*R*)-**179**. It should be

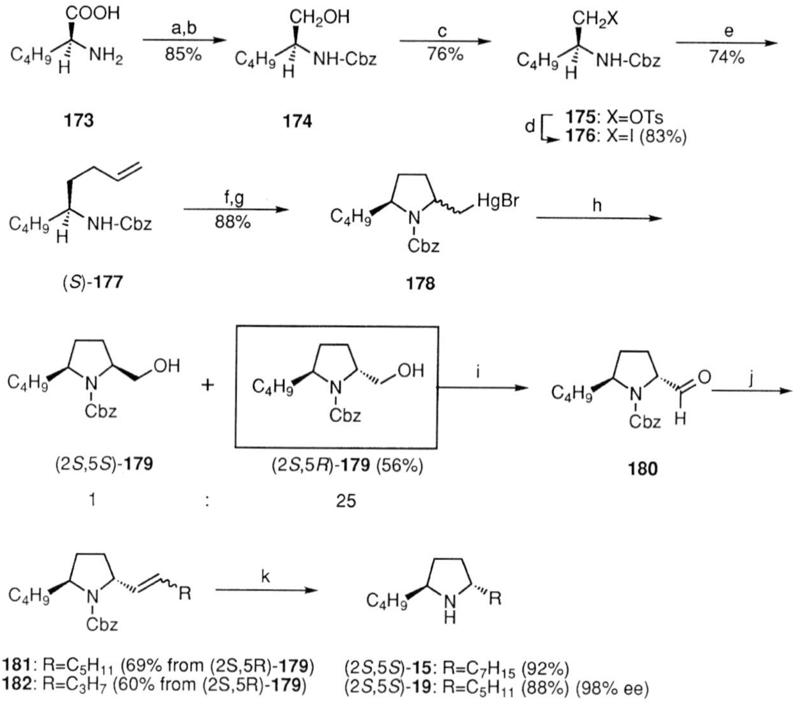

(a) LiAlH$_4$; (b) CbzCl, NaOH; (c) TsCl, NEt$_3$; (d) NaI, acetone; (e) allyl magnesium chloride, CuI; (f) Hg(OAc)$_2$, THF; (g) NaHCO$_3$, KBr; (h) O$_2$, NaBH$_4$, DMF; (i) (COCl)$_2$, DMSO; (j) Ph$_3$P$^+$CH$_2$R Br$^-$, *n*-BuLi; (k) H$_2$, Pd(OH)$_2$.

Scheme 14. Asymmetric synthesis of (2*S*,5*S*)-**15** and (2*S*,5*S*)-**19** from L-norleucine (*73*)

mentioned that diastereoselective electrophilic heterocyclization reactions, which proceed with high asymmetric induction, are commonly employed to control relative stereochemistry in cyclic compounds, and are increasingly recognized as an attractive method for stereoselective synthesis of biologically active heterocycles. The pure diastereoisomer *trans*-(2*S*,5*R*)-**179**, isolated by chromatography, underwent Swern oxidation to provide aldehyde **180**. This latter was subjected to a Wittig reaction using *n*-hexylidenetriphenylphosphorane, generated *in situ* from the appropriate phosphonium bromide and *n*-BuLi, to afford olefin **181** in 69% yield from *trans*-**179**. Finally, **181** underwent simultaneous hydrogenation of the double bond and hydrogenolysis of the Cbz group over Pd(OH)$_2$ under an hydrogen atmosphere to give the desired (2*S*,5*S*)-*trans*-5-butyl-2-heptylpyrrolidine **15**. (2*S*,5*S*)-*Trans*-5-butyl-2-pentylpyrrolidine (**19**), a component of the venom of *Solenopsis punctaticeps* was synthesized from *trans*-**179** in a similar manner.

b) From L-Proline

Reaction of the *N*-acyliminium ion precursor **184** (derived from (*S*)-proline) with C$_7$H$_{15}$Cu in the presence of BF$_3$·Et$_2$O gives preferentially the *trans* adduct **185**. Using such a procedure, a general synthetic route to (2*R*,5*R*)-*trans*-2,5-dialkylpyrrolidines has been developed by WISTRAND and SKRINJAR (*75*), as exemplified by enantioselective syntheses of the ant venom pheromones (2*R*,5*R*)-*trans*-5-butyl-2-heptylpyrrolidine (**15**), (2*R*,5*R*)-*trans*-5-ethyl-2-heptylpyrrolidine (**163**) and (2*R*,5*R*)-trans-5-butyl-2-(5-hexenyl)-pyrrolidine (**16**) (Scheme 15). **184** Was prepared

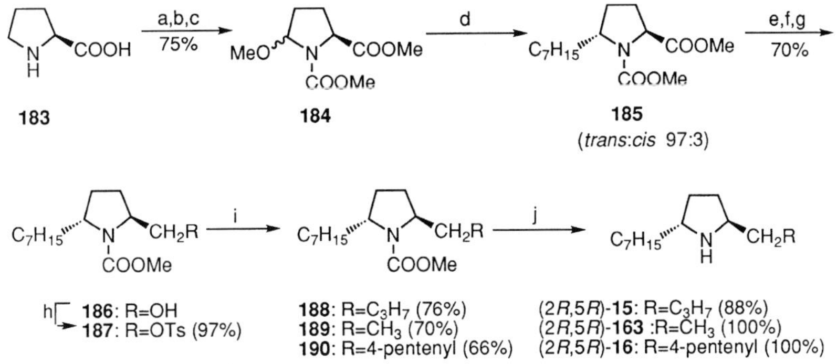

(a) ClCOOMe; (b) MeOH, HCl; (c) -2e⁻, MeOH; (d) C$_7$H$_{15}$Cu, BF$_3$.Et$_2$O, CuBr.Me$_2$S; (e) KOH, MeOH; (f) ClCOCOCl; (g) NaBH$_4$; (h) TsCl; (i) R$_2$CuLi; (j) TMSI.

Scheme 15. Asymmetric synthesis of (2*R*,5*R*)-**15**, (2*R*,5*R*)-**16**, and (2*R*,5*R*)-**163** from L-proline (*75*)

from natural (S)-proline in 75% yield via N-protection, esterification and anodic methoxylation following a procedure published by SHONO (76). This author had shown that reaction of **184** with p-nucleophiles in the presence of $TiCl_4$ preferably gives the *cis*-substituted product, whereas reaction of **79** with RCu in the presence of $BF_3 \cdot Et_2O$ yields the *trans* compound with a high degree of stereoselectivity. In this context, WISTRAND and SKRINJAR (75) optimized the *trans* stereoselectivity (*trans* : *cis* 97 : 3) by using two equivalents each of $C_7H_{15}Cu$, $BF_3 \cdot Et_2O$ and $CuBr \cdot Me_2S$ as the source of Cu(I). Functional group interconversion yielded tosylate **187** from **185**. Unlike bromides, tosylate **187** reacted very cleanly with R_2CuLi to yield the alkylated *trans* compounds **188–190** with no evidence of epimerization. Conversion to the free amines (2R,5R)-**15**, (2R,5R)-**16** and (2R,5R)-**163** was then carried out with Me_3SiI in very good yields.

c) From (S)-Pyroglutamic Acid

(S)-Pyroglutamic acid is cheap and readily available in very high ee. His potential as a chiral template for the asymmetric syntheses of alkaloids was practically neglected up to 1985.

The first synthesis of nonracemic pyrrolidine alkaloids using (S)-pyroglutamic acid as starting material was reported by RAPOPORT and SHIOSAKI in 1985 (77) (Scheme 16). They converted it to thiolactam **191** and then introduced the seven-carbon side chain as a single unit through a sulfide-contraction reaction. Thus, alkylation of thiolactam **191** with triflate **192**, prepared from 2-hydroxyoctanoic acid, proceeded smoothly at room temperature to thioimidate salt **193**. First triphenylphosphine then triethylamine were introduced at room temperature to form the vinylogous carbamate **194** as a 5:1 mixture of geometrical isomers in 70% yield from **191**. Conversion of **194** to pyrrolidine ester **195** was accomplished in one-pot using ammonium formate as the hydrogen source with >99% stereoselectivity. N-protection and hydrolysis of the ester group afforded *cis*-pyrrolidine **197** which was oxidized with $POCl_3$ and treated with KCN to afford amino nitrile **199** in 82% yield from **195**. On trapping iminium salt **198** with cyanide, the kinetic addition product **199** was a 1:3 mixture of the *cis* and *trans* isomers which was then equilibrated in a silica gel slurry to produce a 1:9 *cis*:*trans* mixture of amino nitriles **199**. Hydrolysis of the CN group with strong mineral acid led to the same 1:9 isomeric mixture of *cis*- and *trans*-pyrrolidines, and *trans*-pyrrolidine **197** was isolated by crystallization (>99% de) in 50% yield in three steps from *cis*-**197**. Elaboration of the butyl side chain at C-2 was carried out by reaction with an excess of propyl lithium.

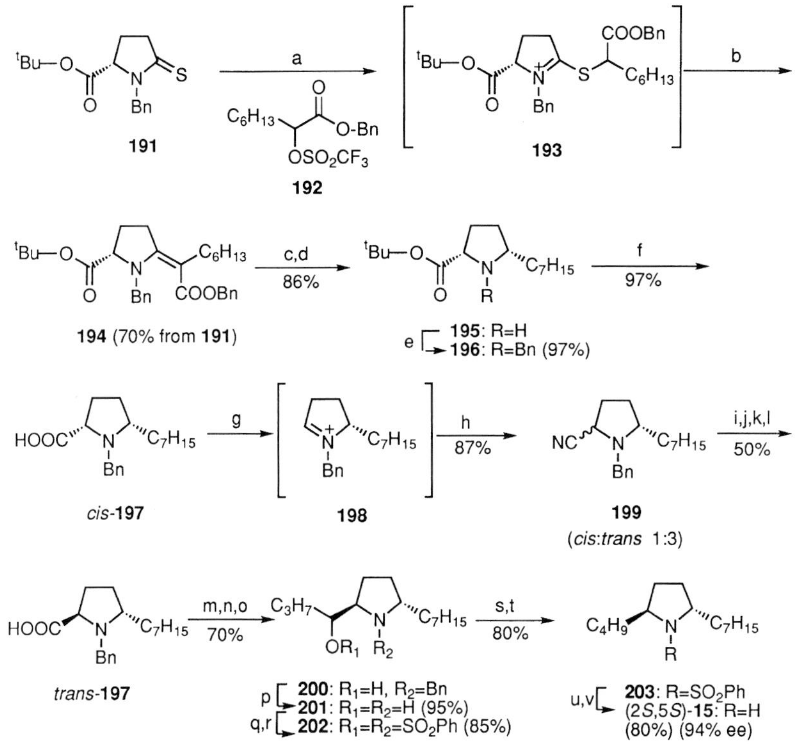

Scheme 16. Asymmetric synthesis of (2S,5S)-**15** from (S)-pyroglutamic acid (77)

(a) CH₃CN; (b) Ph₃P, NEt₃,CH₂Cl₂; (c) 10% Pd/C, ammonium formate, MeOH, AcOH; (d) NaHCO₃; (e) K₂CO₃, CH₃CN,PhCH₂Br; (f) propanol, H₂O, AcOH; (g) POCl₃; (h) isopropyl alcohol, KCN; (i) silica gel, isooctane, AcOEt; (j) conc. HCl; (k) NaHCO₃; (l) two crystallizations; (m) PrLi, ether; (n) acetone; (o) NaBH₄, EtOH; (p) H₂,10%Pd/C, AcOH; (q) phenylsulfonyl chloride, CHCl₃, NaOH; (r) H₃PO₄; (s) phenol, HBr; (t) NaOH.

Subsequent protic acid workup cleanly released the amino ketone which was immediately reduced with NaBH₄ to the diastereoisomeric amino alcohols **200**. Addition of organolithium reagents gave a single ketone, the absence of any epimeric ketone demonstrating that stereochemical integrity at C-2 was maintained despite the absence of an adjacent protecting nitrogen anion. Deoxygenation of amino alcohol **200** required hydrogenolysis of the secondary amino alcohol **201** and conversion to the bi-sulfonated product **202**. The latter was isolated in 90% yield and the sulfonate readily displaced with NaBH₄ in DMSO to provide **203**. Deprotection of **203** gave the final *trans*-(2S,5S)-dialkylpyrrolidine [(2S,5S)-**15**] (94% ee) by using HBr with phenol as bromine scavenger.

Capitalizing on the chemistry used in the racemic synthesis described in Scheme 11, LHOMMET *et al.* have developed general routes to non-racemic *trans*-2,5-dialkylated pyrrolidines and nonracemic pyrrolines, especially with unsaturated substituents. Enantioselective syntheses of natural pyrrolines will be discussed in Section 2.2.3. In 1991, these authors reported an enantioselective synthesis of (2*R*,5*R*)-*trans*-pyrrolidines **13** (Scheme 17) (*78*). It uses as starting material (*S*)-pyroglutamic acid [(*S*)-**204**] which was first transformed in six steps into a key intermediate, the β-enaminoester **211**. Treatment of **211** with sodium hydride, addition of 1-bromo-4-pentene and decarboxylation using boric acid, provided the natural imine (*R*)-**26** in 30% yield. Finally, (*R*)-**26** was reduced with sodium borohydride in acetic acid to provide (2*R*,5*R*)-*trans*-2-(5-hexenyl)-5-nonylpyrrolidine [(2*R*,5*R*)-**13**] and its *cis*-isomer in a 65:35 ratio.

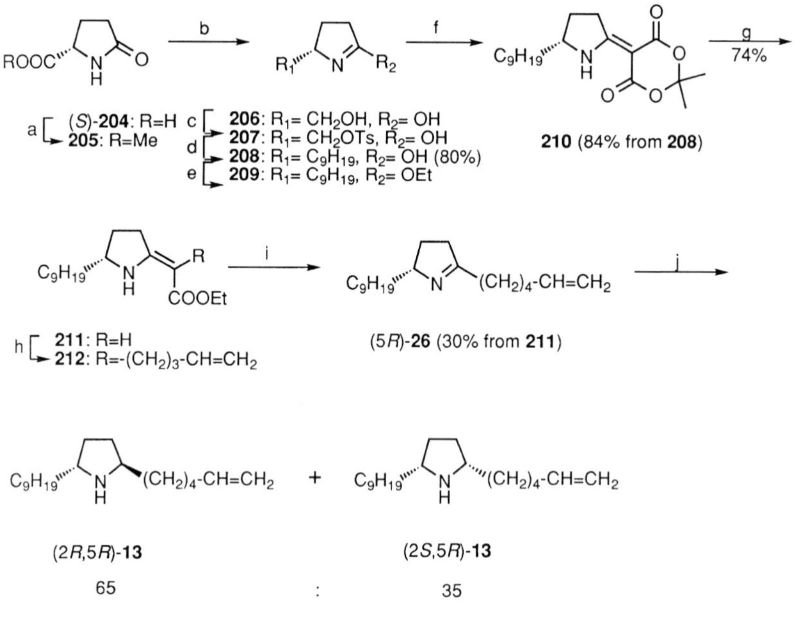

(a) MeOH, SOCl$_2$; (b) NaBH$_4$, EtOH; (c) TsCl, NEt$_3$; (d) (C$_8$H$_{17}$)$_2$CuLi; (e) Et$_3$OBF$_4$, CH$_2$Cl$_2$; (f) Meldrum's acid, Ni(acac)$_2$, CHCl$_3$; (g) EtOH, Δ; (h) NaH, Br-(CH$_2$)$_3$-CH=CH$_2$; (i) H$_3$BO$_3$, Δ; (j) NaBH$_4$, AcOH.

Scheme 17. Asymmetric synthesis of (2*S*,5*R*)-**13** from (*S*)-pyroglutamic acid (*78*)

Another route to nonracemic (2*R*,5*R*)-*trans*-pyrrolidine **13** was also described in this paper (Scheme 18) (*78*). Alcohol **206** was acetylated to give acetate **213** which was treated with Lawesson's reagent, affording

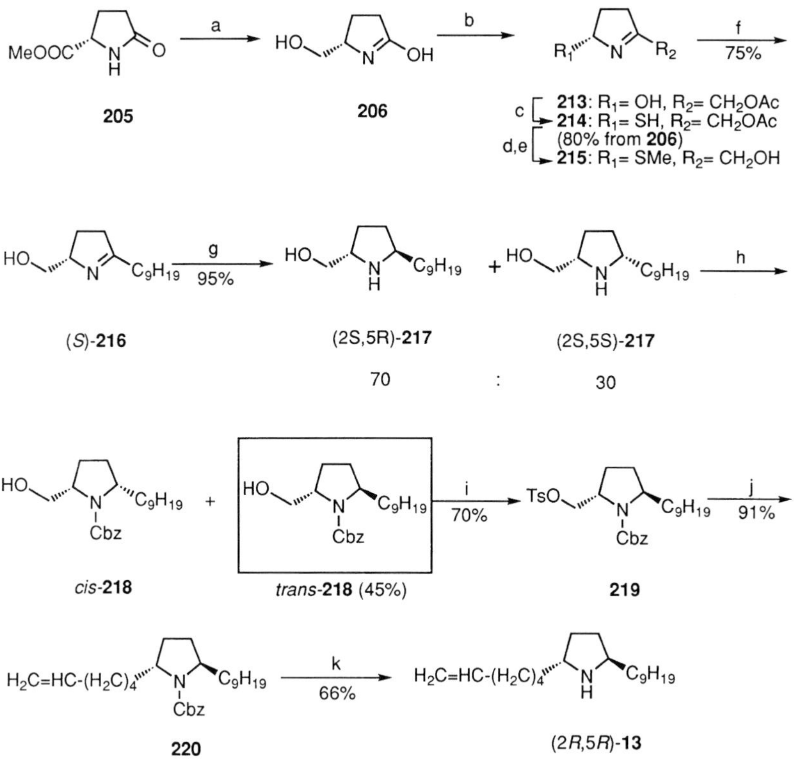

Scheme 18. Asymmetric synthesis of (2R,5R)-13 from (S)-pyroglutamic acid (78)

(a) NaBH₄, EtOH; (b) Ac₂O, pyridine; (c) Lawesson's reagent; (d) MeI; (e) MeONa, MeOH; (f) n-C₉H₁₉MgBr, ether, CH₂Cl₂; (g) NaBH(OAc)₃, toluene; (h) CbzCl, NaHCO₃, H₂O; (i) TsCl, pyridine; (j) (CH₂=CH-(CH₂)₃)₂CuLi; (k) Me₃SiI, CHCl₃.

the cristalline thiolactam **214** in 80% yield. *S*-methylation and alkylation with nonylmagnesium bromide led to (*S*)-imino alcohol **216** which was then reduced using NaBH(OAc)₃ to give a mixture of *trans*- and *cis*-hydroxymethylpyrrolidines **217** in a 70:30 ratio. The *trans* isomer, isolated as its carbamate **218** by flash chromatography, was converted into tosylate **219** in 70% yield. Chain elongation was carried out by reaction with lithium dipentenyl cuprate to yield pyrrolidine carbamate **220** in very good yield. Finally, treatment of **220** with Me₃SiI gave the natural (2*R*,5*R*)-*trans*-2-(5-hexenyl)-5-nonylpyrrolidine [(2*R*,5*R*)-**13**] in 66% yield.

d) From (−)-Phenylglycinol

The synthesis of (2S,5S)-*trans*-pyrrolidines **15** and **142** by HUSSON *et al.* (*79, 80*) is based on their earlier procedure for the synthesis of enantiopure solenopsins (*81*). It uses as starting material 2-cyano-5-oxazolopyrrolidine **223** which is obtained by a Robinson-Schöpf type condensation between formaldehyde and (−)-phenylglycinol [(R)-**221**] in the presence of KCN, followed by addition of 3-bromopropanal (**222**) (Scheme 19). Synthon **223** reacts chemo- and stereoselectively at C-2 and C-5 thus allowing control of the absolute configuration at these centers. Therefore, the anion of **223** generated by treatment with LDA was alkylated on C-2 with heptyl bromide to afford **224** as a 1:1 mixture of two diastereoisomers. This was followed by selective removal of the cyano group without opening of the oxazolidine ring by treatment with a modest excess of lithium in liquid ammonia, which fixed the (S)-configuration at C-2. The ethyl side chain was introduced at C-5 of **225** by reaction with C_2H_5MgBr giving a 72:28 mixture of *trans*:*cis*

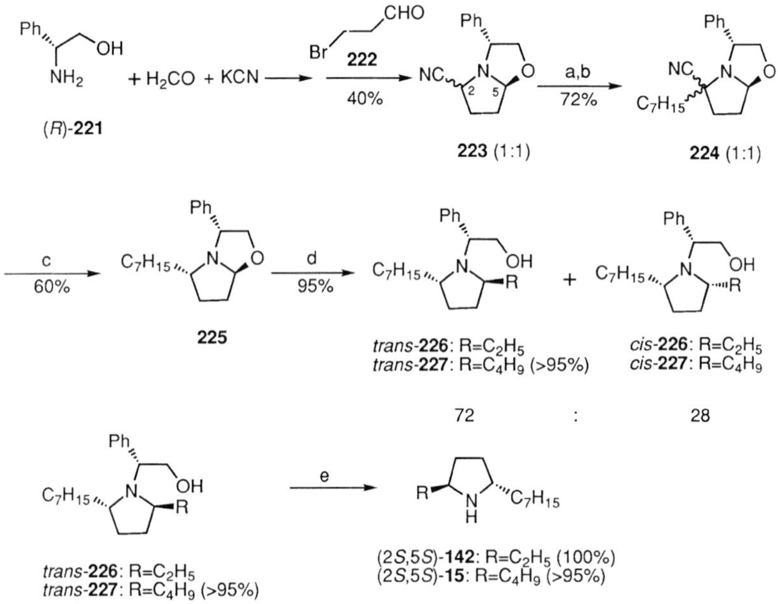

(a) LDA, TMEDA; (b) $C_7H_{15}Br$; (c) Li, NH_3; (d) C_2H_5MgBr, ether; (e) Pd/C, H_2, AcOH.

Scheme 19. Asymmetric synthesis of (2S,5S)-**15** and (2S,5S)-**142** from (−)-phenylglycinol (*79, 80*)

epimers **226** in a 95% yield. The observed *trans* stereoselectivity was explained by a mechanism which involves prior formation of an iminium ion by opening of the oxazolidine ring and preferential addition of the nucleophile from the less hindered face of the molecule. Compound *trans*-**226** was separated from the *cis* isomer by flash chromatography, and the chiral auxiliary was cleaved in quantitative yield to afford the corresponding secondary amine (+)-(2*S*,5*S*)-**142**. Alkaloid (2*S*,5*S*)-**15** was synthesized in a similar manner.

e) From D-Mannitol

In 1991, MACHINAGA and KIBAYASHI reported a 14 step synthesis of (2*S*,5*S*)-*trans*-2-butyl-5-pentylpyrrolidine [(2*S*,5*S*)-**19**] and its enantiomer, starting from D-mannitol (Scheme 20) (*82*). This synthesis is based on using the optically active diepoxide **230** as a C_2 symmetrical building block. Both the (*S*,*S*) and (*R*,*R*)-diepoxides **230** were easily prepared in two and four steps, respectively, using (*S*,*S*)-1,2,5,6-hexanetetraol [(*S*,*S*)-**228**] prepared from D-mannitol as a single common chiral synthon. Thus tosylation of the primary alcohol functions of tetraol **228** followed by alkaline treatment of ditosylate **229** provided (*S*,*S*)-**230** in 58% overall yield from **228**. On the other hand, **228** was converted to dimesylate **232** by protection of the primary alcohol functions by silylation, followed by mesylation of **231**, in 77% yield. Removal of the silyl groups of **232** by acid treatment and subsequent alkaline treatment of the resultant diol **223**, led to (*R*,*R*)-**230** in 51% overall yield from **231**. Treatment of (*S*,*S*)-**230** with Grignard reagent **234** resulted in epoxide ring opening to give diol (*R*,*R*)-**235** which was converted to **237** by silylation followed by debenzylation in 47% yield from (*S*,*S*)-**230**. Ditosylate **238** was treated with one equivalent of Super-Hydride, and, in this way, the requisite monotosylate (5*R*,8*R*)-**239** was obtained in 60% yield. Coupling with methylmagnesium bromide catalysed by Li_2CuCl_4 and removal of both silyl groups by acid treatment gave the unsymmetrical diol (5*R*,8*R*)-**241**. This latter compound reacted with thionyl chloride and triethylamine and then with RuO_4 to afford cyclic sulfate (5*R*,8*R*)-**242** in 80% yield from (5*R*,8*R*)-**240**. Nucleophilic displacement on (5*R*,8*R*)-**242** with lithium azide and subsequent acidic hydrolysis resulted in an unseparable mixture of the two azides (6*R*,9*S*)-**243** and (5*R*,8*S*)-**244** in a 1:1 ratio, which was treated with methanesulfonyl chloride to give a 1:1 mixture of the corresponding mesylates (6*R*,9*S*)-**245** and (5*R*,8*S*)-**246** in a 92% yield. The mixture, without separation was subjected to hydrogenation over Pd/C to produce (2*S*,5*S*)-*trans*-2-butyl-5-pentylpyrrolidine [(+)-**19**] as a single product.

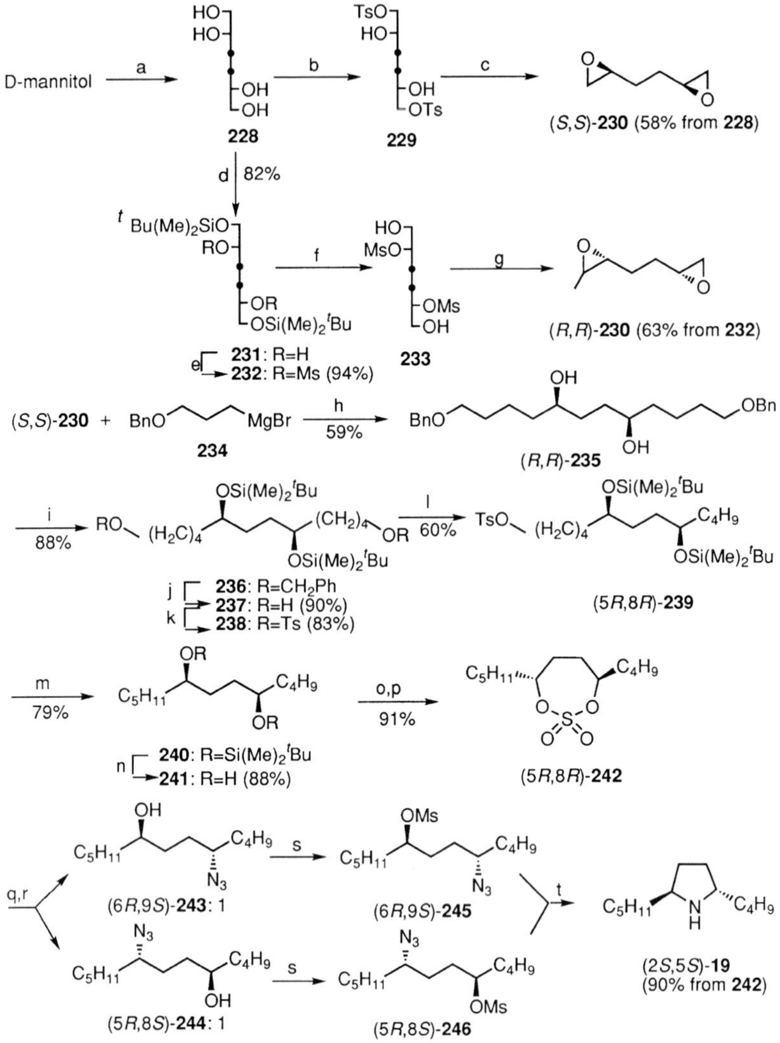

(a) *J. Chem. Soc. Chem. Com.*, 813 (1969); (b) TsCl, pyridine; (c) aq. NaOH, THF; (d) ^{t}Bu(Me)$_2$SiCl; (e) MsCl; (f) HCl; (g) aq. KOH, THF; (h) CuI cat., THF; (i) ^{t}Bu(Me)$_2$SiCl, imidazole; (j) H$_2$, Pd/C; (k) NaH, PhCH$_2$Br; (l) LiEt$_3$BH, THF; (m) MeMgBr, Li$_2$CuCl$_4$ cat; (n) HCl; (o) SOCl$_2$, NEt$_3$; (p) RuO$_4$; (q) LiN$_3$; (r) H$_2$SO$_4$; (s) MsCl; (t) H$_2$, Pd/C.

Scheme 20. Asymmetric synthesis of (2*S*,5*S*)-**19** from D-mannitol (*82*)

2.2.3. Pyrrolines

Several syntheses of racemic 2,5-dialkylpyrroline alkaloids have been reported.

The synthetic scheme developed by LHOMMET et al. (*26*) is based on lactam chemistry, a procedure the authors have also used for the synthesis of racemic 2,5-dialkylpyrrolidines (see Scheme 11, section 2.2.2). This scheme allowed them to regiospecifically synthesize imines (±)-**25** and (±)-**26** without any double bond transposition (Scheme 21). Overall yields of **25** and **26** from lactams **255** and **256** were 13% and 10% respectively. Capitalizing on this chemistry, LHOMMET and co-workers (*78*) reported three years later an enantioselective synthesis of (5*R*)-**26** (*S*)-pyroglutamic acid in 15% yield (Scheme 17, section 2.2.2).

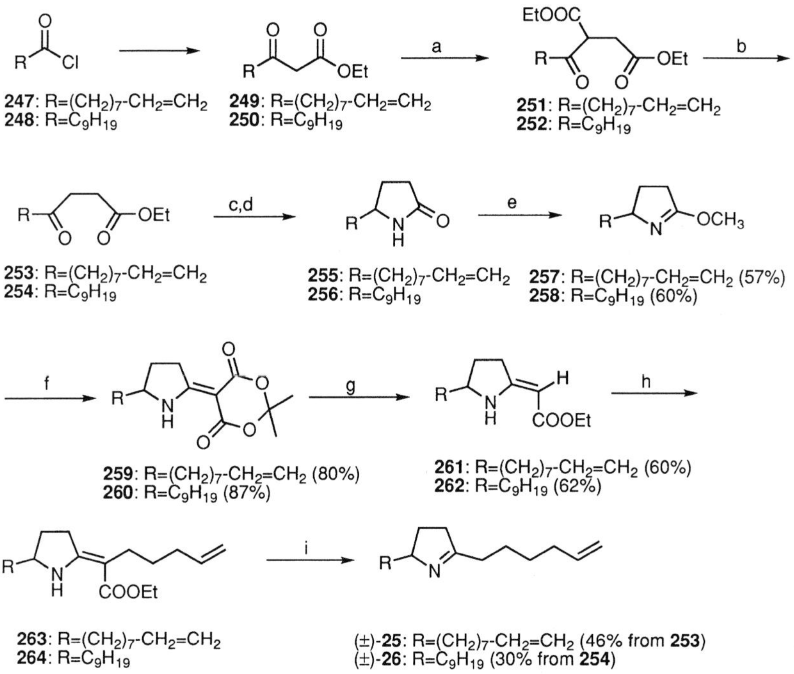

247: R=$(CH_2)_7$-CH=CH_2
248: R=C_9H_{19}

249: R=$(CH_2)_7$-CH=CH_2
250: R=C_9H_{19}

251: R=$(CH_2)_7$-CH=CH_2
252: R=C_9H_{19}

253: R=$(CH_2)_7$-CH=CH_2
254: R=C_9H_{19}

255: R=$(CH_2)_7$-CH=CH_2
256: R=C_9H_{19}

257: R=$(CH_2)_7$-CH=CH_2 (57%)
258: R=C_9H_{19} (60%)

259: R=$(CH_2)_7$-CH=CH_2 (80%)
260: R=C_9H_{19} (87%)

261: R=$(CH_2)_7$-CH=CH_2 (60%)
262: R=C_9H_{19} (62%)

263: R=$(CH_2)_7$-CH=CH_2
264: R=C_9H_{19}

(±)-**25**: R=$(CH_2)_7$-CH=CH_2 (46% from **253**)
(±)-**26**: R=C_9H_{19} (30% from **254**)

(a) Br-CH_2-COOEt; (b) H_3BO_3; (c) NH_3, H_2, Raney Ni; (d) $NaBH_3CN$; (e) Me_2SO_4; (f) Meldrum's acid, Ni(acac)$_2$, $CHCl_3$; (g) EtOH, Δ; (h) Br-$(CH_2)_3$-CH=CH_2, NaH, toluene; (i) H_3BO_3, Δ.

Scheme 21. Synthesis of (±)-**25** and (±)-**26** (*26*)

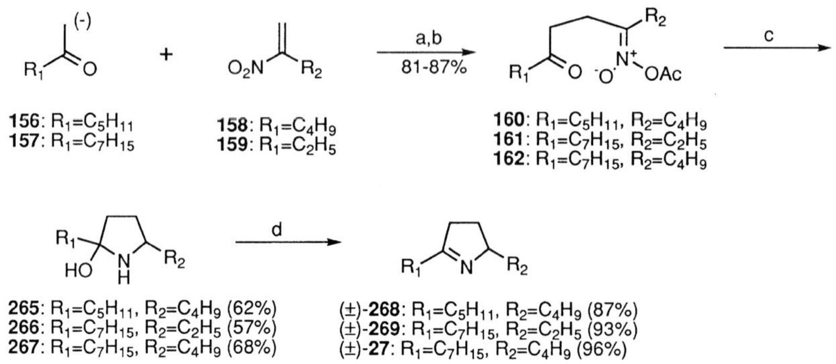

156: $R_1=C_5H_{11}$
157: $R_1=C_7H_{15}$

158: $R_1=C_4H_9$
159: $R_1=C_2H_5$

160: $R_1=C_5H_{11}$, $R_2=C_4H_9$
161: $R_1=C_7H_{15}$, $R_2=C_2H_5$
162: $R_1=C_7H_{15}$, $R_2=C_4H_9$

265: $R_1=C_5H_{11}$, $R_2=C_4H_9$ (62%)
266: $R_1=C_7H_{15}$, $R_2=C_2H_5$ (57%)
267: $R_1=C_7H_{15}$, $R_2=C_4H_9$ (68%)

(±)-**268**: $R_1=C_5H_{11}$, $R_2=C_4H_9$ (87%)
(±)-**269**: $R_1=C_7H_{15}$, $R_2=C_2H_5$ (93%)
(±)-**27**: $R_1=C_7H_{15}$, $R_2=C_4H_9$ (96%)

(a) LDA, THF; (b) Ac$_2$O; (c) H$_2$, PtO$_2$, AcOH; (d) PPTS, CHCl$_3$.

Scheme 22. Synthesis of (±)-**27**, (±)-**268** and (±)-**269** *(71)*

A short synthesis of non-symmetrical racemic 2,5-dialkylpyrrolines has been reported by YOSHIKOSHI *et al.* (Scheme 22) *(71)*. With an acyl nitronate as starting material it has been employed to synthesize (±)-**27**, (±)-**268** and (±)-**269**, three pyrrolines isolated from the genera *Solenopsis* and *Monomorium*. Acyl nitronates **160–162** were prepared as before (cf. Scheme 12, section 2.2.2). Hydrogenation of **160–162** was found to give, as already mentioned, different results depending on the catalyst employed. Thus hydrogenation of **160–162** over 5% Rh on Al$_2$O$_3$ in methanol directly produced 2,5-dialkylpyrrolidines while hydrogenation over PtO$_2$ in acetic acid yielded 2,5-dialkyl-2-hydroxypyrrolidines **265–267** which are suitable precursors for the synthesis of 2,5-dialkylpyrrolines. Indeed, conversion of **265–267** into (±)-**268**, (±)-**269** or (±)-**27** respectively was easily performed by treatment of **265-267** with PPTS in CHCl$_3$ to give excellent yields of the corresponding pyrroline derivatives.

In 1988, JONES *et al.* identified two novel dialkylpyrrolines **21** and **22** from an Australian species of *Monomorium* *(25)*. In order to confirm this assignment, pyrrolines (±)-**21** and (±)-**22** were synthesized as shown in Scheme 23, which is based on reductive amination of 1,4-diketones and has been widely used by these authors to synthesize ant alkaloids (see Scheme 5, section 2.2.2). The long chain 1,4-diketones **97**, prepared by condensation of 1-tetradecen-3-one with propanal, underwent reductive amination to provide the *cis*- and *trans*-2,5-disubstituted pyrrolidines (±)-**10** in a 1:1 ratio in 80% GC-yield. Treatment of this mixture with NaOCl and NaOH yielded an unseparable mixture of the isomeric 2-ethyl-5-undecylpyrrolines (±)-**21** and (±)-**22** in 68% GC-yield.

References, pp. 221–229

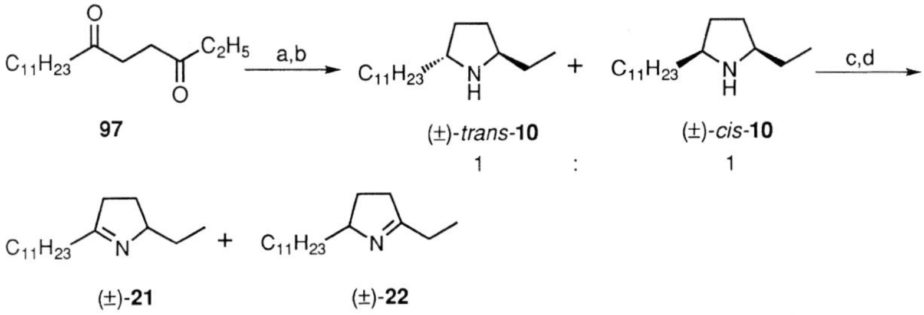

(a) NH₄OAc, NaBH₃CN, MeOH, KOH; (b) NaBH₄; (c) NaOCl, MeOH; (d) NaOH.

Scheme 23. Synthesis of (±)-**21** and (±)-**22** (*25*)

Three years later JONES and co-workers studied the venom of the ant *Megalomyrmex foreli* from Costa Rica and reported the presence of two new pyrrolines, **28** and **29**, whose structures were assigned from their spectral and chemical behavior, and unambiguous syntheses (Scheme 24) (*30*). Nitro ketal **272**, formed by addition of 1-nitropentane to methylvinylketone, was hydrogenated to pyrroline **273** in nearly 60% yield from 1-nitropentane. Compound **273** was deprotonated with LDA, and kinetic azaenolate trapping with diethyl chlorophosphate afforded pyrroline phosphonate **274** in a good yield. Condensation of the anion of

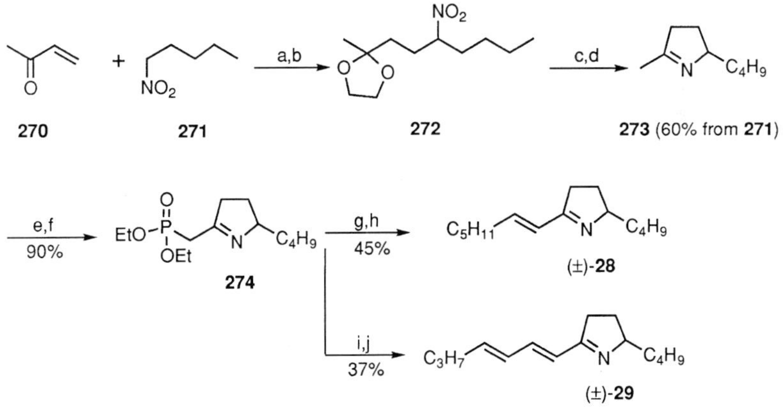

(a) tributylphosphine, benzene; (b) ethylene glycol, *p*-TsOH; (c) H₂, 10% Pd/C, EtOH; (d) HCl 10%, HClO₄ 60%; (e) diisopropylamine, BuLi, THF; (f) Diethylchlorophosphate, THF; (g) BuLi, THF; (h) hexenal; (i) BuLi, THF; (j) (*E*)-2-hexenal.

Scheme 24. Synthesis of (±)-**28** and (±)-**29** (*30*)

274 with hexanal or (E)-2-hexenal in a Wadsworth-Emmons reaction formed (±)-**28** and (±)-**29**, respectively, in moderate yield.

2.2.4. Pyrrolizidines

2.2.4.1. 3,5-Dialkylpyrrolizidines

In order to enhance the clarity of the discussion, we here mention that the Z and E descriptors for 3,5-dialkylpyrrolizidines refer to the relative stereochemistry of the hydrogens at C-5 and C-8 relative to the hydrogen at C-3, as illustrated in Scheme 25 (*83*).

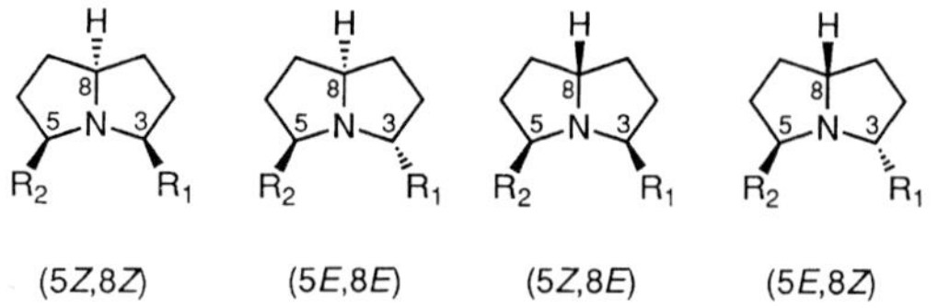

(5Z,8Z)   (5E,8E)   (5Z,8E)   (5E,8Z)

Scheme 25. E,Z nomenclature for specifying the relative configuration of 3,5-dialkylpyrrolizidines (*83*)

Two different 3,5-dialkylpyrrolizidine alkaloids, xenovenine [(5Z,8E)-**36**] and (5E,8E)-3-butyl-5-hexylpyrrolizidine **42**, have been isolated from the poison gland of ants. Several syntheses of racemic and nonracemic xenovenine, an important component of *Solenopsis xenovenum* venom, have been described.

A. Xenovenine (**36**)

Nitrones were utilized by LATHBURY and GALLAGHER (*84*) for the diastereoselective synthesis of (±)-xenovenine [(±)-**36** (Scheme 26)]. The key step was the cyclization of the (E)-allenic oxime **276** to produce the substituted nitrone **277**, which was trapped with methylvinylketone as 1,3-dipolarophile to give **278** as a 1:1 mixture of diastereoisomers in 48% overall yield. The hydrogenation of this mixture afforded hydroxypyrrolizidine **279** by reduction of the double bond, cleavage of the N-O bond, and intramolecular reductive amination. Removal of the hydroxyl group in **279** by Jones oxidation gave a single ketone which was immediately converted into 1,3-dithiolane **280** in 48% yield from **279**. Desulphurisation of **280** gave (±)-**36** in 61% yield.

Another synthesis of (±)-xenovenine by HESSE and co-workers (*85*) used 5-nitropentadecane-2,8-dione (**282**) as key intermediate

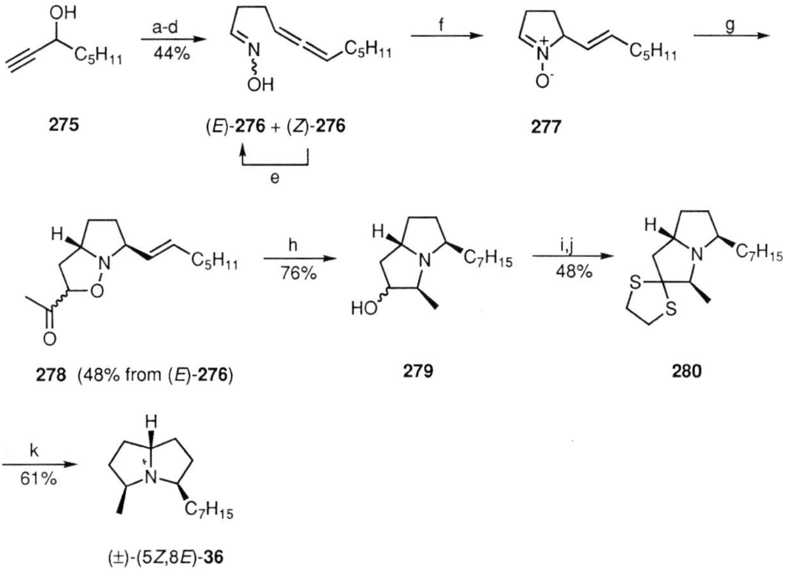

Scheme 26. Synthesis of (±)-xenovenine using nitrone cycloaddition (84)

(a) MeC(OEt)₃, pivalic acid; (b) LiAlH₄, ether; (c) *p*-MeC₆H₄-SO₂Cl, pyridine, then KCN, Me₂SO; (d) Bu₂AlH, ether and quenched with NH₂OH.HCl, NaOAc, H₂O; (e) CHCl₃, overnight, silica gel chromatography; (f) AgBF₄, CH₂Cl₂; (g) MeC(=O)CH=CH₂, THF; (h) H₂, PdCl₂, EtOH; (i) Jones' reagent, acetone; (j) HSCH₂CH₂SH, BF₃.Et₂O, CH₂Cl₂; (k) W-2 Raney Ni, EtOH.

(Scheme 27). The synthesis of **282** by a Michael reaction between nitromethane and α,β-unsaturated ketones was found to be more difficult than expected. Finally, reaction of nitromethane with methylvinylketone, catalysed with tributylphosphine, afforded **281** which, after reaction with dec-1-en-3-one (**98**), yielded **282**. Reduction of the latter, first with NaBH₃CN in the presence of NH₄OAc, KOH and then with NaBH₄, provided all four isomers of (±)-**36**, the natural (5Z,8E) isomer being the major one. The overall yield of the synthesis cannot be calculated since no yields were quoted in the original paper (85).

Recently, LHOMMET and co-workers reported synthesis in eight steps of a diastereoisomer of xenovenine, presumably (±)-(5Z,8Z)-**36** (Scheme 28) (86). β-Enaminolactone **284**, obtained by condensation of lactim ether **144** with acetylbutyrolactone **283**, was easily hydrolyzed with aqueous hydrochloric acid, leading to cyclic iminoalcohol **285** in 92% yield. Catalytic reduction of the latter led to the *cis*-2,5-disubstituted pyrrolidine **286** in high diastereoselectivity. Protection of the secondary amine of **286** with benzylchloroformate and oxidation of

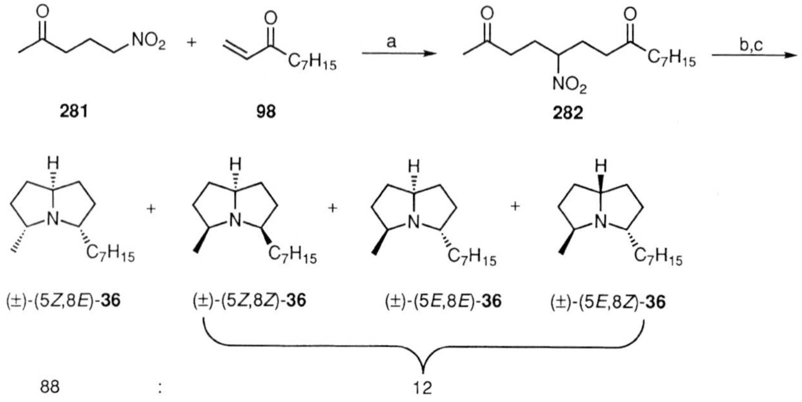

154   S. Leclercq, J. C. Braekman, D. Daloze, and J. M. Pasteels

281 + **98** →[a] **282** →[b,c]

(±)-(5Z,8E)-**36** + (±)-(5Z,8Z)-**36** + (±)-(5E,8E)-**36** + (±)-(5E,8Z)-**36**

88 : 12

(a) PBu₃, benzene; (b) NH₄OAc, KOH, MeOH, NaBH₃CN; (c) NaBH₄

Scheme 27. Synthesis of (±)-xenovenine using nitroketone reductive cyclisation (*85*)

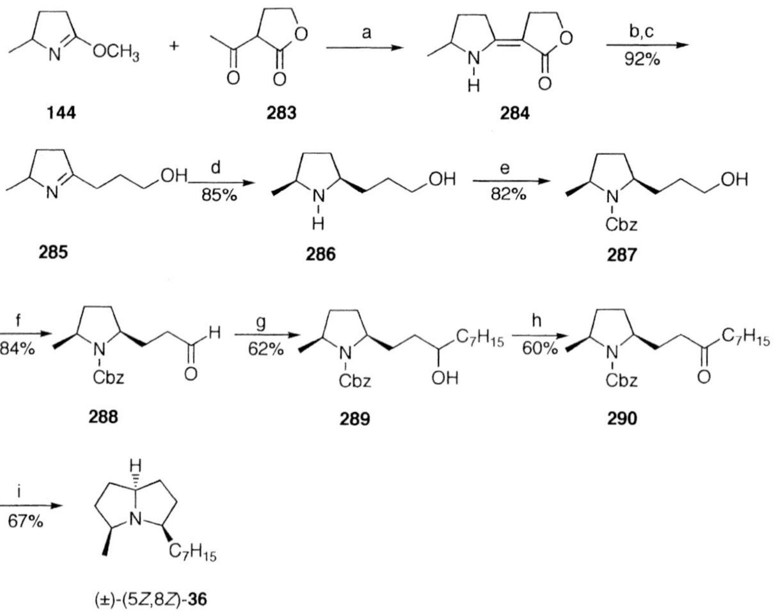

144 + **283** →[a] **284** →[b,c] 92%

285 →[d] 85% **286** →[e] 82% **287**

→[f] 84% **288** →[g] 62% **289** →[h] 60% **290**

→[i] 67% (±)-(5Z,8Z)-**36**

(a) G. Lhommet et al., *Synthesis*, 69 (1993). (b) HCl 3N; (c) K₂CO₃; (d) H₂, Raney Ni, HCl, MeOH; (e) CbzCl, Na₂CO₃, H₂O; (f) PCC, CH₂Cl₂; (g) C₇H₁₅MgBr, ether; (h) PCC, CH₂Cl₂; (i) H₂, Pd/C, AcOH.

Scheme 28. Synthesis of (±)-xenovenine from lactim ether **144** (*86*)

the hydroxy group by PCC in CH_2Cl_2 led to γ-aminoaldehyde **288** in 69% yield from **286**. Subsequently introduction of the heptyl chain was carried out by a Grignard reaction using heptylmagnesium bromide. The resulting secondary aminoalcohol *cis*-**289** was oxidized to ketone **290**. Finally, the pyrrolizidine skeleton was formed as a result of three consecutive chemical transformations, fast debenzylation of pyrrolidine **290**, iminium formation and reduction of the latter, to provide ($\pm$)-(5Z,8Z)-**36** in 67% yield.

Since 1985, five syntheses of optically active xenovenine (**36**) have been published. Four used the chirality of L-alanine or (S)-pyroglutamic acid, whereas the fifth employed 2-cyano-5-oxazolopyrrolidine as chiral auxiliary.

Enantioselective syntheses of both enantiomers of xenovenine were devised by MOMOSE *et al.* (Scheme 29) (*87*). D-alanine (**291**) was converted to alkene (*R*)-**292** in 56% yield. Intramolecular amidomercuration led to the organomercury compound **293**, which was oxidized to provide only the *trans* diastereoisomer (2*R*,5*S*)-**294** in 75% yield from alkene **292**. Parikh-Doering oxidation (DMSO/pyridine-SO_3 complex) of **294** gave aldehyde **295**, which, without purification, was chain-elongated through a Wittig-Horner reaction affording α,β-unsaturated ketone (2*R*,5*S*)-**296** in 49% overall yield from **294** (>99% ee). Hydrogenation of (2*R*,5*S*)-**296** produced the desired pyrrolizidine (3*S*,5*R*,8*S*)-**36** as a result of double bond reduction, removal of the protecting group, ring formation, and reduction of the imine formed on

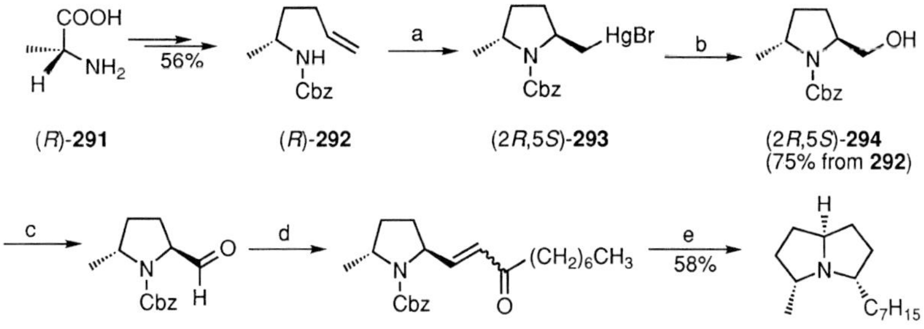

(a) $Hg(OAc)_2$, THF, $NaHCO_3$, NaBr; (b) O_2, $NaBH_4$, DMF; (c) pyridine-sulfur trioxide complex, DMSO; (d) $(MeO)_2POCH_2CO(CH_2)_6CH_3$, NaH, THF; (e) H_2, $Pd(OH)_2$, MeOH.

Scheme 29. Asymmetric synthesis of (3*S*,5*R*,8*S*)- and (3*R*,5*S*,8*R*)-xenovenine from D- and L-alanine (*87*)

cyclization. The enantiomer (3R,5S,8R)-**36** was similarly prepared from L-alanine.

Two syntheses of LHOMMET and co-workers exploit the innate chirality of (S)-pyroglutamic acid (88,89). In the first, (R)-5-methylpyrrolidinone (**300**) prepared from the readily available natural (S)-pyroglutamic acid [(S)-**204**] was treated with dimethyl sulfate to give lactim ether (R)-**144** (Scheme 30) (88). The key feature of their strategy was the condensation reaction between this lactim ether and 2-acetylbutyrolactone catalyzed by Ni(acac$_2$) which afforded the chiral

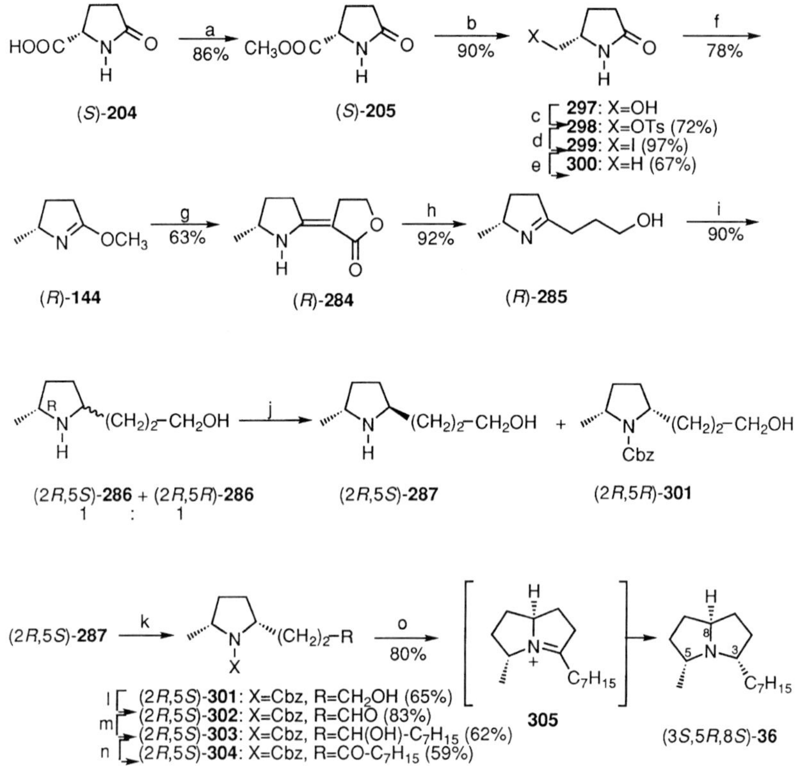

(a) MeOH, SOCl$_2$; (b) NaBH$_4$, EtOH; (c) TsCl, NEt$_3$, CH$_2$Cl$_2$; (d) NaI, CH$_3$CN; (e) H$_2$ (1 atm), PtO$_2$, NEt$_3$, MeOH; (f) MeSO$_4$; (g) 2-acetylbutyrolactone, Ni(acac)$_2$; (h) HCl 3N; (i) NaBH$_4$, EtOH; (j) CbzCl, NaHCO$_3$, H$_2$O, 0°C; (k) CbzCl, NaHCO$_3$, H$_2$O, 80°C; (l)PCC, CH$_2$Cl$_2$; (m) n-C$_7$H$_{15}$MgBr, ether; (n) PCC, CH$_2$Cl$_2$; (o) H$_2$ (1 atm), Pd/BaSO$_4$, MeOH.

Scheme 30. Asymmetric synthesis of (3S,5R,8S)-xenovenine from (S)-pyroglutamic acid (88)

References, pp. 221–229

β-enamino lactone (R)-**284**, a reaction, which, allows direct introduction of a protected alcohol as a carbonyl equivalent that can be transformed into iminium **305**. Thus, hydrolysis and decarboxylation of β-enaminolactone (R)-**284** gave imino alcohol (R)-**285** which was reduced with sodium borohydride to afford a mixture of *cis*- and *trans*-pyrrolidines **286** in a 1 : 1 ratio [17% yield from (R)-**300**]. Separation of the two isomers was achieved by treatment of the mixture with benzyl chloroformate at 0°C which gave only the carbamate of the *cis*-isomer. The unreacted *trans*-**287** was readily separated, treated with benzyl chloroformate at 80°C, and oxidized to aldehyde (2R,5S)-**302**. The latter was converted by a Grignard reaction to the secondary alcohol (2R,5S)-**303**. Oxidation of this alcohol and final reductive cyclisation gave (3S,5R,8S)-**36** (de = 98%). Thus the iminium reduction reaction allowed introduction of the asymmetric carbon C-3 in high diastereoisomeric excess (98%) while on the contrary, no selectivity was observed during the reduction of imine **285** which introduces the stereochemistry at C-8.

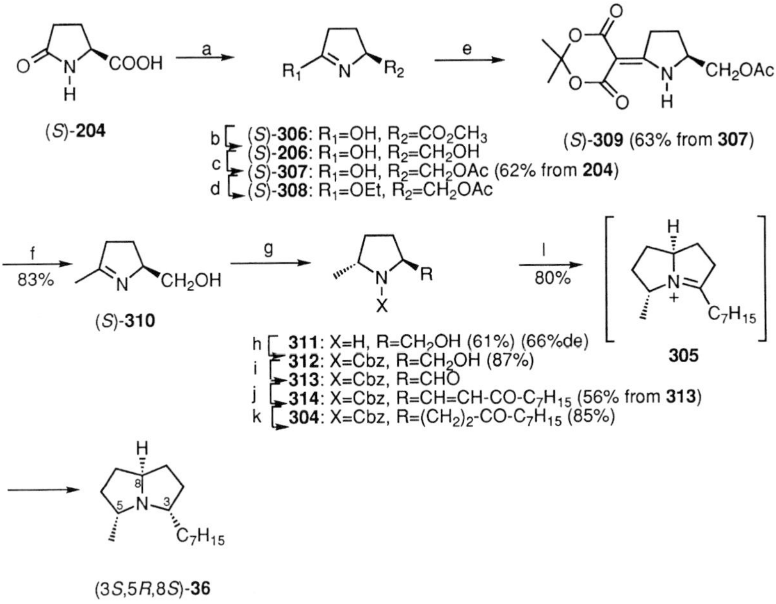

(a) MeOH, SOCl$_2$; (b) NaBH$_4$, EtOH; (c) Ac$_2$O, pyridine; (d) Et$_3$OBF$_4$, CH$_2$Cl$_2$; (e) Meldrum's acid, Ni(acac)$_2$, CHCl$_3$; (f) HCl 5N; (g) NaBH(OAc)$_3$, toluene; (h) CbzCl, NaHCO$_3$, H$_2$O; (i) (COCl)$_2$, DMSO, NEt$_3$; (j) Ph$_3$P=CH-CO-C$_7$H$_{15}$, toluene; (k) H$_2$, PtO$_2$, MeOH; (l) H$_2$, Pd/BaSO$_4$, MeOH.

Scheme 31. Asymmetric synthesis of (3S,5R,8S)-xenovenine from (S)-pyroglutamic acid (89)

As a consequence, the same authors reported a stereoselective synthesis of *trans*-pyrrolidine (2*R*,5*S*)-**311** (de=66%) by reduction of imino alcohol (*S*)-**310** using NaBH(OAc)$_3$ in 80% yield (Scheme 31). It is though that the participation of the hydroxy group of **310** directs the delivery of hydride ion from the *si* face of the imino group by forming a boronate intermediate (*89*). Lactim ether (*S*)-**308**, prepared from (*S*)-**204**, was condensed with Meldrum's acid to give compound (*S*)-**309** in 39% overall yield from (*S*)-**204**. Further hydrolysis and decarboxylation led to imino alcohol (*S*)-**310** in 83% yield. After reduction, the pure *trans*-amino alcohol (2*R*,5*S*)-**311** was isolated by selective acylation of the *cis*-pyrrolidine (61%). (3*S*,5*R*,8*S*)-**36** was obtained from alcohol **311** by a procedure similar to the one used in Scheme 30.

HUSSON and co-workers have extended their CN(*R,S*) method (*90*) to the synthesis of pyrrolizidine alkaloids such as (3*S*,5*R*,8*S*)-**36** (*91*) (Scheme 32). Nitrile **223**, prepared from (*R*)-phenylglycinol, was alkylated to yield a mixture of diastereoisomers **315** whose nitrile groups were stereospecifically removed using Li-NH$_3$ reduction to give compound **316** in 63% yield as a single isomer. The *S* configuration at

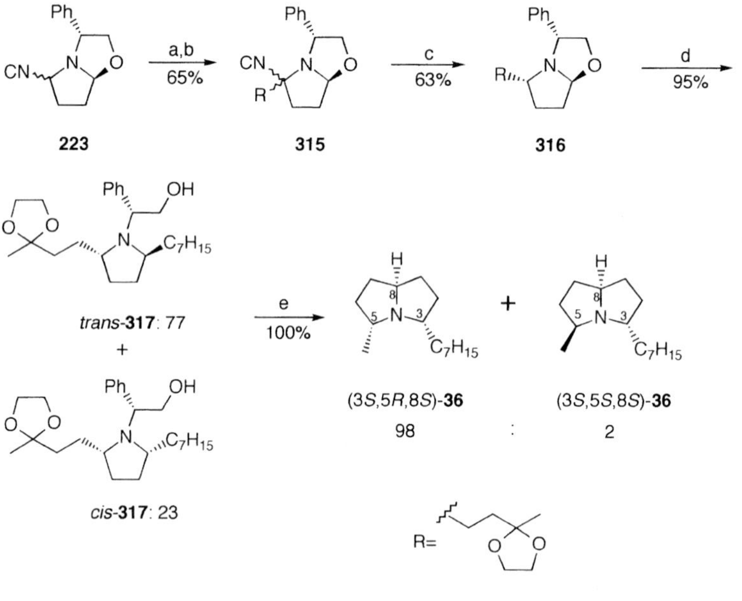

(a) LDA, THF, HMPA; (b) RBr; (c) Li, NH$_3$, THF, EtOH; (d) C$_7$H$_{15}$MgBr, ether; (e) H$_2$, 10% Pd/C, MeOH, 3% HCl 1N.

Scheme 32. Asymmetric synthesis of (3*S*,5*R*,8*S*)-xenovenine from (−)-phenylglycinol (*91*)

the new asymmetric center was established by X-ray analysis and NMR spectroscopy. The required heptyl chain was then introduced by a Grignard reaction to afford a mixture of *trans*-isomer **317** and the *cis*-form in a ratio of 77:23, which were separated by column chromatography. *Trans*-**317** was hydrogenated in the presence of acid to yield the target alkaloid (3S,5R,8S)-**36**. The overall yield of this four-step sequence was 30%.

B. (5E,8E)-3-Butyl-5-hexylpyrrolizidine

In 1991, a new pyrrolizidine alkaloid, (5E,8E)-3-butyl-5-hexylpyrrolizidine **42**, was isolated from the Venezuelan ant *Megalomyrmex modestus*. To establish structure and relative configuration, JONES *et al.* developed a short two step non-stereoselective synthesis of the four racemic diastereoisomers of **42** in 38% overall yield (Scheme 33) (*29*). Condensation of 4-oxooctanal (**318**) and 1-nonen-3-one (**319**) in the presence of a thiazolium salt catalyst provided triketone **320** in 60% yield. Reductive amination of **320** with ammonium acetate and NaBH$_3$CN afforded the four diastereoisomers of ($\pm$)-**42**, with (5Z,8E)-**42** being the major product and the naturally-occurring

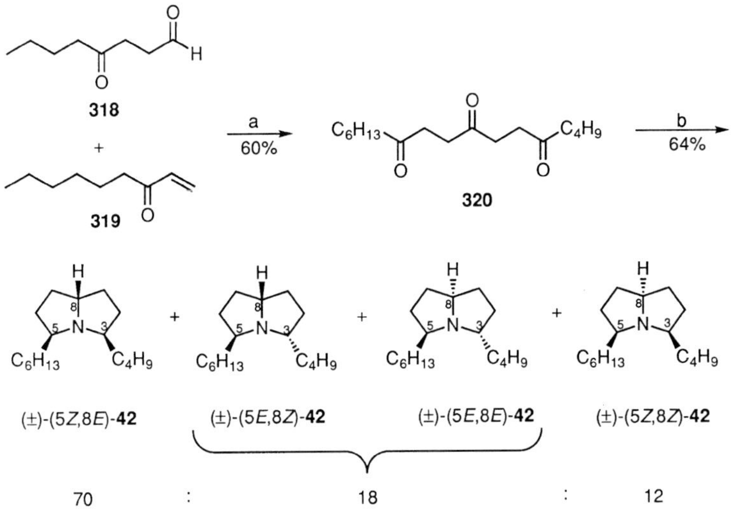

(a) NEt$_3$, 5-(2'-hydroxyethyl)-4-methyl-3-benzylthiazolium chloride; (b) NH$_4$OAc, NaBH$_3$CN, MeOH, KOH.

Scheme 33. Synthesis of ($\pm$)-**42** (*29*)

(5*E*,8*E*)-**42** one of the minor ones. An attempt to improve the yield of (5*E*,8*E*)-**42** by changing the reductive amination conditions failed.

2.2.4.2. 3-Methyl-5-alkenylpyrrolizidines and 3,5-Dialkenylpyrrolizidines

The venoms of ants of the genera *Monomorium, Solenopsis* and *Chelaner* are well-known as a source of a variety of pyrrolizidines containing an unsaturated alkyl chain at C-3 and/or C-5. Some of these alkaloids have been the goal of several syntheses which afforded either racemic or nonracemic end products.

For instance, the structure and relative configuration of (5*Z*,8*E*)-3-methyl-5-(8-nonenyl)pyrrolizidine (**39**) isolated by BLUM *et al.* in 1988 were confirmed by synthesis of the racemate outlined in Scheme 34 (*27*). The necessary triketone, 16-heptadecene-2,5,8-trione (**322**), was formed in good yield by condensation of 6-hepten-2,5-dione (**321**) and 9-decenal (**102**) in the presence of triethylamine and a thiazolium salt catalyst. Reductive amination of **322** produced all four pyrrolizidine isomers in the ratio 40:7:3:1, with (5*Z*,8*E*)-**39** being the major one. No yields are mentioned in the original paper.

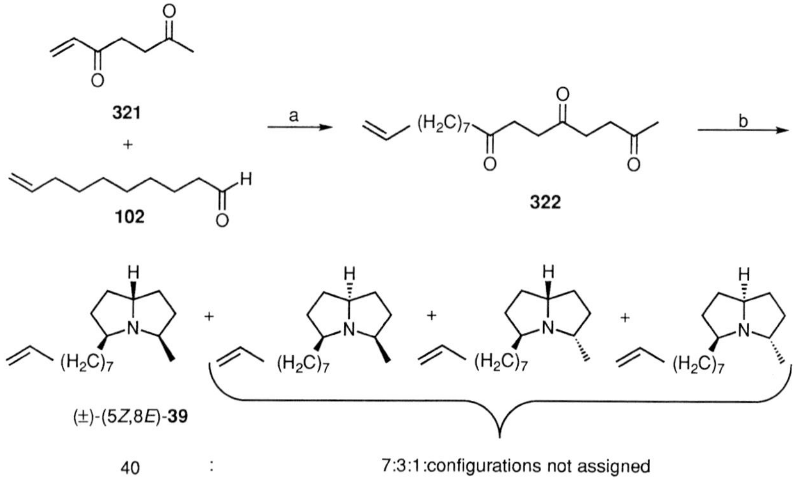

(a) NEt$_3$, 5-(2'-hydroxyethyl)-4-methyl-3-benzylthiazolium chloride; (b) NH$_4$OAc, NaBH$_3$CN, MeOH, KOH

Scheme 34. Synthesis of (±)-**39** (*27*)

A few years later, MOMOSE and co-workers (92) reported a short and highly enantioselective synthesis of (3*R*,5*S*,8*S*)-**39** by a procedure similar to their synthesis of xenovenine (**36**) (see Scheme 29). Pyrrolidine (2*R*,5*S*)-**295**, available from D-alanine, was submitted to a Horner-Wadsworth-Emmons elongation to afford the α,β-unsaturated ketone **323** (Scheme 35). Construction of the bicyclic ring by catalytic hydrogenation of **323** gave pyrrolizidine **324** as a single diastereoisomer in 65% yield. Debenzylation of **324** in acidic medium afforded primary alcohol **325** quantitatively. The last step of the synthesis called for a terminal olefination at the C-5 substituent of **325**. Swern oxidation of the primary hydroxyl group and subsequent Wittig olefination of the resulting aldehyde gave (3*R*,5*S*,8*S*)-**39** in 58% yield.

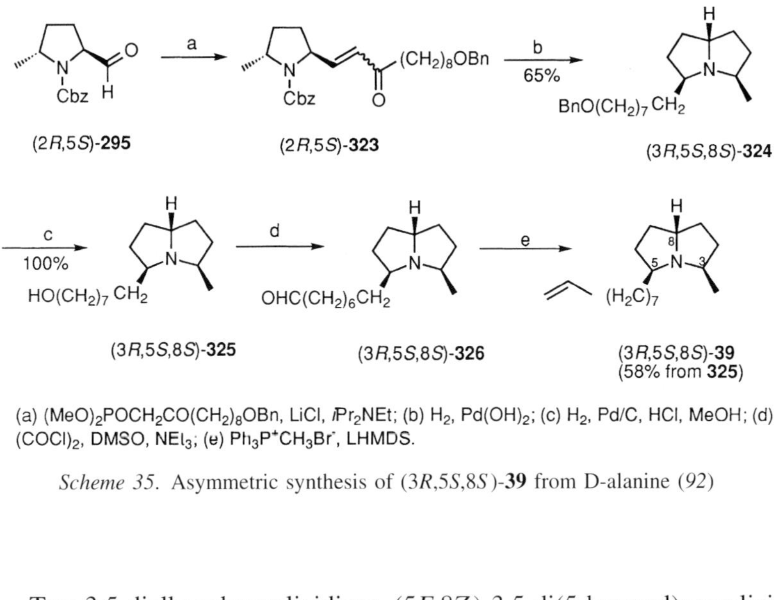

(a) (MeO)$_2$POCH$_2$CO(CH$_2$)$_8$OBn, LiCl, *i*Pr$_2$NEt; (b) H$_2$, Pd(OH)$_2$; (c) H$_2$, Pd/C, HCl, MeOH; (d) (COCl)$_2$, DMSO, NEt$_3$; (e) Ph$_3$P$^+$CH$_3$Br$^-$, LHMDS.

Scheme 35. Asymmetric synthesis of (3*R*,5*S*,8*S*)-**39** from D-alanine (*92*)

Two 3,5-dialkenylpyrrolizidines, (5*E*,8*Z*)-3,5-di(5-hexenyl)pyrrolizidine [(5*E*,8*Z*)-**41**] and its (5*Z*,8*E*) isomer, have been synthesized according to the procedure already described for the preparation of (±)-**39** and (±)-**42** (*27*). Divinyl ketone **327** was condensed with two equivalents of 6-heptenal (**101**) to form 1,18-nonadecadien-7,10,13-trione **328** in 55% yield (Scheme 36). Reductive amination of triketone **328** gave three diastereoisomeric 3,5-di(5-hexenyl)pyrrolizidines eluting from a SP-1000 column in the order (5*Z*,8*Z*)-**41**, (5*Z*,8*E*)-**41**, (5*E*,8*Z*)-**41**, in 1:9:3 ratio.

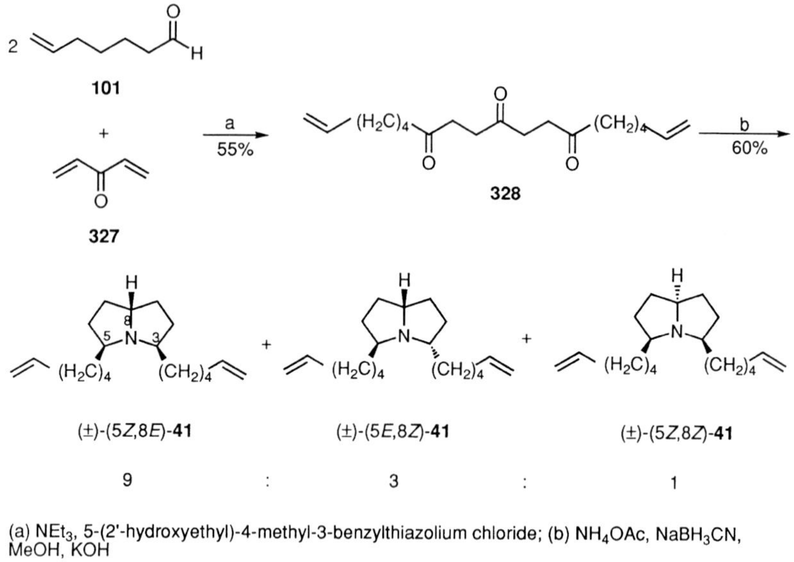

Scheme 36. Synthesis of (±)-(5*E*,8*Z*)-**41** and (±)-(5*Z*,8*E*)-**41** (*27*)

(a) NEt₃, 5-(2'-hydroxyethyl)-4-methyl-3-benzylthiazolium chloride; (b) NH₄OAc, NaBH₃CN, MeOH, KOH

Finally, (5*E*,8*Z*)-3-(1-non-8-enyl)-5((*E*)-1-prop-1-enyl)pyrrolizidine [(±)-(5*E*,8*Z*)-**40**], the major alkaloid produced by a species of *Chelaner* ants, was prepared by JONES and co-workers in 1986 through a non-stereoselective synthesis (*35*). The approach was based on reductive amination of triketone **335**, obtained from furan (Scheme 37). Treatment of 2-furyllithium sequentially with propylene oxide and acetic anhydride gave **329** in 59% yield. Oxidative methoxylation, selective hydrogenation of the ring double bond and subsequent hydrolysis with 50% acetic acid provided a mixture of **332** and **333**. This freshly prepared mixture was then condensed with vinyl ketone **334** in the presence of a thiazolium salt catalyst to give the 19-carbon triketone **335** in 50% yield from **329**. The conjugated double bond of **335** was epoxidized and reductive amination yielded the unstable epoxy-pyrrolidine **336** as a mixture of isomers. Removal of the epoxide on the three-carbon side chain to form the olefin was achieved by treatment of **336** with potassium selenocyanate. GLC analysis showed the presence of the four diastereoisomers (±)-(5*Z*,8*Z*)-**40**, (±)-(5*Z*,8*E*)-**40**, (±)-(5*E*,8*E*)-**40** and (±)-(5*E*,8*Z*)-**40** in the ratio 3:7:10:1. No yield were reported.

References, pp. 221–229

2.2.5. Indolizidines

2.2.5.1. Monomorine I [(5Z,9Z)-**346**]

A. Syntheses of Racemic Monomorine I

No fewer than ten syntheses of the popular target (±)-monomorine I [(5Z,9Z)-**346**] have appeared since Numata and Ibuka's 1987 review (8).

The synthetic strategy of KIBAYASHI et al. (93, 94) was based on a highly regio- and stereoselective intramolecular acyl nitroso Diels-Alder cycloaddition leading to bicyclic oxazinolactam **339** (Scheme 38). Hydroxamic acid **337** was obtained from ethyl (E)-2-heptenoate by a

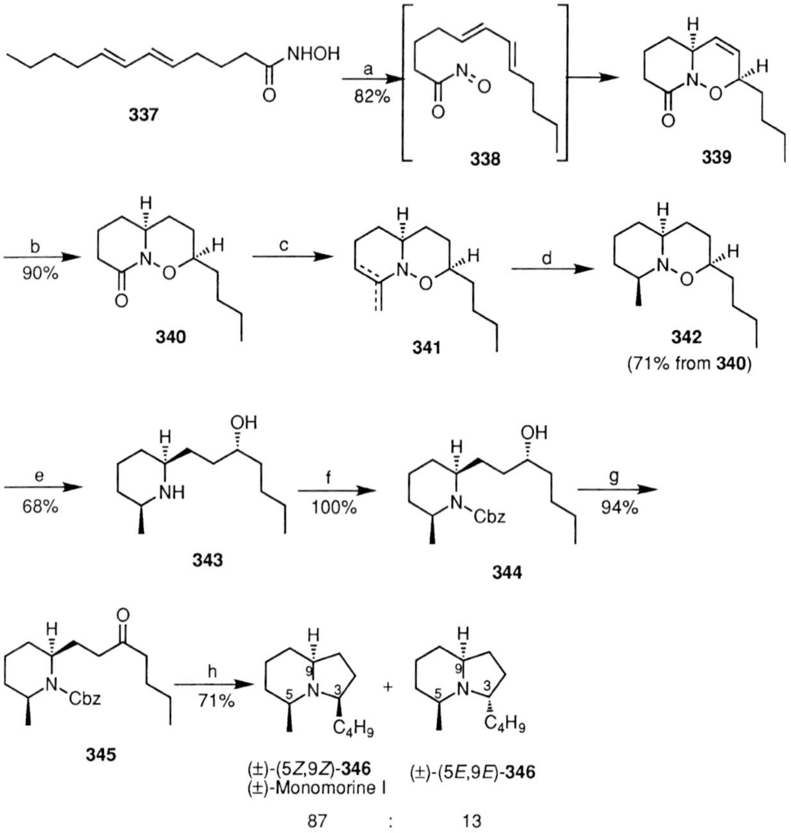

(a) n-$Pr_4N(IO_4)$, $CHCl_3$; (b) H_2, 5% Pd/C, MeOH; (c) MeMgBr, Et_2O; (d) H_2, 5% Pd/C, MeOH; (e) Zn, AcOH-H_2O; (f) $PhCH_2OCOCl$, aqueous Na_2CO_3, CH_2Cl_2; (g) CrO_3.2Py, CH_2Cl_2; (h) H_2, 5% Pd/C, MeOH.

Scheme 38. Synthesis of (±)-monomorine I through acylnitroso Diels-Alder cycloaddition (*93, 94*)

somewhat lengthy synthesis in 23% overall yield. Oxidation to the acyl nitroso compound **338** and Diels-Alder cycloaddition afforded **339**. Subsequent introduction of the C-8 methyl group was achieved by means of a stereocontrolled process involving a Grignard reaction followed by dehydration and catalytic reduction. The resulting bicyclic oxazine **342** was subjected to reductive N-O bond cleavage, affording the *cis*-2,6-dialkylpiperidine **343**. After deprotection, the latter was converted to ketone **345** by Collins oxidation in 94% yield. On reductive cyclization of **345**, (±)-monomorine I [(±)-(5*Z*,9*Z*)-**346**] was formed along with its C-3 epimer (±)-(5*E*,9*E*)-**346** in a 87 : 13 ratio.

McGrane and Livinghouse (95) have reported the synthesis of (±)-monomorine I based on a [2+2] cycloaddition between a monocyclopentadienyltitanium imido complex and an alkyne. The key γ-aminoalkyne **353** required for the imidotitanium [2+2] cycloaddition was prepared by the highly convergent approach detailed in Scheme 39. Reaction of the acetylide of **347** with paraformaldehyde and sequential

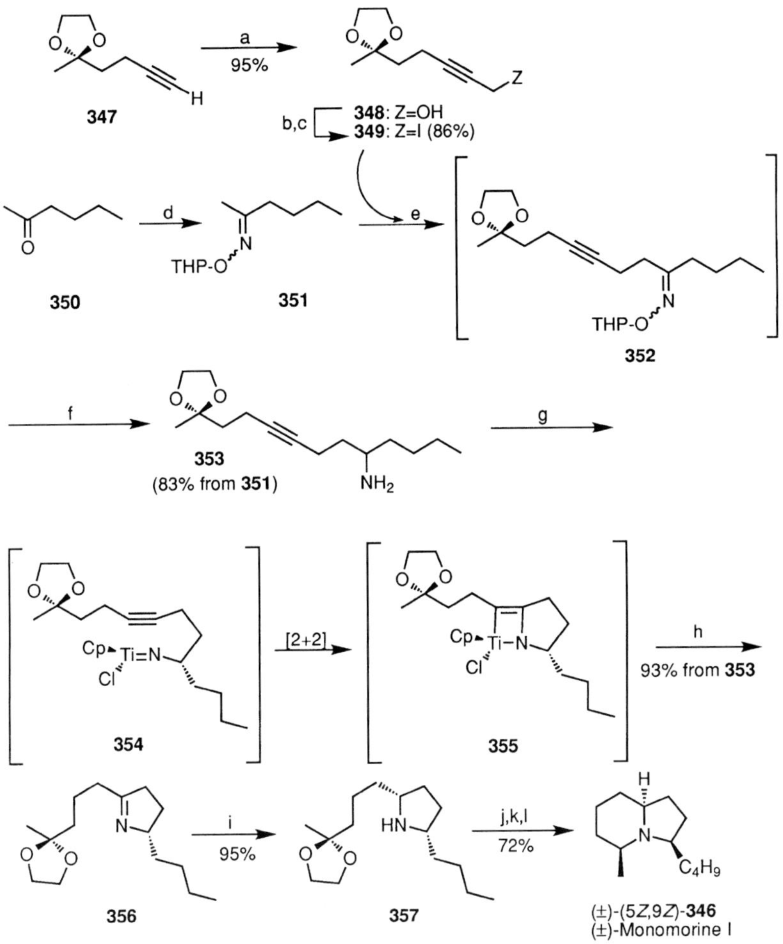

(a) *n*-BuLi, CH₂O; (b) MsCl, Et₃N; (c) NaI, CH₃CN; (d) H₂N-OTHP; (e) LDA, **349**; (f) LiAlH₄; (g) CpTiCl₃, Et₃NH⁺Cl⁻; (h) Et₃NH⁺Cl⁻; (i) DIBALH, THF; (j) aqueous HCl; (k) K₂CO₃; (l) NaBH₃CN.

Scheme 39. Synthesis of (±)-monomorine I through [2+2] cycloaddition (95) mesylation-iodide-mediated displacement of the resulting alcohol **348** afforded iodide **349** in 81% overall yield. Condensation of 2-hexanone (**350**) with H$_2$N-OTHP provided the corresponding oxime **351** as a mixture of *syn* and *anti* isomers. Lithiation of **351** followed by alkylation with iodide **349** secured the thermally labile oxime **352** which was immediately reduced to γ-aminoalkyne **353**. Exposure of **353** to a catalytic amount of CpTiCl$_3$ at room temperature provided pyrroline **356** in 93% yield by way of a [2+2] cycloaddition sequence involving the transient imido complex **354** and the titanetine **355**. Stereoselective reduction of the pyrroline to the *cis*-pyrrolidine **357** and cyclization by the procedure of STEVENS and LEE (*96*) furnished (±)-monomorine I in a 53% overall yield from **352**.

HAMAGUCHI *et al.* reported a stereoselective formal synthesis of (±)-monomorine I starting with 6-methylpiperidin-2-one (**358**) (Scheme 40) (*97*). Reaction of **358** with benzylchloroformate in the presence of (MeSi)$_2$NLi gave *N*-benzyloxycarbonyllactam **359** in 74% yield. Ring opening of **359** with acetylene **360** by the method of NOZOE and co-workers (*98–100*) gave the desired product **361** only in very low yield (8%). Catalytic hydrogenation-cyclization of the acyclic alkynone **361** gave exclusively the 2,6-*cis*-disubstituted piperidine **362**. Catalytic hydrogenation under acidic conditions, described elsewhere by YAMA-GUCHI (*101*), converted **362** into (±)-monomorine I with a stereoselectivity close to 100% (60% yield).

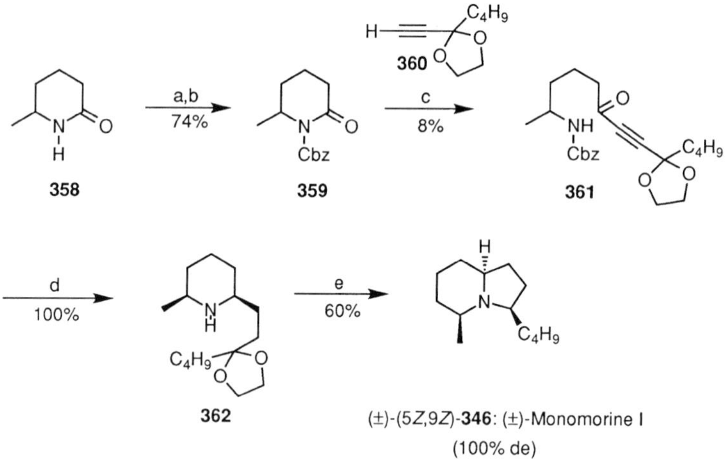

(a) (Me$_3$Si)$_2$NLi, THF; (b) CbzCl; (c) **360**, LDA, ether (d) H$_2$, 5%Pd/C, EtOH; (e) H$_2$, Pd/C, aqueous HCl-MeOH.

Scheme 40. Synthesis of (±)-monomorine I from 6-methylpiperidin-2-one (*97*)

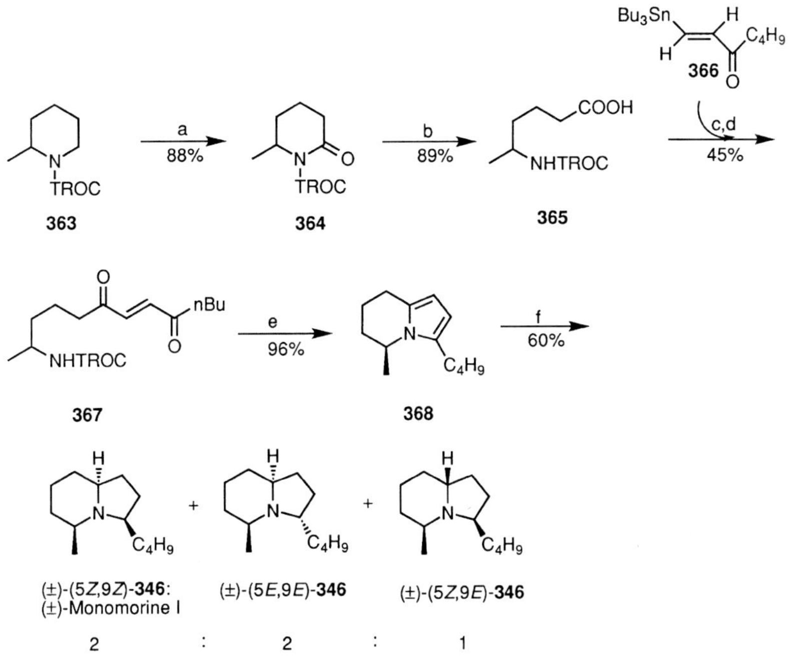

TROC: 2,2,2-trichloroethoxycarbonyl-

(a) $RuCl_3$, $NaIO_4$; (b) H_2O Δ; (c) $SOCl_2$; (d) **366**, $Pd(PPh_3)_4$, dioxane; (e) Cd, AcOH-DMF; (f) H_2, Rh/C .

Scheme 41. Synthesis of (±)-monomorine I from 6-methylpiperidin-2-one (*102*)

Another synthesis starting with a 6-methylpiperidin-2-one derivative has been described by Echavarren *et al.* in 1994 (*102*). Its novel feature was the palladium-mediated coupling of the acid chloride derived from **365** with (*E*)-vinylstannane **366** to form 1,4-diketone **367** subsequently used in a modified Paal-Knorr pyrrole synthesis to give pyrrole **368** (Scheme 41). Catalytic hydrogenation of the latter over rhodium on charcoal yielded a 2 : 2 : 1 mixture of (±)-monomorine I and its two diastereoisomers (±)-(5*E*,9*E*)-**346** and (±)-(5*Z*,9*E*)-**346**, both of which are known alkaloids from the skin of dendrobatid frogs.

Several syntheses of 3,5-disubstituted indolizidine alkaloids exploit reductive alkylations in which the cyclization of keto pyrrolidines such as **375** (Scheme 42) is followed by the well-precedented and highly diastereoselective reduction of iminium ion intermediates such as **376**. For instance, Shawe and co-workers (*103*) reported a successful application of this strategy to the synthesis of (±)-monomorine I in

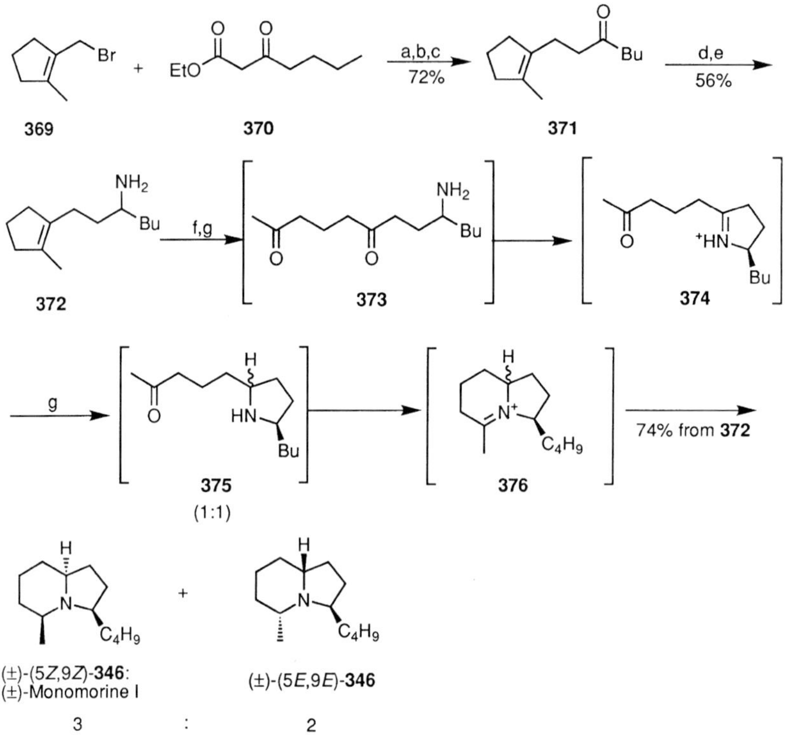

369 + **370** →(a,b,c, 72%) **371** →(d,e, 56%) **372** →(f,g) **373** → **374** →(g) **375** (1:1) → **376** → (±)-(5Z,9Z)-**346**: (±)-Monomorine I + (±)-(5E,9E)-**346**

3 : 2

74% from **372**

(a) NaOEt; (b) NaOH; (c) H$_3$O$^+$; (d) NH$_2$OH; (e) LAH, THF; (f) O$_3$; (g) NaBH$_3$CN.

Scheme 42. Synthesis of (±)-monomorine I through cyclization of ketopyrrolidines (*103*)

four steps with 18% overall yield (Scheme 42). Precursor **372** was prepared from allylic bromide **369** and ethyl 3-oxoheptanoate **370** in three steps in 56% yield. Ozonolysis, removal of excess ozone and treatment with sodium cyanoborohydride afforded a mixture of (±)-monomorine I and its (±)-(5E,9E)-diastereoisomer in a 3:2 ratio. This reductive alkylation, while unselective in the formation of intermediate **375**, showed complete stereochemical control during the reduction of the bicyclic iminium ion intermediate with sodium cyanoborohydride.

Recently, SOMFAI *et al.* (*104*) reported a total synthesis of (±)-monomorine I by an aza-[2,3]-Wittig rearrangement of vinylaziridine **380** to tetrahydropyridine **381** as a key step (Scheme 43). Protection of crotyl alcohol as its *tert*-butyldiphenylsilyl ether followed by epoxidation gave **377** in 88% yield. Exposure of **377** to sodium azide gave the corresponding vicinal azido alcohol which was converted into aziridine **378** by Ph$_3$P in refluxing toluene. Subsequent *N*-alkylation of **378**, in

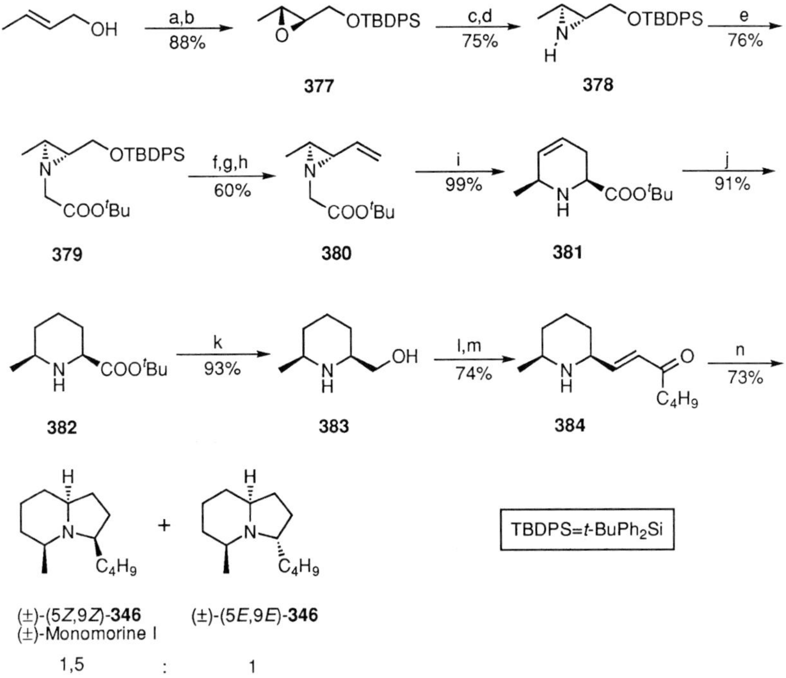

(a) TBDPSCl, NEt_3, DMAP, CH_2Cl_2; (b) *m*-CPBA, CH_2Cl_2; (c) NaN_3, NH_4Cl, $MeOCH_2CH_2OH$, H_2O; (d) Ph_3P, PhMe; (e) *tert*-butyl bromoacetate, K_2CO_3, 18-crown-6, CH_3CN; (f) Bu_4NF, THF; (g) DMSO, $(COCl)_2$, NEt_3, CH_2Cl_2; (h) Ph_3PCH_3Br, KHMDS, THF; (i) LDA, THF; (j) H_2, 5% Rh/C, MeOH; (k) $LiAlH_4$, THF; (l) DMSO, $(COCl)_2$, NEt_3, CH_2Cl_2; (m) $(MeO)_2P(O)CH_2COBu$, LiCl, *i*-Pr_2NEt, CH_3CN; (n) H_2, 5% Pd/C, MeOH.

Scheme 43. Synthesis of (±)-monomorine I through aza-[2,3]-Wittig rearrangement (*104*)

order to install the necessary anion-stabilizing group for the projected aza-[2,3]-Wittig rearrangement, was carried out by treatment with *tert*-butyl bromoacetate and gave ester **379** as a 2:1 mixture of *N*-invertomers. Removal of the silyl group, Swern oxidation of the resulting primary alcohol, and Wittig olefination yielded the key intermediate vinylaziridine **380**. Subjecting **380** to LDA resulted in the rapid formation of tetrahydropyridine **381** as the only detectable diastereoisomer in 99% yield. Hydrogenation of **381** and subsequent reduction of the ester group gave amino alcohol **383** which was oxidized to the corresponding aldehyde under Swern conditions. This aldehyde proved to be labile and was directly cannulated into a slurry of dimethyl(2-oxohexyl)phosphonate, LiCl and *i*-Pr_2NEt to give the *E*-configurated α,β-unsaturated ketone **384**. Finally, hydrogenation and concomitant intramolecular reductive amination of **384** gave a separable 1.5:1 mixture of (±)-monomorine I [(±)-(5Z,9Z)-**346**] and its (5E,9E) diastereoisomer, (±)-indolizidine 195B earlier isolated from dendrobatid frogs.

A short route to (±)-monomorine I described by JEFFORD et al. (*105*) exploits rhodium (II)-catalyzed decomposition of diazoketone **386**, the bicyclic system resulting from an intramolecular carbene insertion into the C(2)-H bond of the pyrrole ring of **386** (Scheme 44). Catalytic hydrogenation was also used to ensure excellent stereocontrol at the four stereogenic centers of the resulting alcohol **388**, which was deoxygenated to the target compound (±)-(5Z,9Z)-**346** *via* the imidazole carbothioate **389**. The overall yield for this six-step synthesis was 26%.

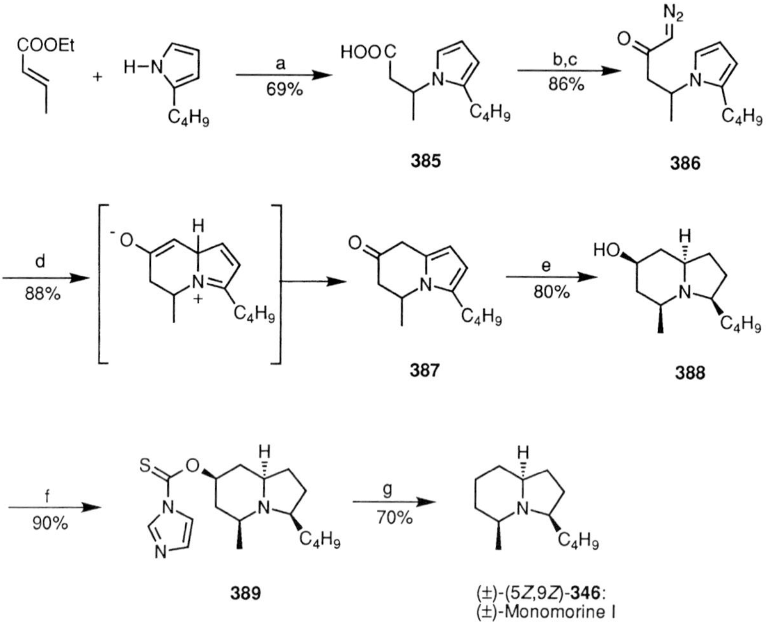

(a) KOH, CH$_3$CN; (b) t-BuOCOCl, N-methylmorpholine; (c) CH$_2$N$_2$, Et$_2$O; (d) Rh$_2$(OAc)$_4$, CH$_2$Cl$_2$; (e) H$_2$, PtO$_2$, EtOH, AcOH; (f) N,N-thiocarbonyldiimidazole, ClCH$_2$CH$_2$Cl; (g) Bu$_3$SnH, toluene.

Scheme 44. Synthesis of (±)-monomorine I through Rh-catalyzed insertion of diazoketones (*105*)

ZELLER and GRIERSON (*106*) have reported a synthesis of (±)-monomorine I in seven steps from aminonitrile **390** (Scheme 45) which initially involved replacement of the cyano group of **390** by a methyl

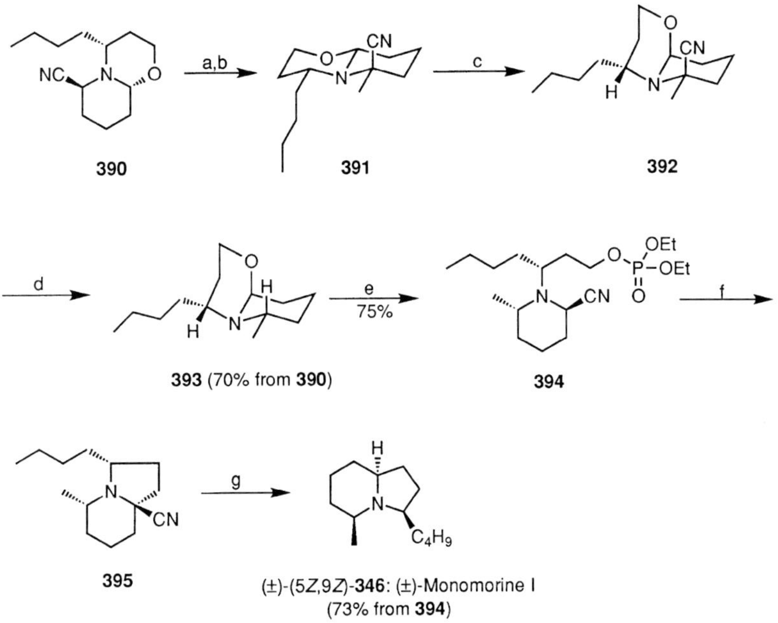

(a) *s*-BuLi, THF-HMPT; (b) CH_3I; (c) $ZnBr_2$, CH_2Cl_2; (d) Na/NH_3; (e) $ZnBr_2$, $(EtO)_2POCN$; (f) LDA, THF; (g) Na/NH_3.

Scheme 45. Synthesis of (±)-monomorine I from aminonitrile **390** (*106*)

substituent, ring opening of intermediate **393** and ring closure. Reaction of synthon **390** was *sec*-BuLi followed by addition of methyl iodide produced the C-10 methylated product **391**, which was completely converted into the more stable **392** in refluxing dichloromethane containing $ZnBr_2$. Reductive decyanation was stereospecific, producing only **393** in 70% yield from **390**. Ring opening of the tetrahydrooxazine using diethylcyanophosphonate and cyclization provided bicyclic compound **395** which was then decyanated by Na/NH_3 to selectivity yield (±)-monomorine I (38% overall yield).

Recently, Hesse and co-workers have reported a racemic synthesis of (±)-monomorine I based on reductive cyclization of 4-nitroalkanone **397** to a 2,5-*cis*-disubstituted pyrrolidine **398** (Scheme 46) (*107*). The preparation of **397** was accomplished by ethanolysis of nitrocyclohexanone derivative **396**, which was synthesized from 2-nitrocyclohexanone and 1-hepten-3-one (**99**). Further cyclization by way of α-iodoester **398** produced a mixture of two indolizidines (5*E*,9*Z*)-**399** and (5*Z*,9*Z*)-**399**;

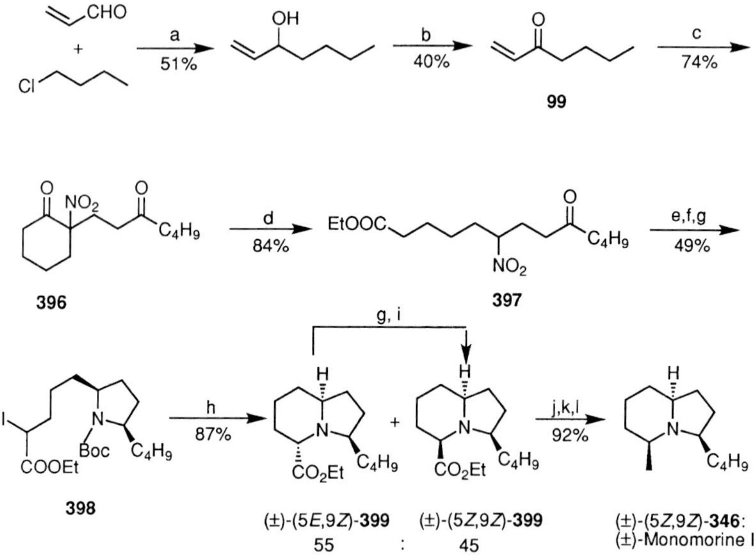

(±)-(5E,9Z)-**399** : (±)-(5Z,9Z)-**399**
55 : 45

(±)-(5Z,9Z)-**346**:
(±)-Monomorine I

(a) Mg, THF; (b) MnO$_2$, CH$_2$Cl$_2$; (c) THF, PPh$_3$, 2-nitrocyclohexanone; (d) EtOH, EtONa; (e) H$_2$, 10% Pd/C, EtOH, HCl; (f) BOC-ON, NEt$_3$, THF; (g) Lithiumcyclohexylisopropyl amide, THF, then I$_2$, THF, then NEt$_3$, THF; (h) CF$_3$COOH, then NEt$_3$, THF; (i) AcOH, NaBH$_3$CN, MeOH; (j) LiAlH$_4$, THF; (k) SOCl$_2$; (l) Bu$_3$SnH, AIBN, toluene.

Scheme 46. Synthesis of (±)-monomorine I through reductive cyclization of nitroalkanone **397** (*107*)

the undesired (5E,9Z) epimer could be isomerized to the (5Z,9Z) isomer by iodination α to the ester followed by reduction of the iminium salt produced therefrom. Subsequent reduction of the carboethoxy group of (5Z,9Z)-**399** gave racemic monomorine I. The overall yield for this sequence was 6%.

The last synthesis of (±)-monomorine I described up to now is based on atmospheric nitrogen fixation (Scheme 47) (*108*). Ozonolysis of **371**, obtained from allylic bromide **369** and ethyl-3-oxoheptanoate (**370**), followed by treatment with Me$_2$S gave triketone **400** in 57% yield from starting materials. The desired indolizine derivative **401** was obtained from **400** by reaction with titanium nitrogen complexes prepared from TiCl$_4$, Li and TMSCl under dry air. Hydrogenation of **401** catalyzed by Rh on alumina afforded monomorine I [(±)-(5Z,9Z)-**346**] as the main product in 32% yield, along with (5E,9E)-**346** (4%) and two other stereoisomers (10%).

References, pp. 221–229

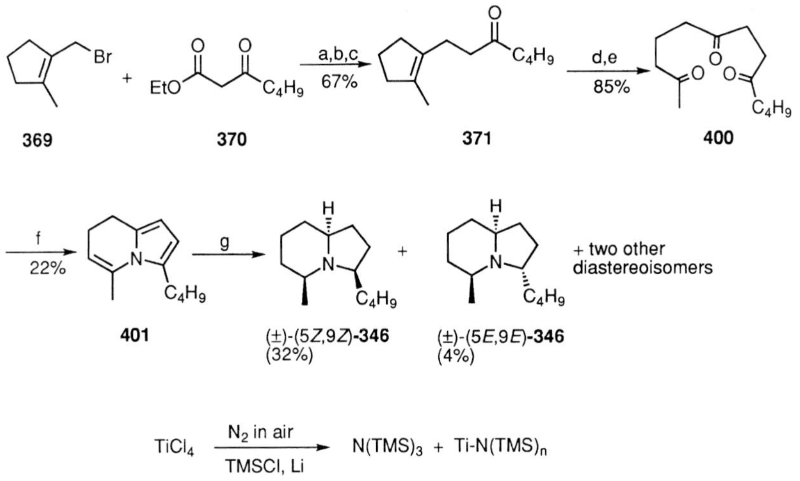

Scheme 47. Synthesis of (±)-monomorine I through atmospheric nitrogen fixation (*108*)

(a) NaOEt; (b) NaOH; (c) H_3O^+; (d) O_3; (e) Me_2S; (f) dry air, $TiCl_4$, Li, TMSCl, THF; (g) Rh/Al_2O_3, H_2, EtOH.

B. Syntheses of Nonracemic Monomorine I

Since 1987, fourteen syntheses of nonracemic monomorine I have been published. Ten of them are based on the chiron approach, whereas the other four use chiral auxiliaries.

Syntheses Using Chirons

Nine of the ten monomorine I syntheses based on this approach exploit the innate chirality of readily available α-amino-acids or tartaric acid. In most cases, the strategy requires the transformation of the chosen α-amino-acid into a chiron that can undergo cyclization and chain elongation.

a) From L-Alanine

JEFFORD et al. (*109*) used L-alanine [(S)-**291**] as the point of departure for a short synthesis of natural (+)-(3R,5S,9S)-monomorine I (Scheme 48) which paralleled their previously published synthesis of the racemic alkaloid (*105*). The key step was the decomposition of diazoketone **402** in the presence of rhodium (II) acetate, which gave the desired bicyclic product **404** along with the unexpected carbene insertion-rearrangement

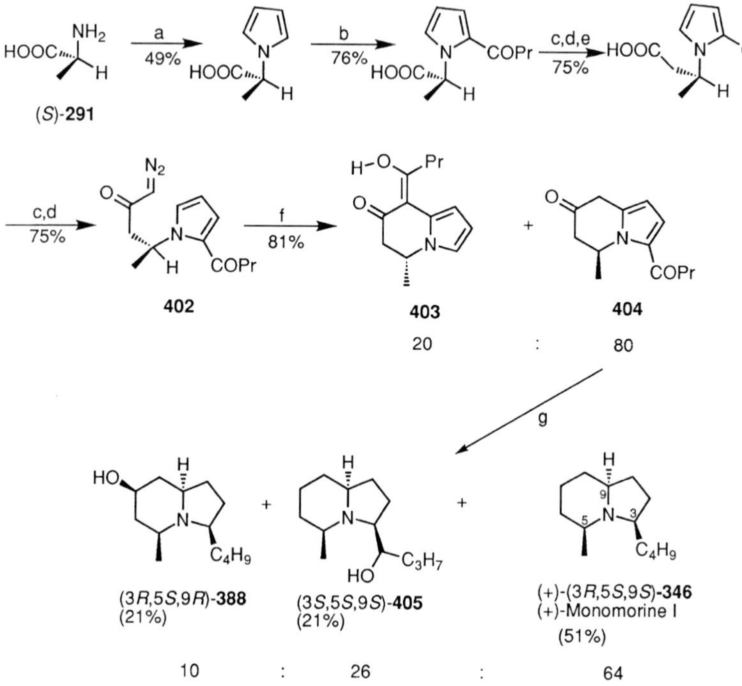

(a) 2,5-dimethoxytetrahydrofuran, AcONa, AcOH; (b) PrCOCl, N-methylmorpholine, ether, then AlCl$_3$; (c) Me$_2$CHCH$_2$OCOCl, N-methylmorpholine, Et$_2$O; (d) CH$_2$N$_2$, Et$_2$O; (e) AgOAc, THF-H$_2$O; (f) Rh(OAc)$_4$, CH$_2$Cl$_2$; (g) H$_2$, 10% Pd/C, HCl, AcOH.

Scheme 48. Asymmetric synthesis of (+)-(3R,5S,8S)-monomorine I from L-alanine (*109*)

product **403** in 80:20 ratio. Catalytic hydrogenation of the former in the presence of HCl yielded (+)-monomorine I in 51% yield accompanied by alcohols (3R,5S,9R)-**388** and (3S,5S,9S)-**405**, which can in principle be deoxygenated to the desired product through their thiocarbonylimidazole derivatives. This procedure has already been shown to work well with (±)-**388**, which gave (±)-monomorine I in 63% yield (see Scheme 44).

(+)-Monomorine I has also been prepared by ANGLE and BREITENBUCHER (*110*) by a route which features as key reaction the conformationally restricted Claisen rearrangement of lactone **410** to piperideine ester **412** (Scheme 49). (+)-Monomorine I was obtained in ten steps from N-Boc-L-alanine ethyl ester (**406**) in 5% overall yield. Allyl alcohol **407** was prepared from Boc-alanine ethyl ester **406** via YAMAMOTO'S one pot reduction-alkylation procedure (*111*) in 59% yield.

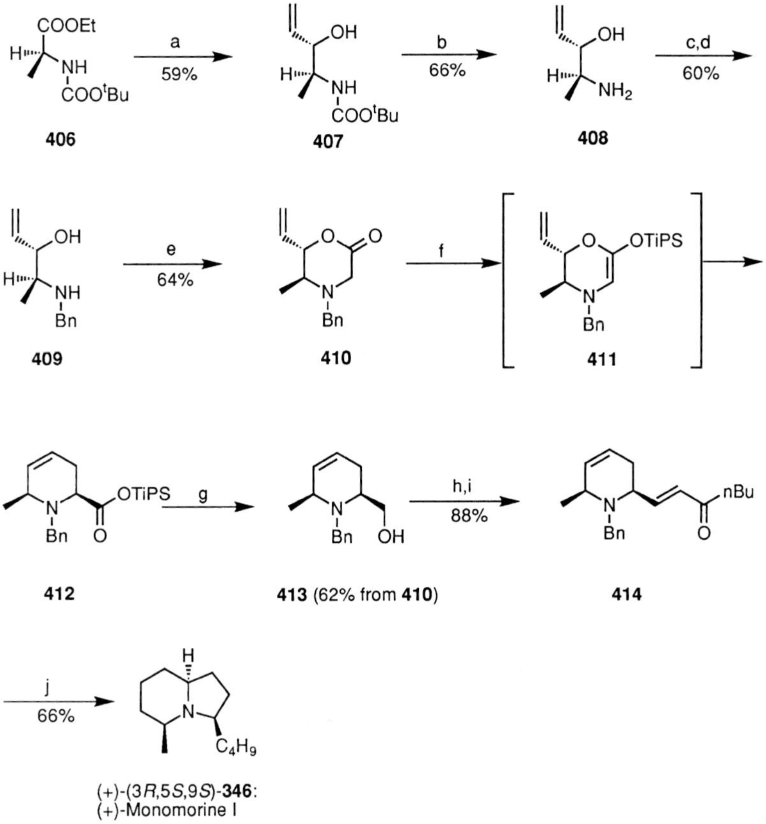

Scheme 49. Asymmetric synthesis of (+)-(3R,5S,8S)-monomorine I from L-alanine (110)

(a) DIBALH, CH$_2$=CHMgCl; (b) CF$_3$COOH; (c) PhCO$_2$Cl, C$_5$H$_5$N; (d) LiAlH$_4$; (e) BrCH$_2$CO$_2$Ph, (iPr)$_2$NEt; (f) TiPS-OTf, Et$_3$N; (g) LiAlH$_4$; (h) (COCl)$_2$, DMSO; (i) (EtO)$_2$POCH$_2$COnBu; (j) 10% Pd/C, H$_2$, MeOH, HCl.

Removal of the Boc protecting group of **407**, reaction with benzoyl chloride and further reduction of the resulting amidoester provided *N*-benzyl amine **409** as a 8:1 mixture of diastereoisomers in 25% yield from **407**. Formation of lactone **410** was accomplished in 64% yield by treatment of **409** with α-bromophenylacetate in the presence of Hünig's base. Addition of triethylamine and triisopropylsilyltrifluoromethane-sulfonate (TiPS-OTf) to the mixture of lactone diastereoisomers **410** resulted in the immediate formation of silyl ketene acetals **411**. Claisen rearrangement of the major ketene acetal diastereoisomer, with the vinyl and methyl groups in a *trans*-orientation, proceeded at room temperature, whereas the minor *cis* isomer failed to undergo the rearrangement under these conditions. Reduction of TiPS ester **412** obtained as a single diastereoisomer, and one pot Swern oxidation-Horner-Wittig olefination afforded enone **414**. Exposure of **414** to an atmosphere of hydrogen in the presence of Pd/C as catalyst caused simultaneous reduction of its double bond, debenzylation leading to iminium formation, and reduction of the iminium intermediate to give stereoselectively the desired (+)-monomorine I in 66% yield.

A few months later, Momose *et al.* (*112*) reported another synthesis of (+)-(3R,5S,9S)-monomorine I starting from L-alanine, which is very similar to the synthesis described by Angle and Breitenbucher (*110*) (Scheme 50). The synthesis began with an intramolecular amidomercuration of (*S*)-*N*-(benzyloxycarbonyl)-1-methyl-5-hexenylamine **415**, readily available from L-alanine [(*S*)-**291**], followed by treatment with sodium bromide to afford the organomercurial bromide **416**. The latter was then oxidatively demercurated to provide a 5.6:1 mixture of diastereoisomeric *cis*- and *trans*-2,6-disubstituted piperidines **417**. Transformation of *cis*-**417** to (+)-monomorine I was accomplished by

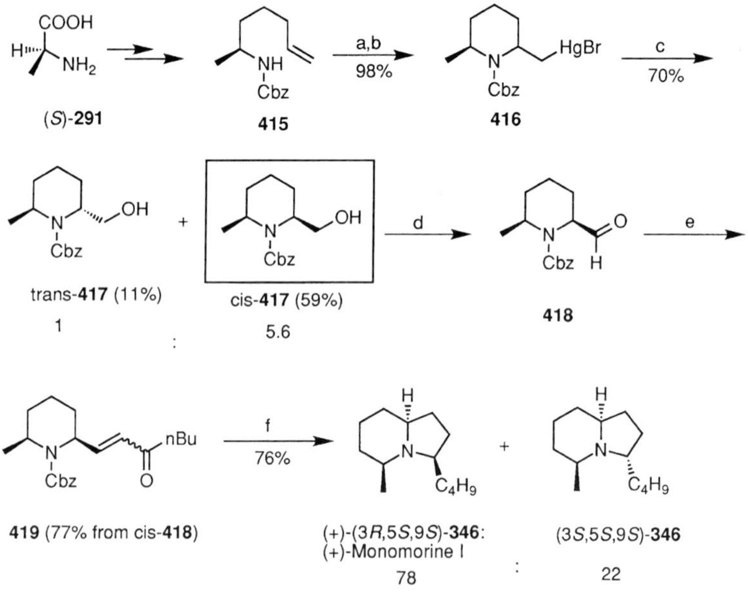

(a) Hg(OCOCF$_3$)$_2$, CH$_3$NO$_2$; (b) NaBr, NaHCO$_3$; (c) O$_2$, NaBH$_4$, DMF; (d) (COCl)$_2$, DMSO, EtN$_3$; (e) (H$_3$CO)$_2$POCH$_2$COC$_4$H$_9$, NaH, THF; (f) H$_2$, Pd(OH)$_2$, MeOH.

Scheme 50. Asymmetric synthesis of (+)-(3R,5S,8S)-monomorine I from L-alanine (*112*) the procedure already described by ANGLE and BREITENBUCHER (*110*) (see above).

b) From L-Glutamic Acid and Derivatives

Three syntheses of nonracemic monomorine I use glutamic acid and derivatives as chiral starting material.

The first by JEFFORD *et al.* (*113*) in 1994 describes an improved procedure for constructing enantiomerically pure indolizidines by intramolecular acylation of a suitable *N*-substituted pyrrole followed by a substituent-directed stereoselective hydrogenation of the resulting bicyclic intermediate (Scheme 51). In the present instance, the source of chirality was provided by diethyl L-glutamate hydrochloride (*S*)-**420**. The *N*-substituted pyrrole derivative **422**, synthesized from (*S*)-**420** and 2,5-dimethoxytetrahydrofuran and subsequent acylation in 42% yield, underwent reductive deoxygenation of the ketone carbonyl with sodium cyanoborohydride and zinc iodide to provide the butylated derivative

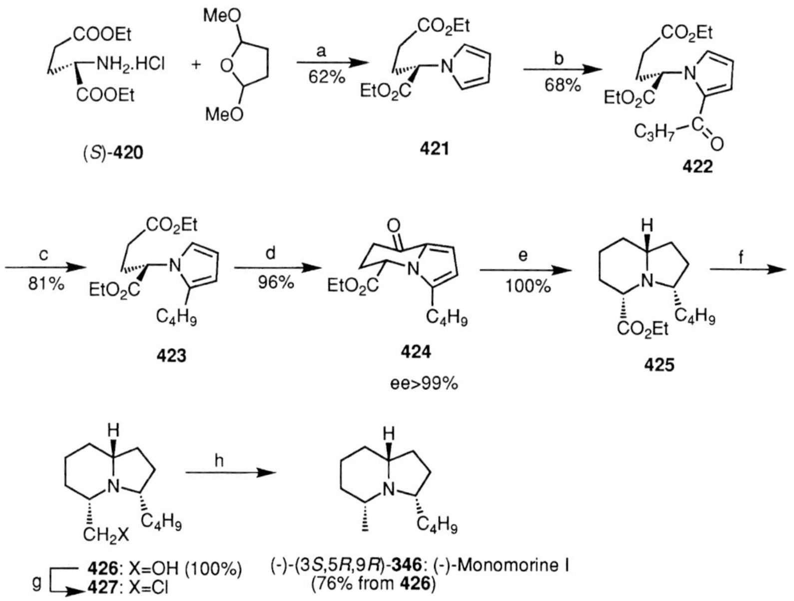

(a) H_2O; (b) PrCOCl, PhMe; (c) $NaBH_3CN$, ZnI_2, $Cl(CH_2)_2Cl$; (d) BBr_3, CH_2Cl_2; (e) H_2, Pd/C, EtOH, H_2SO_4; (f) LAH, THF; (g) $SOCl_2$; (h) Bu_3SnH, AIBN, PhMe.

Scheme 51. Asymmetric synthesis of (−)-(3*S*,5*R*,8*R*)-monomorine I from diethyl L-glutamate (*113*)

423. Intramolecular acylation to the bicyclic keto pyrrole **424** occurred regioselectivity with complete retention of configuration in 96% yield. Catalytic hydrogenation of **424** over palladium on charcoal in acidic ethanol proceeded with remarkable stereocontrol to give **425** which could be converted into (−)-(3S,5R,9R)-monomorine I, the unnatural enantiomer of the alkaloid, in three steps in a 76% yield.

LHOMMET and co-workers (*114*) started from the observation that reduction of the bicyclic iminium ion **435** is totally stereoselective and provides only (+)-(3R,5S,9S)-monomorine I (Scheme 52). Thus, tosylate (S)-**298**, prepared from readily available (S)-pyroglutamic acid **204** in three steps (*115*), reacted with lithium dipropyl cuprate to give the alkylated product **428** in 61% yield. Treatment of **428** with dimethylsulfate, condensation with Meldrum's acid and monodecarboxylating transesterification led to the key β-enaminoester intermediate **429**. This latter was hydrogenated to a *cis:trans* mixture of 2,5-disubstituted pyrrolidines **430** in a 96:4 ratio. LiAlH$_4$ reduction converted the diastereoisomeric esters **430** into the alcohols **431**, which were then

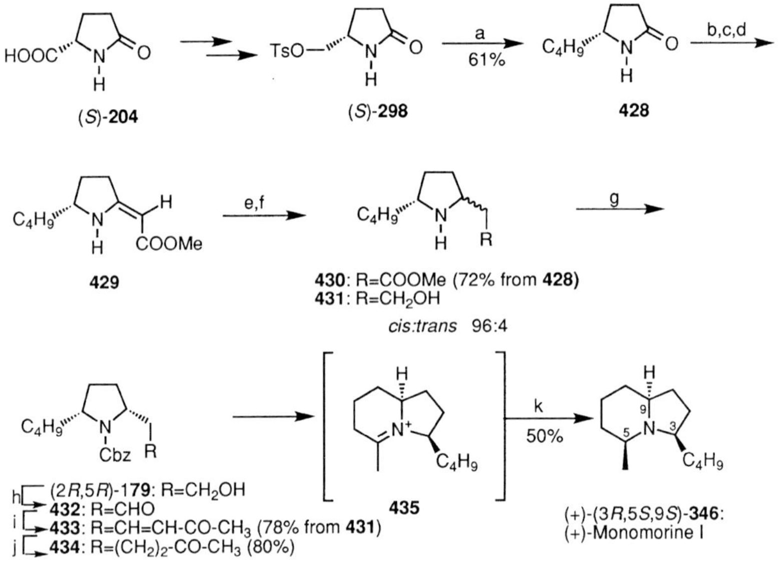

(a) *n*Pr$_2$CuLi, ether-CH$_2$Cl$_2$; (b) Me$_2$SO$_4$; (c) Meldrum's acid, Ni(acac)$_2$, CHCl$_3$; (d) MeONa, MeOH; (e) H$_2$ (100 atm), Raney Ni, HCl, MeOH; (f) LiAH$_4$, Et$_2$O; (g) ClCOCH$_2$Ph; (h) PCC, CH$_2$Cl$_2$; (i) CH$_3$COCH=PPh$_3$, THF; (j) H$_2$ (1 atm), PtO$_2$; (k) H$_2$ (1 atm), 10% Pd/C.

Scheme 52. Asymmetric synthesis of (+)-(3R,5S,8S)-monomorine I from (S)-pyroglutamic acid (*114*)

transformed into the carbamate *cis*-(2*R*,5*R*)-**179**. Successive oxidation, chain elongation and partial hydrogenation provided ketone **434**. Finally, reduction of **434** with Pd/C gave only one stereoisomer, (+)-monomorine I, in 50% yield, presumably through iminium **435**.

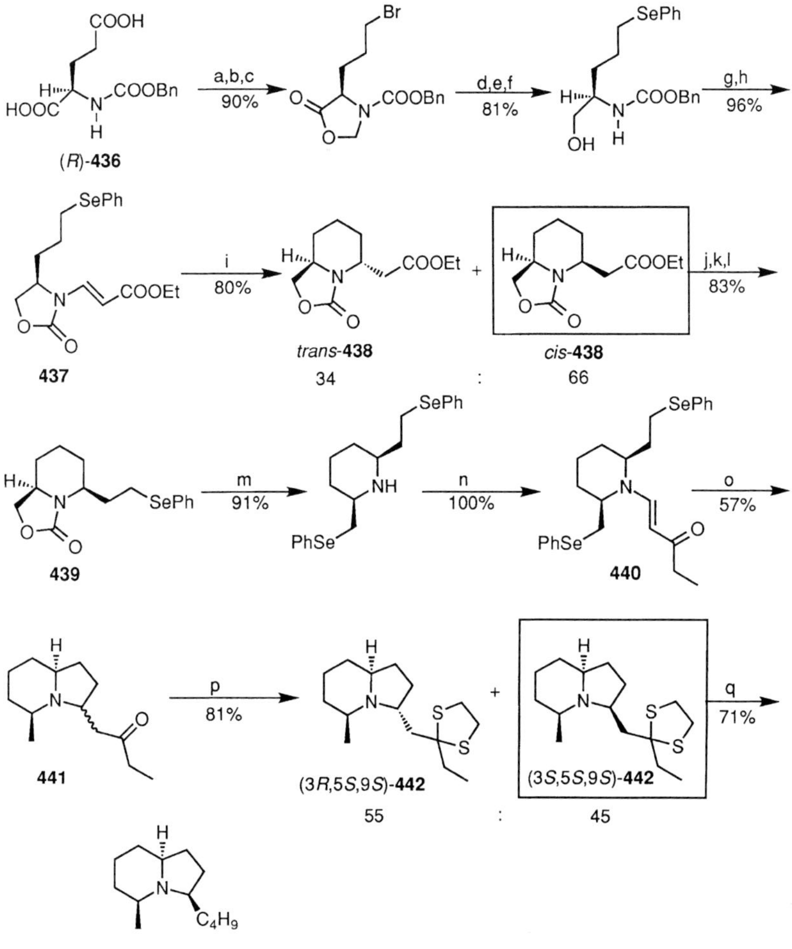

(+)-(3*R*,5*S*,9*S*)-**346**: (+)-Monomorine I

(a) $(CH_2O)_n$, *p*-TsOH, toluene; (b) BH_3.THF, THF; (c) CBr_4, PPh_3, THF; (d) $LiBH_4$, THF, (e) MeOK, MeOH; (f) $(PhSe)_2$, $NaBH_4$, EtOH; (g) MeOK, MeOH-THF; (h) HCCCOOEt, *N*-methylmorpholine, DCM; (i) Bu_3SnH, AIBN, benzene; (j) $LiAlH_4$, THF; (k) CBr_4, PPh_3, CH_2Cl_2; (l) $(PhSe)_2$, $NaBH_4$, EtOH; (m) $PhSeSiMe_3$, ZnI_2, toluene; (n) HCCCOOEt, *N*-methylmorpholine, CH_2Cl_2; (o) Bu_3SnH, AIBN, benzene; (p) $HSCH_2CH_2SH$, $BF_3.OEt_2$, CH_2Cl_2, N_2; (q) Raney Ni, EtOH.

Scheme 53. Asymmetric synthesis of (+)-(3*R*,5*S*,8*S*)-monomorine I from D-glutamic acid (*116*)

The last synthesis, recently described by LEE et al. (*116*), exemplifies the construction of both the six- and the five-membered rings of the indolizidine system by stepwise radical cyclizations of a β-aminoacrylate **437** and a β-amino vinyl ketone **441**, respectively (Scheme 53). Starting from the Cbz-protected D-glutamic acid (*R*)-**436**, which was transformed in eight steps into the β-amino acrylate **437** in 70% yield, cyclization of **437** to a 2:1 mixture of the *cis*- and *trans*-piperidine derivatives **438** could be achieved with tributyltin hydride under high dilution conditions. The major *cis* isomer, *cis*-**438**, was then transformed into selenide **439** *via* the corresponding alcohol and bromide. Cleavage of the cyclic carbamate and treatment with 1-pentyn-3-one gave β-amino vinyl ketone **440** in 75% from *cis*-**439**. A second radical cyclization under high dilution conditions gave an inseparable mixture of indolizidine derivatives **441** in 57% yield. The dithioketal derivatives (3*R*,5*S*,9*S*)-**442** and (3*S*,5*S*,9*S*)-**442** were obtained in 44 and 37% yield from the mixture of ketones **442** and the dithioketal (3*S*,5*S*,9*S*)-**442** reacted with Raney Ni to afford (+)-monomorine I in a good yield.

c) From L-Tartaric Acid

An enantioselective total synthesis of (+)-monomorine I, starting from chiral precursor **444**, itself available in six steps from diethyl L-tartrate (L-**443**) (*117*), was devised in 1988 by KIBAYASHI and YAMAZAKI (*118*) (Scheme 54). Reduction of **444** with zinc borohydride yielded alcohol **445** with high anti-selectivity (>99:1). The latter was converted to aldehyde **446** in 41% overall yield by a Mitsunobu reaction with phthalimide followed by debenzylation and Swern oxidation. Subsequent Grignard reaction, Swern oxidation and reduction with L-selectride provided alcohol **447** with high syn-selectivity (syn:anti=98:2). Removal of the phthaloyl group, *N*-benzyloxycarbonylation, mesylation and base-induced cyclization afforded exclusively the (2*S*,5*R*)-pyrrolidine **448**. The 3-pyrroline derivative **450**, obtained from **448** in two steps, underwent selective Wacker oxidation of the terminal olefin to yield **451** which, on hydrogenation over Pd/C in MeOH, gave exclusively (+)-monomorine I in 61% yield from **450**.

Two years later, ITO and KIBAYASHI reported an alternative enantioselective synthesis of (+)-monomorine I by means of an asymmetric 1,3-dipolar cycloaddition of a nitrone with a chiral allylic ether as dipolarophile (*119, 120*). The L-threitol derivative **452**, prepared from L-diethyl tartrate (*121*), was converted into tosylate **453**, which underwent coupling with propylmagnesium bromide to give **454** in 82% yield from **452** (Scheme 55). Removal of the isopropylidene group under

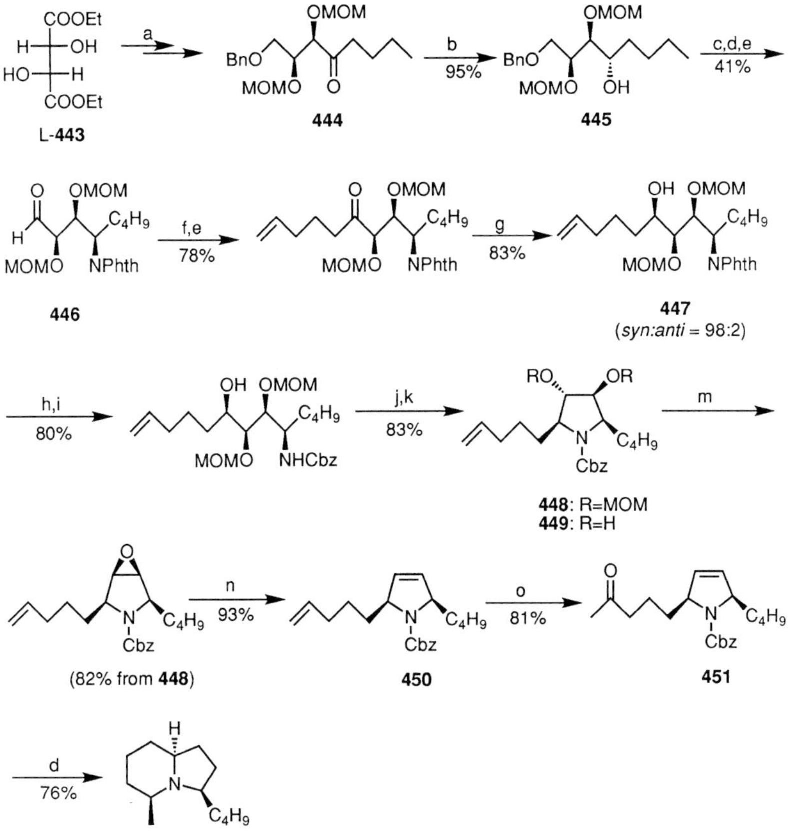

(+)-(3*R*,5*S*,9*S*)-**346**: (+)-Monomorine I

(a) *J. Org. Chem.*, 51, 3769 (1996); (b) $Zn(BH_4)_2$, Et_2O; (c) Phtalimide, PPh_3, EtOOCN=NCOOEt, THF; (d) H_2, Pd/C, MeOH; (e) $(COCl)_2$, Me_2SO, Et_3N, CH_2Cl_2; (f) $CH_2=CH(CH_2)_3MgBr$, THF; (g) $LiBH(sec\text{-}Bu)_3$, THF; (h) $(NH_2)_2.H_2O$, EtOH; (i) $PhCH_2OCOCl$, Na_2CO_3, CH_2Cl_2; (j) MsCl, Et_3N, CH_2Cl_2; (k) t-BuOK, THF; (l) HCl, MeOH; (m) imidazole, triiodoimidazole, PPh_3, toluene; (n) PPh_3, Zn, toluene; (o) $PdCl_2$, $CuCl_2$, DMF-H_2O.

Scheme 54. Asymmetric synthesis of (+)-(3*R*,5*S*,8*S*)-monomorine I from diethyl L-tartrate (*117*)

acidic conditions and treatment of the resulting 1,2-diol **455** with $Me_2NCH(OMe)_2$ afforded dioxolane **456**, which was subsequently converted to the (S)-allyl ether **457**, by heating with acetic anhydride in 68% overall yield from **454**. Nitrone **458**, existing as an *E-Z* equilibrium mixture, was allowed to react with **457** to give a mixture of the adducts (*S*,*S*,*S*)- and (*R*,*R*,*S*)-**459** in a 76% yield and a 1 : 3 ratio. Using a chiral allylic ether as dipolarophile resulted in a reasonable degree of

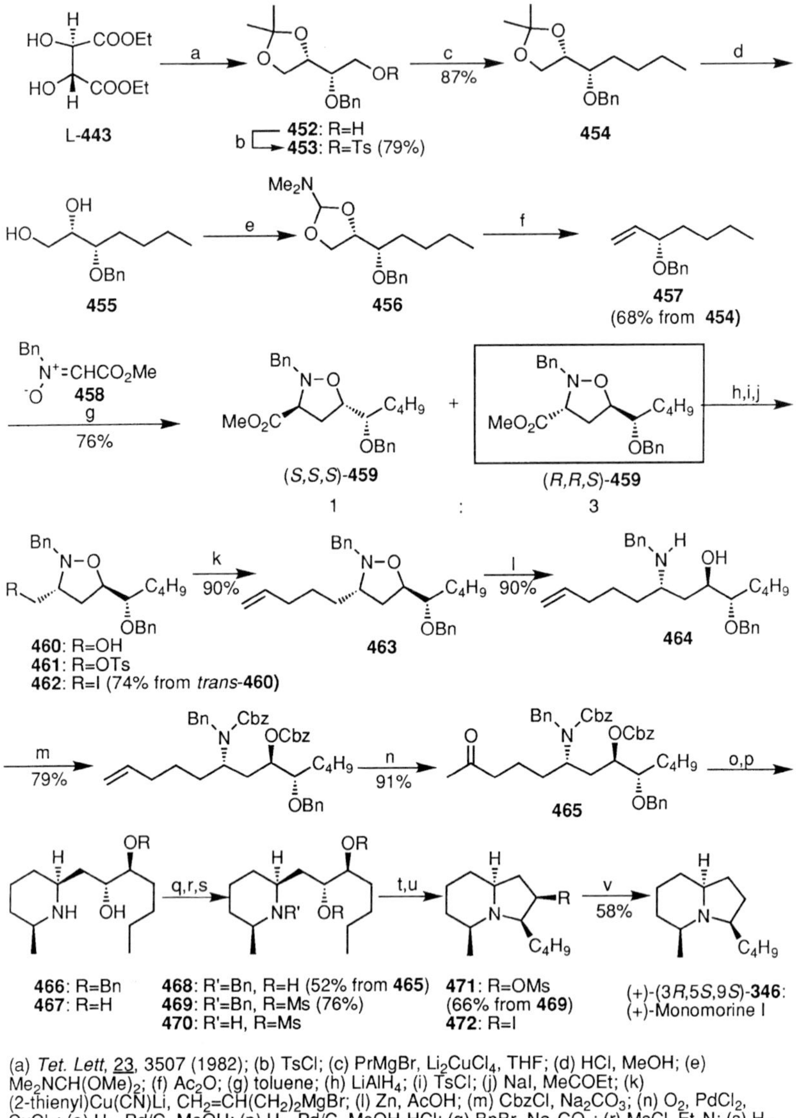

Scheme 55. Asymmetric synthesis of (+)-(3*R*,5*S*,8*S*)-monomorine I from diethyl L-tartrate (*119, 120*)

(a) *Tet. Lett*, **23**, 3507 (1982); (b) TsCl; (c) PrMgBr, Li$_2$CuCl$_4$, THF; (d) HCl, MeOH; (e) Me$_2$NCH(OMe)$_2$; (f) Ac$_2$O; (g) toluene; (h) LiAlH$_4$; (i) TsCl; (j) NaI, MeCOEt; (k) (2-thienyl)Cu(CN)Li, CH$_2$=CH(CH$_2$)$_2$MgBr; (l) Zn, AcOH; (m) CbzCl, Na$_2$CO$_3$; (n) O$_2$, PdCl$_2$, CuCl$_2$; (o) H$_2$, Pd/C, MeOH; (p) H$_2$, Pd/C, MeOH-HCl; (q) BnBr, Na$_2$CO$_3$; (r) MsCl, Et$_3$N; (s) H$_2$, Pd/C, MeOH; (t) Et$_3$N, CH$_2$Cl$_2$; (u) NaI, MeCOEt; (v) H$_2$, Pd/C, Et$_3$N, MeOH.

asymmetric induction. The preferred formation of the major isomer (*R*,*R*,*S*)-**459** can be rationalized according to Houk's concepts (*122*). (*R*,*R*,*S*)-**459** was transformed into iodide **462** which was coupled with a mixed organocuprate to afford **463** in 63% yield from (*R*,*R*,*S*)-**459**. Reductive N-O bond cleavage, treatment of the resultant amino alcohol **464** with PhCH$_2$OCOCl, and oxidation by the Wacker process gave ketone **465**. After simultaneous reductive cyclization and debenzylation, the *cis*-2,6-dialkylpiperidine **467** was subsequently converted to **468** by selective *N*-benzylation in 52% yield from ketone **465**. The di-*O*-mesylate prepared from **468**, was hydrogenolyzed to give **470**, which was immediately cyclized by heating with NEt$_3$ to afford **471**. Finally, **471** was converted into (+)-monomorine I by means of nucleophilic displacement (NaI) followed by reductive deiodination.

d) From D-Norleucinol

In 1997, CRAIG and co-workers synthesized (+)-monomorine I from D-norleucinol using a 5-*endo*-trig cyclization and intramolecular reductive amination as the key ring-forming steps (Scheme 56) (*123*).

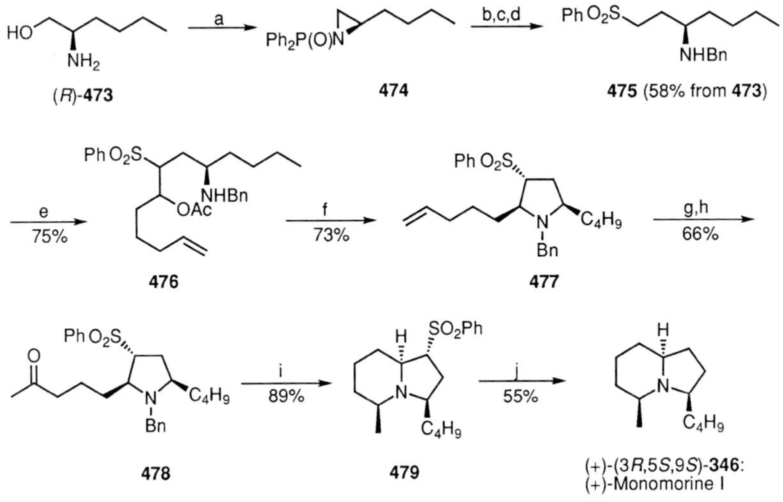

(a) Ph$_2$P(O)Cl, Et$_3$N, THF, then NaH; (b) PhSO$_2$Me, BuLi, THF-Me$_2$N(CH$_2$)$_2$NMe$_2$; (c) BF$_3$.OEt$_2$, CH$_2$Cl$_2$-MeOH; (d) BnCl, pyridine, CH$_2$Cl$_2$; (e) BuLi, THF-Me$_2$N(CH$_2$)$_2$NMe$_2$, hex-5-enal; (f) Bu^{t}OK, Bu^{t}OH, THF; (g) DIBALH, CH$_2$Cl$_2$; (h) Hg(OAc)$_2$, THF-H$_2$O, then PdCl$_2$, CuCl$_2$, THF; (i) 10% Pd/C, cyclohexa-1,4-diene, MeOH; (j) Na$^+$C$_{10}$H$_8^-$, THF.

Scheme 56. Asymmetric synthesis of (+)-(3*R*,5*S*,8*S*)-monomorine I from D-norleucinol (*123*)

D-norleucinol (*R*)-**473** was directly converted into *N*-protected aziridine **474** by treatment with diphenylphosphinic chloride and triethylamine followed by excess sodium hydride according to the method of SWEENEY (*124*). Addition of **474** to lithio(phenylsulfonyl)-methane and proton quench gave the expected product of aziridine ring-opening at the less substituted carbon atom. This was dephosphinylated and reprotected as the benzyl derivative **475** in good overall yield for the three steps from **474**. Exposure of **475** to base followed by hex-5-enal and *in situ* trapping of the intermediate alkoxides gave ester **476**, mostly as one diastereoisomer. Pyrrolidine formation was effected in a single step by treating ester **476** with potassium *tert*-butoxide in the presence of *tert*-butyl alcohol, which effected one-pot elimination and cyclization to give *cis*-**477** in 73% yield. Partial reduction of **477** to the *N*-benzyl analogue using DIBALH, and oxidation of the side chain double bond gave ketone **478**. This was subjected to catalytic transfer hydrogenation, which effected sequential hydrogenolytic debenzylation and intramolecular reductive amination of give exclusively **479** in 89% yield. Finally, indolizidine **479** was treated with sodium naphthalenide to afford (+)-monomorine I.

e) From Other Chirons

In 1994, MUCHOWSKI and co-workers (*125*) described a total synthesis of the unnatural (−) enantiomer of monomorine I based on oxidative radical cyclization of ω-iodoalkylpyrrole **484**, made in seven steps from (*R*)-4-aminopentanoic acid [(*R*)-**480**] (Scheme 57). Condensation of (*R*)-**480** with dimethoxytetrahydrofuran and esterification with diazomethane gave the *N*-substituted pyrrole **481** in 70% yield. Reduction of **481**, conversion of the resulting alcohol to the chloride **482** and subsequent acylation of the pyrrole ring of the latter provided **483** in 35% yield from **481**. Iodide **484**, prepared from **483** by treatment with NaI, cyclized to the enantiomerically pure bicyclic ketone (*R*)-**485**. The latter was converted into (*R*)-**368** which, on catalytic reduction, gave a 33:40:23:3 mixture of (−)-monomorine I [(−)-**346**] and of three of its diastereoisomers, separable by chromatography on alumina.

Syntheses Using Chiral Auxiliaries

a) Using Koga's Base (**487**) (*126*)

In 1990, MOMOSE *et al.* described the asymmetric synthesis of (+)-monomorine I in 24% overall yield, using as key step the cleavage of the

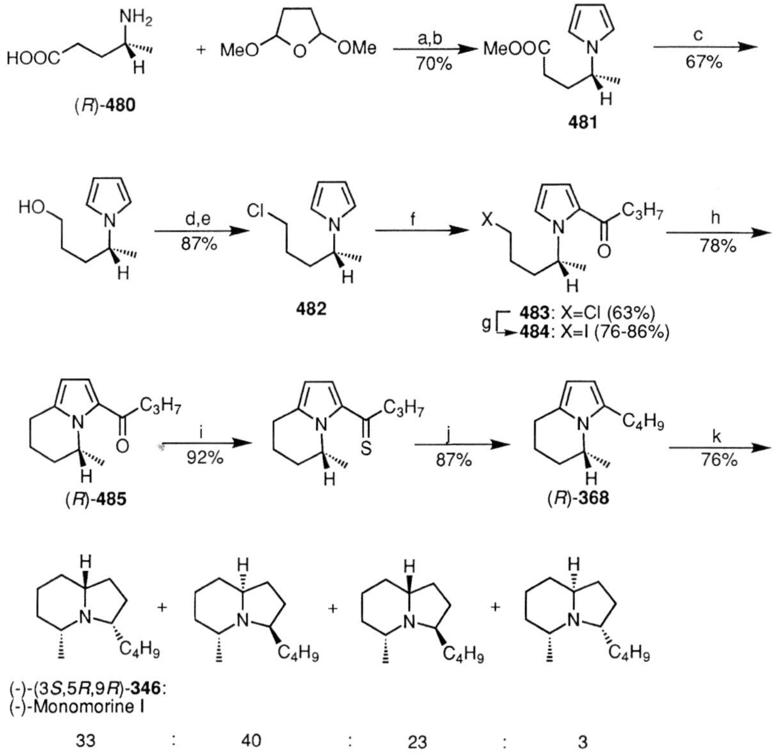

Scheme 57. Asymmetric synthesis of (−)-(3S,5R,8R)-monomorine I from (R)-4-aminopentanoic acid (125)

(a) AcOH; (b) CH$_2$N$_2$; (c) LAH; (d) MeSO$_2$Cl, Et$_3$N; (e) Bu$_4$NCl; (f) PrCOCl; (g) NaI, MeCN; (h) Fe(II), H$_2$O$_2$, DMSO,))); (i) Lawesson's reagent, THF; (j) W-2 Raney Nickel; (k) H$_2$, Rh-Al$_2$O$_3$, MeOH.

optically active silyl enol ether **488**, which was produced in 90% ee by asymmetric kinetic deprotonation of 8-azabicyclo [3.2.1]octan-3-one (**486**) in the presence of Koga's base **487** (Scheme 58) (*127*). Ozonolysis of **488** followed by esterification of the resulting carboxylic acid provided the *cis*-2,5-disubstituted pyrrolidine derivative **489**. Monoprotected diol **490** was obtained in 84% from **489** by protection of the alcohol and reduction of the ester with Super-Hydride. Compound **490** was converted into iodide **492** *via* tosylate **491** and then transformed by a Grignard cross coupling reaction with allylmagnesium chloride into olefin **493** in 70% yield. Further carbon-chain elongation was carried out by Swern oxidation of the deprotected primary alcohol followed by

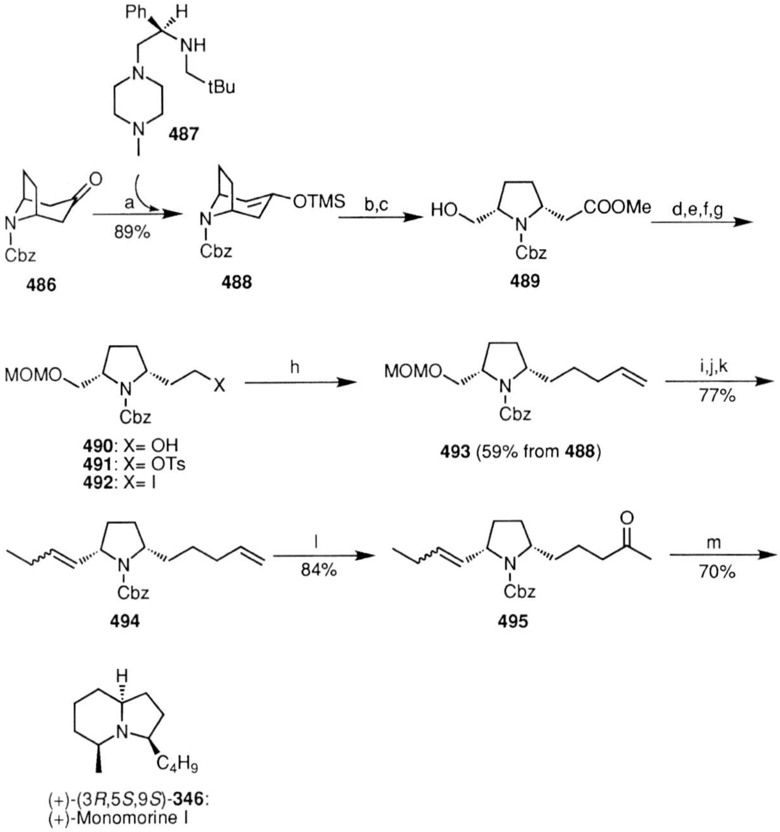

Scheme 58. Asymmetric synthesis of (+)-(3R,5S,8S)-monomorine I using Koga's base (127)

(a) **487**, n-BuLi, TMSCl-HMPA; (b) O_3, CH_2Cl_2-MeOH, then $NaBH_4$; (c) CH_2N_2; (d) MOMCl, (i-Pr)$_2$EtN; (e) Super-Hydride, THF; (f) TsCl, pyridine; (g) NaI, acetone; (h) allylmagnesium chloride, CuI, THF; (i) HCl, MeOH; (j) (COCl)$_2$, DMSO, Et$_3$N; (k) CH_3CH_2CH=PPh$_3$; (l) O_2, PdCl$_2$, CuCl; (m) H_2, 5% Pd/C, MeOH.

Wittig reaction, yielding diolefin **494**. Site-selective oxidation of **494** under Wacker conditions smoothly proceeded to give ketone **495** in 84% yield. Final hydrogenation provided the expected (+)-monomorine I which was obtained in 24% overall yield from starting material **486**.

b) Using Chiral 1,3-Oxazolidines

TAKAHASHI'S route to the same (+)-monomorine I was conceptually quite different, using as key step the highly diastereoselective reaction of

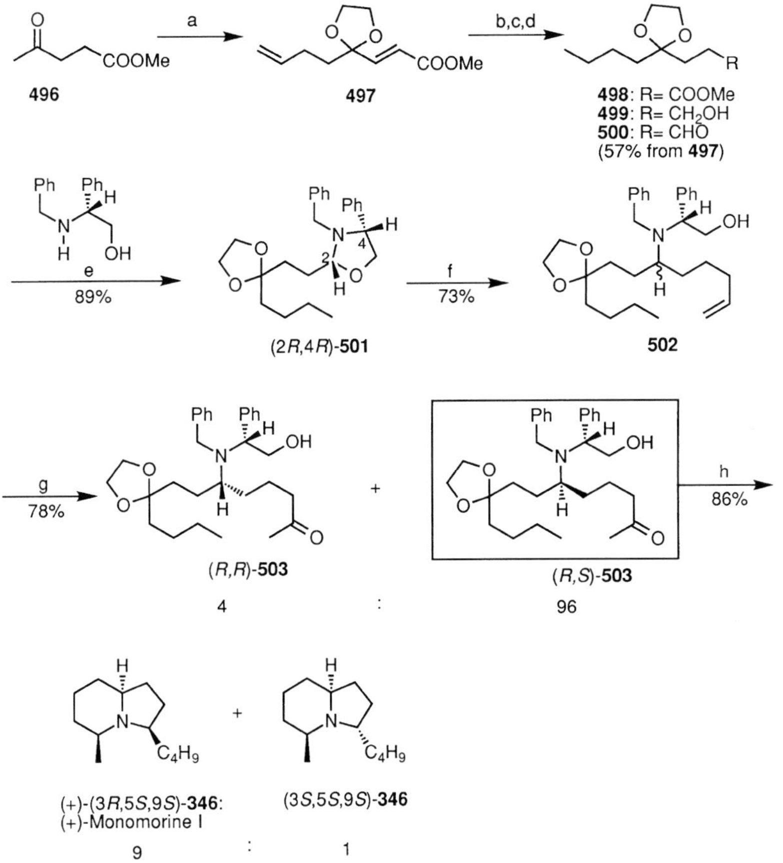

(a) *J. Org. Chem.*, 46, 3137 (1981); (b) H_2, 10%Pd/C, MeOH; (c) $LiAlH_4$, THF; (d) PCC,CH_2Cl_2; (e) (*R*)-*N*-benzylphenylglycinol, $MgSO_4$, CH_2Cl_2; (f) $CH_2=CH(CH_2)_3MgBr$, THF; (g) O_2, $PdCl_2(MeCN)_2$, $CuCl_2$, MeOH; (h) H_2, 10% Pd/C, MeOH-HCl.

Scheme 59. Asymmetric synthesis of (+)-(3*R*,5*S*,8*S*)-monomorine I using chiral 1,3-oxazolidines (*128*)

chiral 1,3-oxazolidine **501** with a Grignard reagent (*128*) (Scheme 59). Thus, dienoate **497**, readily available from methyl levulinate **496** by a known procedure (*129*), was converted into aldehyde **500** by successive catalytic hydrogenation, chemical reduction and oxidation. Condensation of **500** with (*R*)-*N*-benzylphenylglycinol provided (2*R*,4*R*)-1,3-oxazolidine **501** along with its C-2 epimer in a 93 : 7 ratio. Reaction of **501** with pent-4-enylmagnesium bromide furnished alcohol **502** as an unseparable diastereoisomeric mixture. This, when subjected to a Wacker oxidation, afforded a 4:96 ratio of methyl ketones (*R,R*)- and (*R,S*)-**503** in 78% yield. After separation by silica gel chromatography, (*R,S*)-**503** was submitted to catalytic hydrogenation, yielding (+)-monomorine I (78%) along with its C-3 epimer (8%).

c) Using Chiral Sulfoxides

The synthetic strategy recently developed by CHU and SOLLADIÉ (*130*) was based on the formation of the enantiomerically pure *cis*-2,6-disubstituted piperidine intermediate **511** which, after addition of an appropriate substituent, should afford by ring closure the indolizidine skeleton (Scheme 60). In this approach, the first asymmetric center was

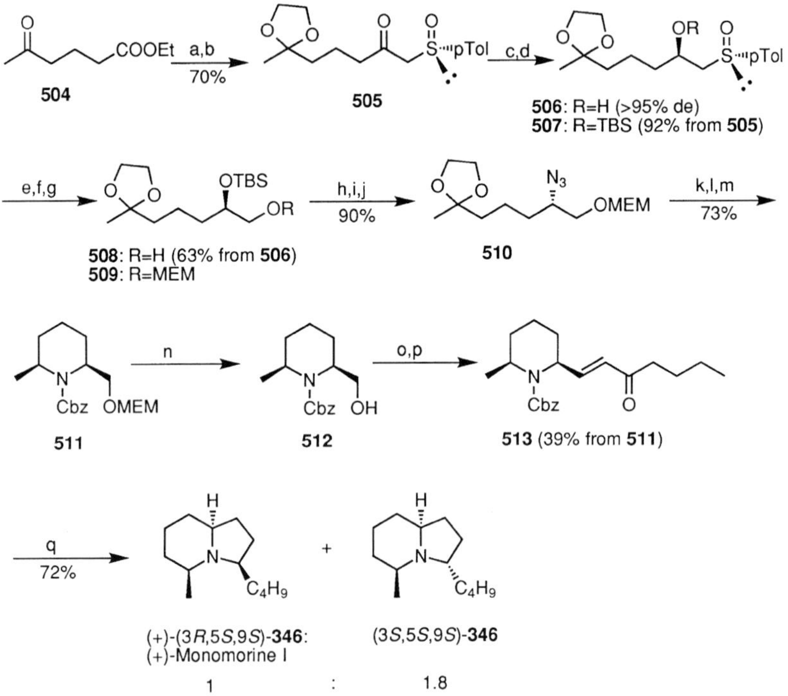

(a) $(CH_2OH)_2$, TsOH, PhH; (b) (*R*)-methyl *p*-tolylsulfoxide, LDA, THF; (c) $ZnCl_2$, DIBALH, THF; (d) TBSCl, imidazole, DMF; (e) Ac_2O, AcONa; (f) $LiAlH_4$, $PhCH_3$; (g) MEMCl, *i*-Pr_2NEt, CH_2Cl_2; (h) TBAF, THF; (i) MsCl, Et_3N, CH_2Cl_2; (j) NaN_3, DMF; (k) TsOH, acetone; (l) H_2, 10% Pd/C, MeOH; (m) CbzCl, 20% K_2CO_3, CH_2Cl_2; (n) $TiCl_4$, CH_2Cl_2; (o) Swern oxidation; (p) LiCl, *i*-Pr_2NEt, $(MeO)_2POCH_2COC_4H_9$; (q) H_2, 10% Pd/C, MeOH.

Scheme 60. Asymmetric synthesis of (+)-(3*R*,5*S*,8*S*)-monomorine I using chiral sulfoxides (*130*)

created by reduction of β-ketosulfoxide **505** while the second one was obtained by reductive cyclization of the amino-ketone derived from **510**. (+)-(R)-β-ketosulfoxide **505**, prepared from 5-oxohexanoate, underwent stereoselective reduction of the carbonyl group to afford (R,R)-β-hydroxysulfoxide **506** in >95% de. Protection of the hydroxyl group of **506**, followed by Pummerer rearrangement and subsequent reduction of the Pummerer intermediate provided compound **508** in 63% overall yield. Compound **510**, obtained from **509** by treatment with sodium azide in 90% yield, was deprotected and cyclized under reducing conditions, giving the (S,S)-benzyl piperidine carbamate **511**. The stereochemistry of the reduction of the imino intermediate was controlled by the chiral center α to the nitrogen atom. Piperidine carbamate **511** was deprotected with $TiCl_4$ and the hydroxyl group oxidized to the corresponding aldehyde which reacted with the appropriate ketophosphonate to provide **513** in 39% yield from **511**. Cyclization of the Wittig adduct under reducing conditions yielded a 1:1:8 mixture of the two diastereoisomers (+)-monomorine I [(+)-**346**] and its C-3 epimer.

d) Using Chiral Bicyclic Thiolactams

In 1995, MUNCHHOF and MEYERS (*131*) reported that thiolactam **515** could be readily transformed in three steps to the chiral non-racemic *cis*-2,6-disubstituted piperidine **519** which is a key intermediate towards the synthesis of (+)-monomorine I (Scheme 61). Thus, thiolactam **515**, prepared from **514** by Belleau's reagent (*37*) was treated with methyl α-bromoacetate and then with trimethylphosphite and triethylamine, the so-called Eschenmoser contraction sequence (*182*) to afford the vinylogous urethane **518** in 80% yield. This transformation of thiolactam **515** was not possible by use of organometallic reagents, sequence **515**→**516**→**517** was therefore crucial to the success of the operation. The transformation of **518** to the *cis*-2,6-disubstituted piperidine **519** as essentially one diastereoisomer (>97%) was accomplished in a single step by hydrogenation using $Pd(OH)_2/C$. The carbobenzyloxy derivative **520**, obtained from **519**, underwent hydrolysis of the ester function and was then subjected to Arndt-Eistert conditions which produced the homologated piperidine ester **521** in 75% from **519**. Utilizing the Weinreb amidation, the *N*-methyl-*N*-methoxyamide **522** was treated with *n*-butylmagnesium bromide, affording the piperidine ketone **523**. Catalytic hydrogenation of **523** yielded (+)-monomorine I in high enantiomeric purity.

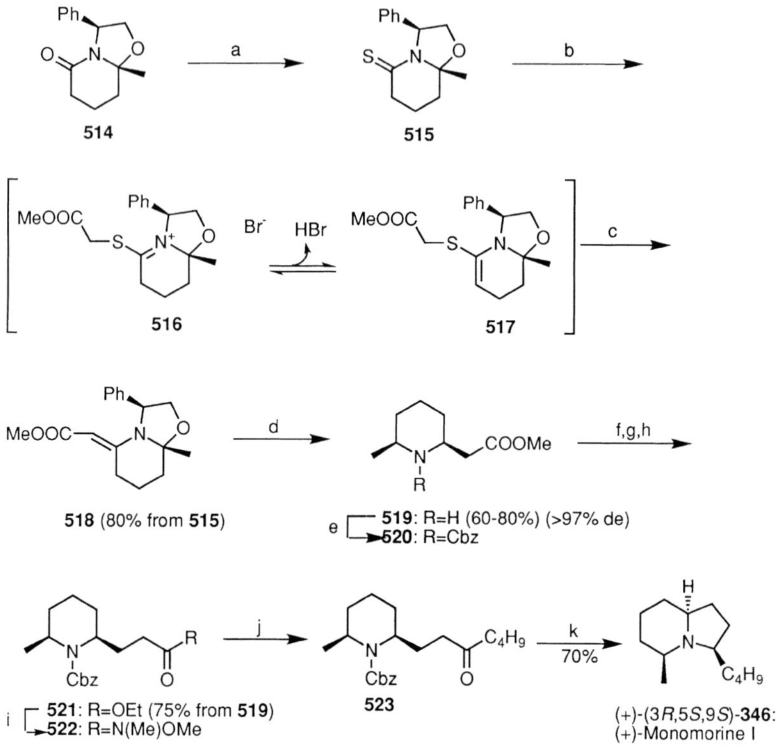

Scheme 61. Asymmetric synthesis of (+)-(3R,5S,8S)-monomorine I using chiral bicyclic thiolactams (131)

(a) Tet. Lett., 24, 3815, (1983); J. Org. Chem., 57, 2818 (1992); (b) BrCH$_2$COOMe, NEt$_3$, CHCl$_3$; (c) P(OMe)$_3$, NEt$_3$, CHCl$_3$; (d) H$_2$, Pd(OH)$_2$/C; (e) benzylchloroformate, (Schotten-Baumann); (f) hydrolysis; (g) (COCl)$_2$; (h) CH$_2$N$_2$, Et$_2$O, then Ag$_2$O, EtOH; (i) Weinreb amidation; (j) n-butylmagnesium bromide; (k) H$_2$, 5% Pd/C, MeOH.

2.2.5.2. 3,5-Dialkylindolizidines

A. 3-Butyl-5(4-penten-1-yl)indolizidine

Alkaloid (5E,9Z)-**51** isolated in 1990 was the first indolizidine from ants which did not possess the all-*cis* (5Z,9Z) relative configuration. Its structure was confirmed by two separate syntheses reported by JONES *et al.* (*28*). The key step of the first was the reductive amination of triketone **524**, synthesized in five steps from 4-chloro-1-butanol in 15% yield, which provided the four possible diastereoisomers of **51** as a

References, pp. 221–229

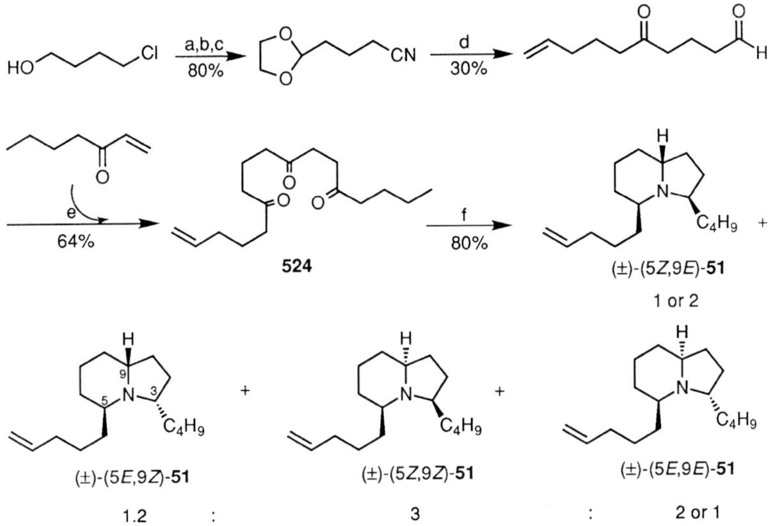

(a) PCC; (b) (CH$_2$OH)$_2$, p-TsOH; (c) NaCN, DMSO; (d) 4-pentenylmagnesium bromide, then HCl; (e) 3-benzyl-5-(2-hydroxyethyl)-4-methylthiazolium chloride, Et$_3$N; (f) NH$_4$OAc, NaBH$_3$CN.

Scheme 62. Non stereoselective synthesis of (±)-**51** (*28*)

mixture (Scheme 62). Alkaloid (±)-(5E,9Z)-**51** was obtained in six steps and 2% overall yield from starting material.

A stereoselective synthesis carried out by the same authors (*28*) provided indolizidine (±)-**51** with the (5E,9Z) geometry (Scheme 63).

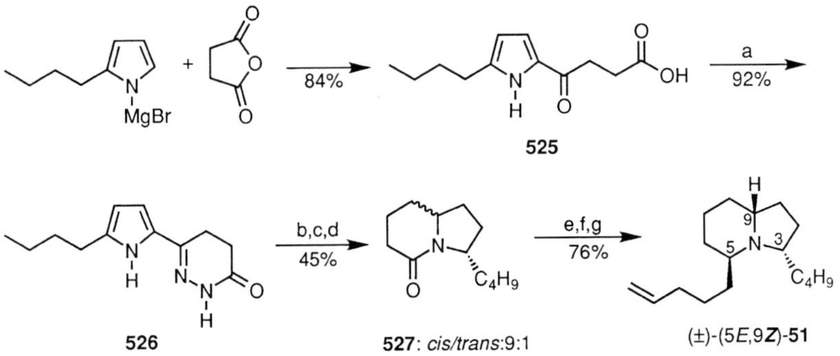

(a) (NH$_2$)$_2$, benzene; (b) KOH, (CH$_2$OH)$_2$; (c) H$_2$, PtO$_2$, AcOH; (d) EtOH; (e) DIBALH, then HClO$_4$; (f) KCN; (g) 4-pentenylmagnesium bromide.

Scheme 63. Stereoselective synthesis of (±)-(5E,9Z)-**51** (*28*)

Thus, pyrrole keto acid **525**, obtained in good yield from 2-butylpyrrolemagnesium bromide and succinic anhydride, was condensed with hydrazine to form tetrahydropyridazine **526** in 92% yield. After Wolff-Kishner reduction of **526**, the relative stereochemistry between positions 5 and 9 was set up by means of a reasonably stereoselective catalytic hydrogenation of the pyrrole ring. The substituent at position C-5 was efficiently introduced by intercepting an intermediate α-cyanoamine, prepared *in situ* from **527**, with 4-pentenylmagnesium bromide. In this way, alkaloid (±)-(5E,9Z)-**51** was obtained in seven steps and 26% overall yield.

In 1996, TAKAHATA *et al.* reported the first asymmetric synthesis of (+)-(3S,5S,9R)-**51** *via* construction of the *cis*-2,5-disubstituted pyrrolidine (S,S)-*cis*-**179** by capitalizing on Sharpless asymmetric

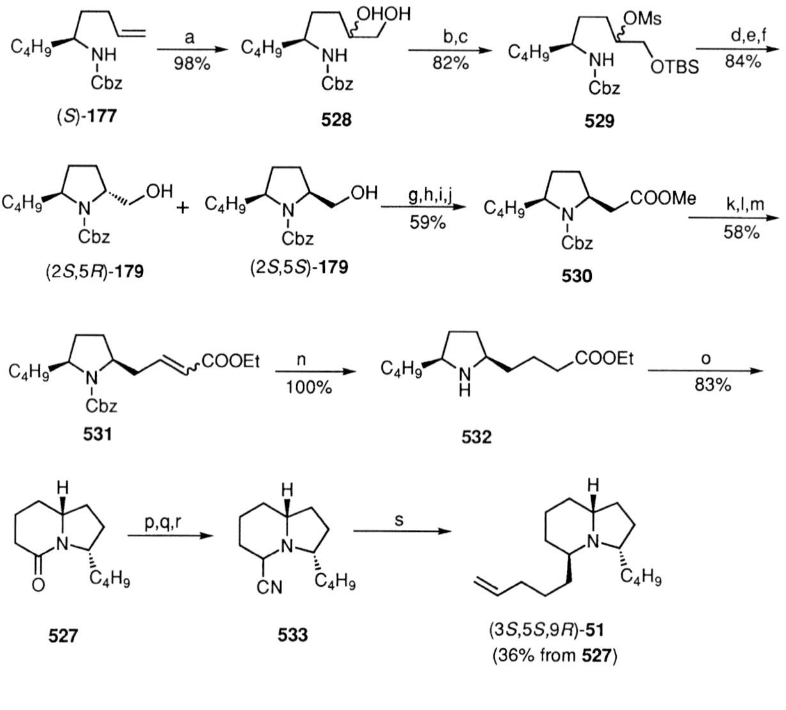

(a) AD-mix-β; (b) TBSCl, imidazole, DMF; (c) MsCl, Et₃N; (d) H₂, Pd(OH)₂; (e) HCl 1%; (f) CbzCl, NaOH; (g) chromatography; (h) CrO₃, H⁺, acetone; (i) CH₂N₂; (j) AgCOOPh, MeOH; (k) LiBHEt₃; (l) (COCl)₂, DMSO, Et₃N; (m) (EtO)₂P(O)CH₂COOEt, NaH; (n) H₂, Pd(OH)₂; (o) Me₃Al; (p) DIBALH; (q) HClO₄; (r) KCN; (s) 1-pentenylmagnesium bromide.

Scheme 64. Asymmetric synthesis of (+)-(3S,5S,9R)-**51** from L-norleucine (*133*)

References, pp. 221–229

dihydroxylation of the homochiral 4-pentenylcarbamate (*S*)-**177** (available from L-norleucine) (Scheme 64) (*133*). Treatment of (*S*)-**177** with AD-mix-β afforded a diastereoisomeric mixture of diols **528**. Selective protection of the primary hydroxyl of **528** with *tert*-butyldimethylsilyl chloride followed by mesylation of the secondary hydroxyl provided mesylate **529**. Exposure of **529** to an atmosphere of hydrogen in the presence of $Pd(OH)_2$ as a catalyst caused concurrent debenzyloxycarbonylation and ring closure to give the pyrrolidine salt, which was converted by a two-step sequence to a separable 4 : 1 mixture of the *cis*-2,5-disubstituted pyrrolidine (*S,S*)-*cis*-**179** and its *trans* isomer. With the requisite **179** in hand, the elongation of the side chain was carried out as follows. Jones oxidation gave the acid, which on the subsequent Arndt-Eistert homologation provided ester **530**. Application of a three-step sequence to **530** gave the α,β-unsaturated ester **531**. Catalytic hydrogenation effected both debenzyloxycarbonylation and olefin reduction to give **532**, but failed to produce further ring closure to an indolizidinone. The intramolecular lactamization of **532** was performed by Weinreb's procedure, and the unsaturated chain at C-5 was introduced by the completely stereoselective sequence already used by JONES (Scheme 63).

B. 3-Ethyl- and 3-Hexyl-5-methylindolizidines

In order to establish the relative stereochemistry of 3-hexyl-5-methylindolizidine **47** isolated from the venom of *Solenopsis diplorhoptrum* from California, JONES *et al.* developed a nonselective synthesis of the four possible isomers (Scheme 65) (*12*). Thus, pyridine alcohol **534** was successfully oxidized to ketone **535** with pyridinium dichromate and, after ketalization, the pyridine ring of **536** was reduced either with sodium in ethanol or by hydrogenation over a rhodium catalyst. The former gave a 5 : 2 mixture of the *cis*- and *trans*-piperidines **537**, while the latter provided only the *cis* isomer. Reductive amination of the *cis* and *trans* mixture of **537**, and removal of the ketal protecting group afforded ($\pm$)-(5Z,9Z)-3-hexyl-5-methylindolizidine [($\pm$)-(5Z,9Z)-**47**] and three stereoisomers in a 3 : 2 : 2 : 1 ratio (Scheme 65).

MOMOSE and co-workers (*112*) reported in 1993 the first synthesis of (+)-3-ethyl- and (+)-3-hexyl-5-methylindolizidines, (+)-(3*R*,5*S*,9*S*)-**46** and (+)-(3*R*,5*S*,9*S*)-**47**, two minor alkaloids isolated from *Solenopsis diplorhoptrum* from Puerto Rico (see Table 1). This synthesis proceeded by the same sequence as the one described by these authors to synthesize (+)-monomorine I (*112*). Aldehyde **418**, prepared from L-alanine by the procedure depicted previously (Scheme 50), underwent Horner-Wittig

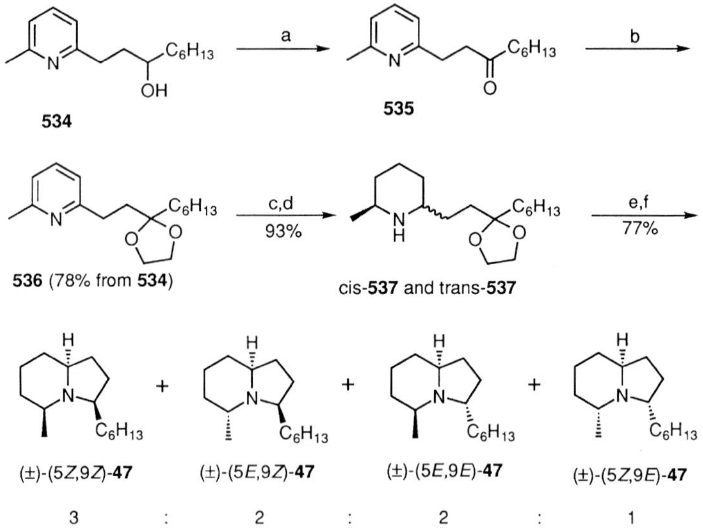

(a) pyridinium dichromate, CH_2Cl_2, pyridine, CF_3COOH; (b) ethylene glycol, benzene, HCl, *p*-TsOH; (c) Na-EtOH; (d) 5% Rh/Al_2O_3, AcOEt; (e) conc. HCl; (f) $NaCNBH_3$.

Scheme 65. Synthesis of (±)-**47** (*12*)

olefinations to afford the ethyl-enone **538** and the *n*-hexyl-enone **539**. Deprotection of the amine, iminium ion formation and stereoselective reduction were accomplished by catalytic hydrogenation (in a single step), affording (+)-(3*R*,5*S*,9*S*)-**46** and (+)-(3*R*,5*S*,9*S*)-**47** in 76% and 82% yield, respectively (Scheme 66).

C. Myrmicarin 237A and 237B

In 1995, two new indolizidine alkaloids with 3,9 *E* relative configuration, (−)-myrmicarin 237A [(−)-(5*E*,9*E*)-**53**] and (+)-myrmicarin 237B [(+)-(5*Z*,9*E*)-**54**] (Scheme 68), were isolated from the poison gland of the African ant *Myrmicaria eumenoides* (*38*). As the assignment of relative and absolute configurations required total synthesis, Francke *et al.* (*38*) developed a first synthesis which, unfortunately, led to diastereoisomers (±)-(5*E*,9*Z*)-**543** and (±)-(5*Z*,9*Z*)-**543** having the wrong 3,9 *Z* relative configuration (Scheme 67). 1-Hepten-3-one **99**, obtained from acrolein in a simple two-step procedure, was allowed to

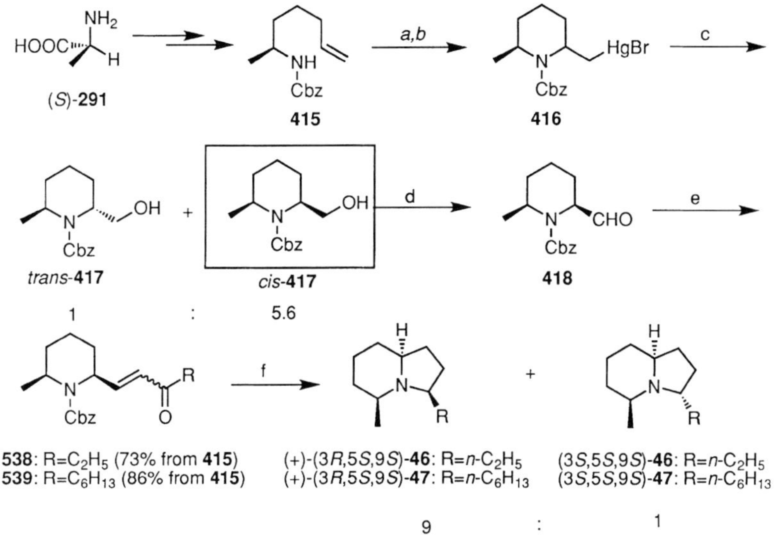

(a) $Hg(OCOCF_3)_2$, CH_3NO_2; (b) NaBr, $NaHCO_3$; (c) O_2, $NaBH_4$, DMF; (d) $(COCl)_2$, DMSO, Et_3N; (e) NaH, THF, $(H_3CO)_2POCH_2COC_2H_5$ for (+)-(3*R*,5*S*,9*S*)-**46**, $(H_3CO)_2POCH_2CO(CH_2)_5CH_3$ for (+)-(3*R*,5*S*,9*S*)-**47**; (f) H_2, $Pd(OH)_2$, MeOH.

Scheme 66. Asymmetric synthesis of (+)-(3*R*,5*S*,9*S*)-**46** and (+)-(3*R*,5*S*,9*S*)-**47** from L-alanine (*112*)

react with ethyl-6-nitrohexanoate **540** prepared from 6-nitrohexanoic acid. The resulting tridecanoate **397** was then hydrogenated to the *cis*-substituted pyrrolidine *cis*-**541** in 98% de. Deprotonation at C-α of the ester group followed by reaction with iodine yielded a mixture of indolizidines (5*E*,9*Z*)-**399** and (5*Z*,9*Z*)-**399** in 16% and 49% yield respectively. These were separated, and the diastereoisomerically pure samples were saponified. Due to epimerization at C-5 under alkaline conditions, mixtures of the diastereoisomeric acids indolizidines (5*E*,9*Z*)-**542** and (5*Z*,9*Z*)-**542** were obtained. Subsequent treatment of these with ethyl lithium furnished (±)-(5*E*,9*Z*)- **543** and (±)-(5*Z*,9*Z*)-**543** which could be separated by column chromatography on aluminium oxide.

As diastereoisomers **543** displayed spectroscopic and chromatographic properties quite different from those of the natural alkaloids (−)-**53** and (+)-**54**, a second synthesis was required to establish their

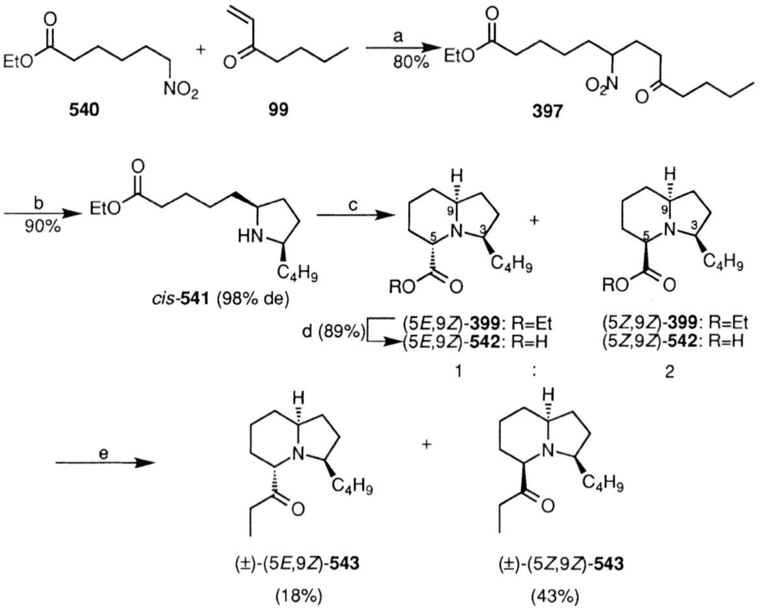

(a) NaOEt, EtOH; (b) H₂, Pd, EtOH; (c) LDA, THF, then I₂; (d) aq. KOH, EtOH; (e) EtLi, Et₂O, then separation on alumina.

Scheme 67. Synthesis of (±)-(5E,9Z)-**543** and (±)-(5Z,9Z)-**543** (*38*)

relative configuration. Therefore, FRANCKE *et al.* (*38*) (Scheme 68) prepared the *trans*-substituted pyrroline **546** by using a method employed by MACDONALD (*134*). N-Boc-pyrroline **544** was alkylated with butyl bromide to afford the corresponding 2-butylpyrroline **545** in 70% yield. Subsequent alkylation with 5-bromopentylbenzylether prepared from 1,5-pentanediol gave exclusively the *trans*-substituted pyrroline **546**. Hydrogenation of the double bond and deprotection of the alcohol using Pd/C catalyst provided the desired *trans*-isomer **547** in 80% de. Acid *trans*-**548**, obtained by oxidation of **547** with pyridinium dichromate, was converted to ester *trans*-**541** in 66% yield from **547**. As described above for the *cis* isomer, *trans*-**541** was transformed into the separable pair of indolizidines (5E,9E)-**399** and (5Z,9E)-**399**. Saponification and treatment of the resulting acids with ethyl lithium afforded two racemic ketones, with spectroscopic properties identical with those of the natural products (−)-**53** and (+)-**54**.

Having established the relative configuration of myrmicarin 237A and 237B by synthesis of the racemates, FRANCKE *et al.* elucidated their

References, pp. 221–229

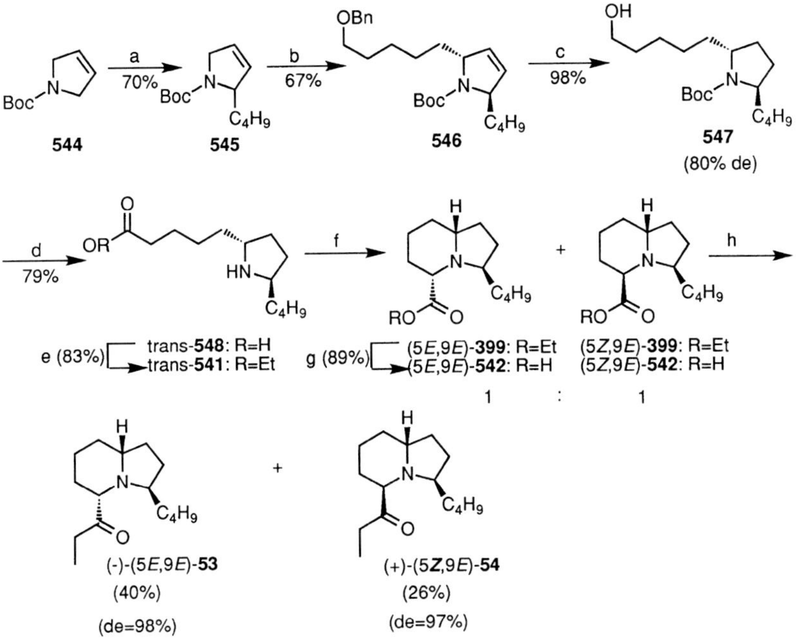

Scheme 68. Synthesis of (±)-(5E,9E)-**53** (myrmicarin 237A) and (±)-(5Z,9E)-**54** (myrmicarin 237B) (38)

(a) LDA, THF, then n-BuBr; (b) LDA, THF, then BnO(CH$_2$)$_5$Br; (c) H$_2$, Pd/C, MeOH; (d) pyridinium dichromate, DMF; (e) HCl, EtOH; (f) LDA, THF, then I$_2$; (g) aq. KOH, EtOH; (h) EtLi, Et$_2$O, then separation on alumina.

absolute configurations by enantioselective synthesis, using (2S)-2-butyloxirane (**552**) as the source of chirality (Scheme 69) (38). The disubstituted pyridine **551**, prepared from 6-methylpyridine-2-carboxaldehyde (**549**) in 60% yield, was allowed to react with (2S)-**552** of 99% ee to give (3S)-**553**. Hydrogenation of the pyridine ring of (3S)-**553** resulted in a 3:2 mixture of (3S,2'R,6'S)-**554** and (3S,2'S,6'R)-**554**. Subsequently, the mixture of these non-protected piperidines was treated with tosyl chloride in the presence of a base to afford the resulting indolizidines (3R,5R,9S)-**555** and (3R,5S,9R)-**555**. After separation on silica gel, these two indolizidines were deprotected to give ketones (3R,5R,9S)-**543** and (3R,5S,9R)-**53** in 98% de and 98% ee [22% overall yield for (3R,5S,9R)-**53** based on **549**]. Under the acidic conditions used for deprotection, the corresponding C-5 epimers were not produced. These were obtained by treating pure (3R,5S,9R)-**53** and (3R,5R,9S)-**543** with silica. Coinjection of natural and synthetic samples on a chiral gas

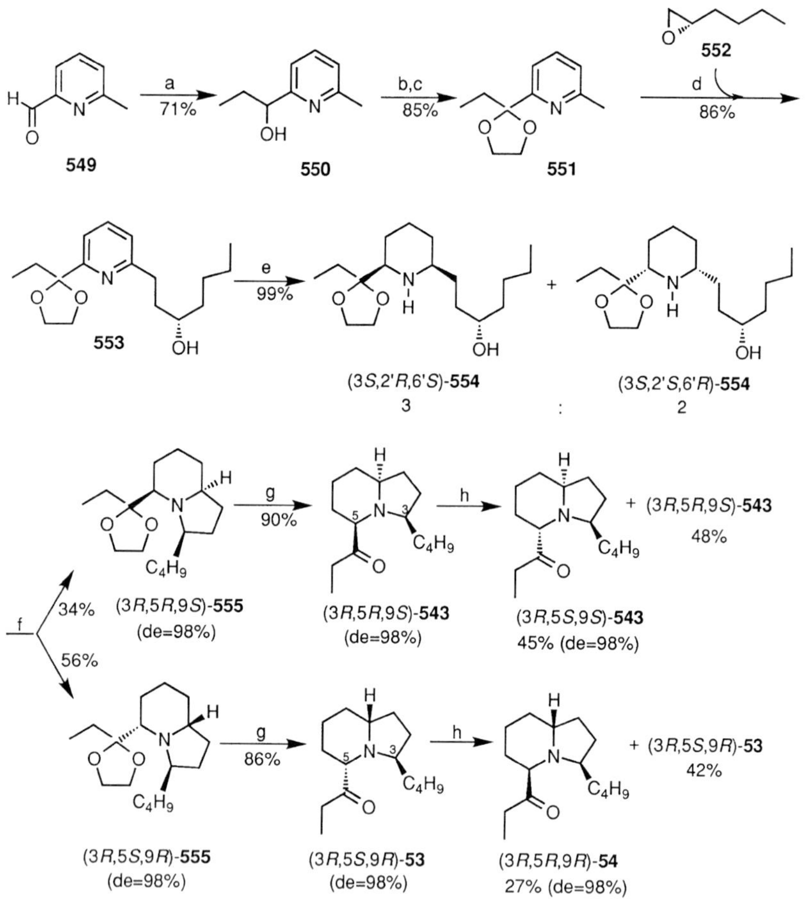

(a) EtMgBr, Et$_2$O; (b) Swern oxidation; (c) (CH$_2$OH)$_2$, TsOH, toluene; (d) n-BuLi, DMPU, THF; (e) H$_2$, Pd, EtOH; (f) TsCl, DMAP, pyridine, then separation on SiO$_2$; (g) aq. HCl, acetone; (h) SiO$_2$, then separation on alumina.

Scheme 69. Asymmetric synthesis of (−)-(3R,5S,9R)-**53** [(−)-myrmicarin 237A] and (+)-(3R,5R,9R)-**54** [(+)-myrmicarin 237B] (*38*)

chromatography column showed the ant alkaloids (−)-myrmicarin 237A and (+)-myrmicarin 237B to be (3R,5S,9R)-**53** and (3R,5R,9R)-**54**, respectively.

D. Myrmicarin 217

In 1998, the pyrrole-indolizidine alkaloid myrmicarin 217 (**57**) was synthezised using a new method for direct cyclization of appropriately

References, pp. 221–229

disubstituted piperidines (Scheme 70) (*40*). The synthetic intermediate (±)-**553** prepared from the available 6-methylpyridine-2-carboxaldehyde (**549**) was converted to the corresponding ketone which was protected as its ketal. Catalytic hydrogenation of the pyridine ring afforded *cis*-piperidine **556** in 81% overall yield from **549** (98% de). Deprotection of **556** with hydrochloric acid yielded the hydrochloride of *cis*-**557**, which under basic conditions was directly converted into myrmicarin 217 (**57**). To elucidate the mechanism leading to the formation of alkaloid **57** from *cis*-**557**, the progress of the reaction was monitored by ^{1}H-NMR spectroscopy. Apparently, the reaction mechanism involves two stereoisomers of indolizidine **558** as intermediates. First, by intramolecular condensation of the piperidine NH-functionality with the 3'-keto group of the heptyl side chain of *cis*-**557**, *cis*-**558** is formed, which rapidly equilibrates to a mixture of both *cis*- and *trans*-isomers. In a much slower process, condensation of the enamine moiety

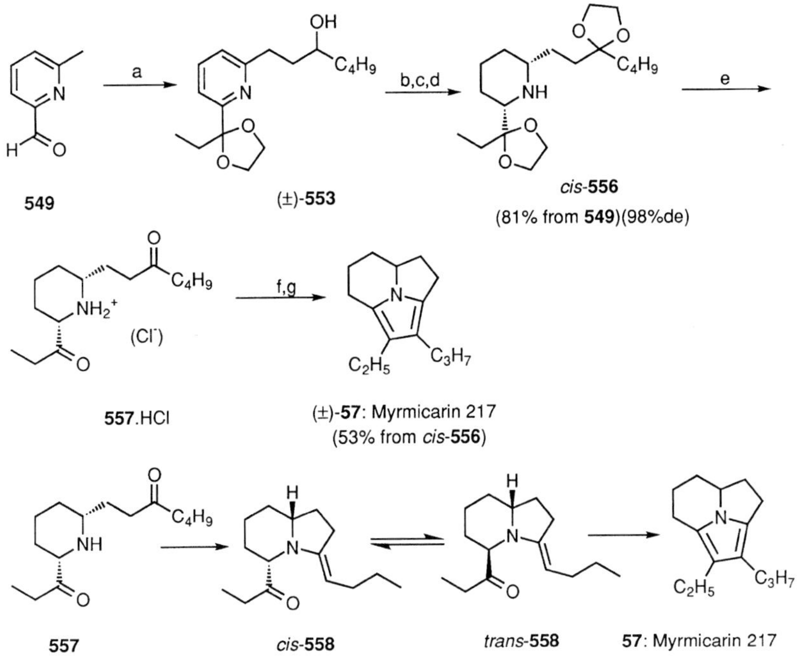

(a) M. Kaib, H. Dittebrand; *Chemoecology*, <u>1</u>, 3-11 (1990); (b)pyridinium dichromate, molecular sieves, CH_2Cl_2; (c) $(CH_2OH)_2$, TsOH, benzene; (d) H_2, Pd/C, EtOH, (e) aq. HCl, acetone; (f) aq. $NaHCO_3$, ether, N_2; (g) benzene, N_2.

Scheme 70. Synthesis of (±)-**57** (*40*)

with the keto group in **558** finally leads to the formation of the pyrrole system of **57**. Interestingly, **53** and **54**, the saturated analogues of **558**, are the major components of the defensive secretion of *Myrmicaria eumenoides*. Furthermore, traces of components showing mass spectra and GC-properties identical with those of the stereoisomers of **558** can be detected in the defensive secretions of *M. striata* and *M. opaciventris*, which both contain myrmicarin 217. Since all the C_{15}-chains in the oligocyclic *Myrmicaria* alkaloids are functionalized in a similar fashion, the authors assume that piperidine derivatives like the synthetic intermediate **558** (Scheme 70) may represent common biosynthetic precursors of the *Myrmicaria* alkaloids.

2.2.6. Tetraponerines

A. Syntheses of Racemic Tetraponerines

In 1987, MERLIN *et al.* reported isolation and structure determination of the tetraponerines, a family of tricyclic alkaloids used for defense by the New Guinean ant *Tetraponera* sp., whose main structural feature is a highly unusual aminal moiety (*44*). Soon afterwards, they published the first total synthesis of ($\pm$)-tetraponerine-8 [($\pm$)-T8, ($\pm$)-**63**], the major constituent of this venom (*47, 48*) (Scheme 71).

Cycloaddition of nitrone **560**, prepared by HgO oxidation of 1-hydroxypiperidine, to 1-heptene afforded isoxazolidine **561** in a 94% yield after chromatography on alumina. Hydrogenolysis by H_2/Raney-Nickel furnished aminoalcohol **562** which was then treated with 3 equivalents of $(CH_3SO_2)_2O$ to provide **563**. The next step, nucleophilic substitution of the mesylate group at C-8 to introduce the nitrogen-containing five membered ring, was accomplished by using succinimide in HMPA with K_2CO_3 as a base (71% yield). All attempts to use stronger nucleophiles and/or stronger bases always led to substantial amounts of elimination products. Attempted transformation of **565** into aminosulfonamide **567** by refluxing with $LiAlH_4$ afforded three compounds, the expected product **567** (30%), diamine **568** (51%) and ($\pm$)-T8 [($\pm$)-**63**] (5%). Compound **567** could be subsequently converted to diamine **568** by treatment with Na/*tert*-BuOH in NH_3/HMPA, and cyclization of **568** into ($\pm$)-T8 was achieved by treatment with *N*-chlorosuccinimide and NEt_3 in ether under UV irradiation.

This synthetic scheme, although stereoselectively affording ($\pm$)-T8 in 6 steps, was flawed by the problems encountered in the reduction of **565** and by the lack of reproducibility of the cyclization reaction. MERLIN *et al.* therefore described a modification of the original synthetic scheme

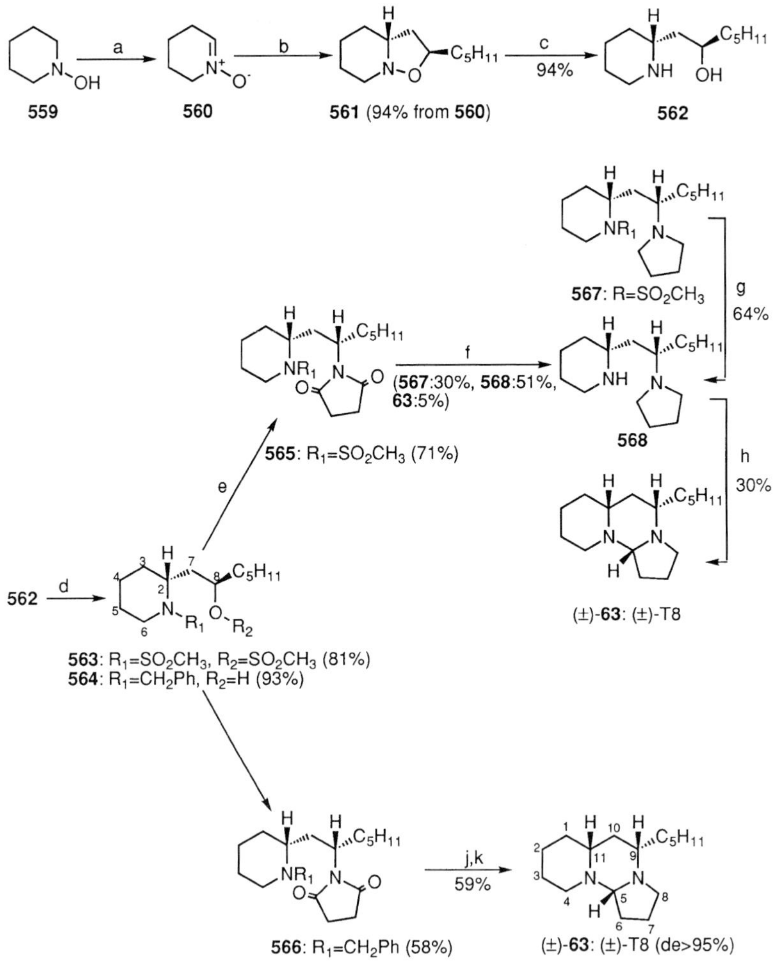

The Defensive Chemistry of Ants 201

(a) HgO, CHCl₃; (b) 1-heptene, CHCl₃; (c) H₂, Raney Ni, MeOH; (d) for **563**: (CH₃SO₂)₂O, NEt₃, CH₂Cl₂; for **564**: PhCH₂Br, KOH, EtOH; (e) succinimide, K₂CO₃, HMPA, THF; (f) LiAlH₄, THF; (g) Na, ᵗBuOH, NH₃-HMPA (h) N-chlorosuccinimide, NEt₃, ether, THF; (i) PPh₃, succinimide, EtOOC-N=N-COOEt, THF; (j) H₂, Pd/C, CH₃OH; (k) LiAlH₄, THF.

Scheme 71. Synthesis of (±)-T8 [(±)-**63**] (*47, 48*)

which increases the overall yield from 11 to 28% (*48*). The second route started with aminoalcohol **562** and was made possible by a selective, more convenient, protection of the nitrogen atom, as shown in Scheme 71. Thus, selective protection of the NH group of **562** was achieved by treatment with 1.1 equivalent of benzyl bromide which afforded *N*-benzylaminoalcohol **564** in a 93% yield. Attempts to mesylate **564** under the conditions developed for **562** were not successful, the mesylated derivative undergoing extensive degradation. To circumvent this problem, the authors turned to a Mitsunobu reaction with succinimide

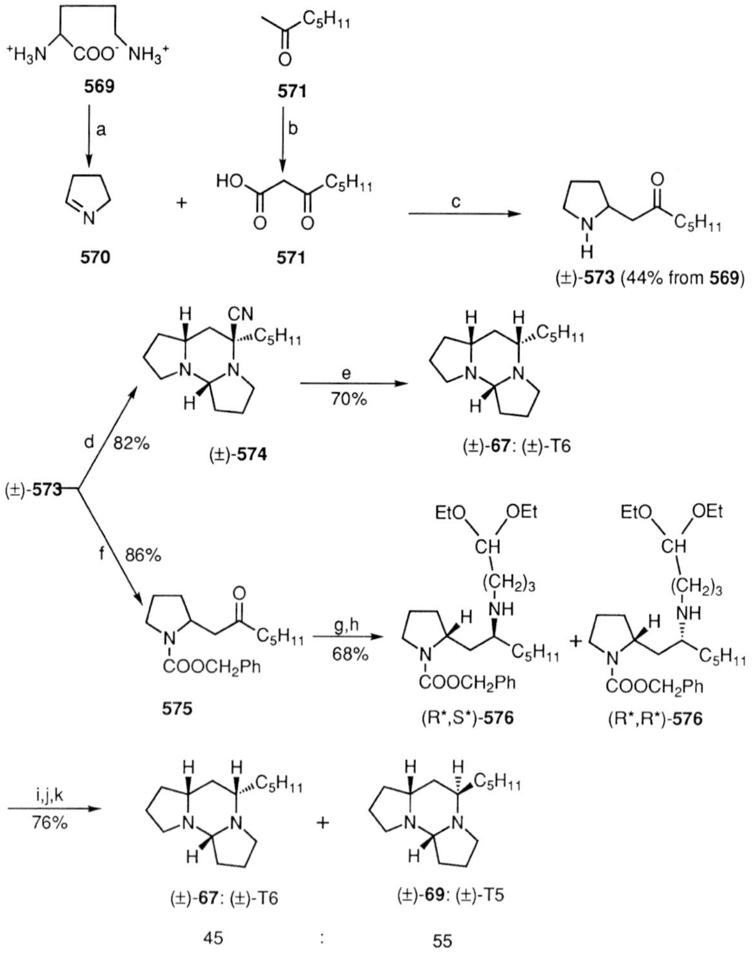

(a) NBS, H_2O; (b) methylmagnesiumcarbonate, DMF; (c) pH=6.9; (d) 1,1-diethoxy-4-aminobutane, KCN, pH=3-4; (e) Na/NH_3; (f) $ClCOOCH_2Ph$, K_2CO_3; (g) 1,1-diethoxy-4-aminobutane, Amberlyst A-15, 3A molecular sieves; (h) $NaBH_4$, MeOH; (i) H_2, Pd/C, MeOH; (j) HCl 1N; (k) NaOH to pH=8.0.

Scheme 72. Synthesis of (±)-T5 [(±)-**69**] and (±)-T6 [(±)-**67**] (*53*)

as the nucleophile, the optimal yield of **566** (58%) was achieved by using 1.5 equivalents of reagent. Reductive cyclization of **566** led to (±)-T8 with a high stereoselectivity and a 59% yield.

Recently, DEVIJVER *et al.* reported a short diastereoselective synthesis of (±)-T5 [(±)-**69**] and (±)-T6 [(±)-**67**], two other tetraponerine alkaloids which both possess a tricyclic 5-6-5 ring system with a pentyl side chain at C-8, instead of the tricyclic 6-6-5 ring present in T3, T4, T7 and T8 (*53*). It uses β-aminoketone (±)-**573** as pivotal intermediate (Scheme 72). Schöpf condensation of the trimer of Δ^1-pyrroline (**570**), obtained by NBS oxidation of L-ornithine hydrochloride (**569**), with 3-oxooctanoic acid (**571**) led in a 44% yield to β-aminoketone (±)-**573**. The latter was easily transformed into (±)-T6 followed the procedure of YUE *et al.* (*135*). Thus treatment of (±)-**573** with 1,1-diethoxy-4-aminobutane in the presence of HCl and KCN stereoselectivity led to the tricyclic diaminonitrile (±)-**574** which was cleanly transformed into (±)-**67** by reduction with Na in liquid ammonia. The next synthetic target, (±)-T5 [(±)-**69**], can also be constructed from β-aminoketone (±)-**573**, by adapting the procedure of JONES (*49*). Thus, **573** was protected as its *N*-benzyloxycarbonyl derivative **575**, which was reacted with 1,1-diethoxy-4-aminobutane. The resulting imine was immediately reduced with NaBH$_4$ to afford a mixture of the two diastereoisomeric aminocarbamates, (±)-(R^*,R^*)-**576** and (±)-(R^*,S^*)-**576**. This mixture was cyclized under the previously described conditions to afford (±)-T5 and (±)-T6 in a 55:45 ratio. These two epimers were easily separated by silica gel chromatography.

In 1990, JONES developed a practical route to (±)-T4 [(±)-**64**] to provide material for his investigation on the repellencies and toxicities of ant venom alkaloids (*49*). It was envisioned that the tetraponerine nucleus could be assembled from the readily available 1-(2-pyridyl)-2-alkanones and a four carbon segment carrying the remaining nitrogen. Thus 1-(2-pyridyl)-2-pentanone **577**, the precursor to tetraponerines T3 and T4, was prepared in 87% yield from *N,N*-dimethylbutyramide and 2-picolyllithium (Scheme 73). The remaining nitrogen and four carbons of the tetraponerine were furnished by 2-(3-aminopropyl)-1,3-dioxolane **578**, prepared in 90% yield by reduction of the corresponding nitrile. Condensation of **577** and **578** in benzene with azeotropic removal of water provided enamine **579** which was chemically reduced to amine **580** in 77% overall yield. Catalytic hydrogenation of **580** provided nearly quantitatively piperidylamine **581** as a mixture of diastereoisomers. Typical acetal hydrolysis conditions produced 90% of a mixture of (±)-T4 [(±)-**64**] and its C-3 epimer (±)-T3 [(±)-**66**]. Catalytic hydrogenation of **580** was investigated under a variety of conditions to

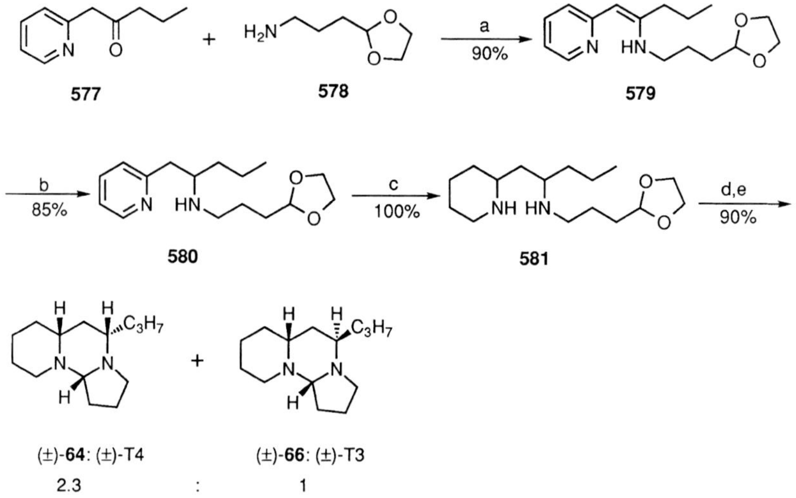

577 + **578** →(a, 90%) **579**

→(b, 85%) **580** →(c, 100%) **581** →(d,e, 90%)

(±)-**64**: (±)-T4   (±)-**66**: (±)-T3
2.3   :   1

(a) pTSA, C$_6$H$_6$; (b) NaBH$_4$, MeOH; (c) H$_2$, PtO$_2$, EtOH, 1 eq. CH$_3$CHO; (d) HClO$_4$; (e) NaOH 10%.

Scheme 73. Synthesis of (±)-T4 [(±)-**64**] (49)

improve the selectivity. While rhodium catalysts provided a 1 : 1 mixture of **64** and **66**, hydrogenation over PtO$_2$ was somewhat more selective. The best results, a 2.3 : 1 mixture of **64** and **66**, were obtained by addition of one equivalent of acetaldehyde to the hydrogenation mixture. Overall yield of (±)-T4 was 48% from **577**.

In the same year, JONES reported another synthesis of (±)-T4 and (±)-T8 which stereoselectively provided the all-*cis* tetraponerines in four operations from commercially available starting materials in 50% overall yield (Scheme 74) (50). Pyridyl amide **583**, prepared in 80% yield by coupling 2-pyridylacetic acid hydrochloride (**582**·HCl) and 2-(3-aminopropyl)-1,3-dioxolane (**578**), was near quantitatively hydrogenated to **584**. Treatment of the latter with aqueous formaldehyde in the presence of a catalytic amount of KOH provided bicyclic aminolactam **585**. The stereochemistry at C-9 was then controlled by the appropriate choice of reducing agent after the addition of an organometallic reagent to the amide carbonyl. In particular, two equivalents of C$_3$H$_7$MgCl and one equivalent of TMEDA were added to **585** and the mixture was subsequently treated with LiAlH$_4$. After acidification with HCl, basic treatment produced 70% of a 98 : 2 mixture of (±)-T4 [(±)-**64**] and its C-9 epimer (±)-T3 [(±)-**66**]. (±)-T7 [(±)-**65**] and (±)-T8 [(±)-**63**] were obtained in the same manner using pentylmagnesium bromide.

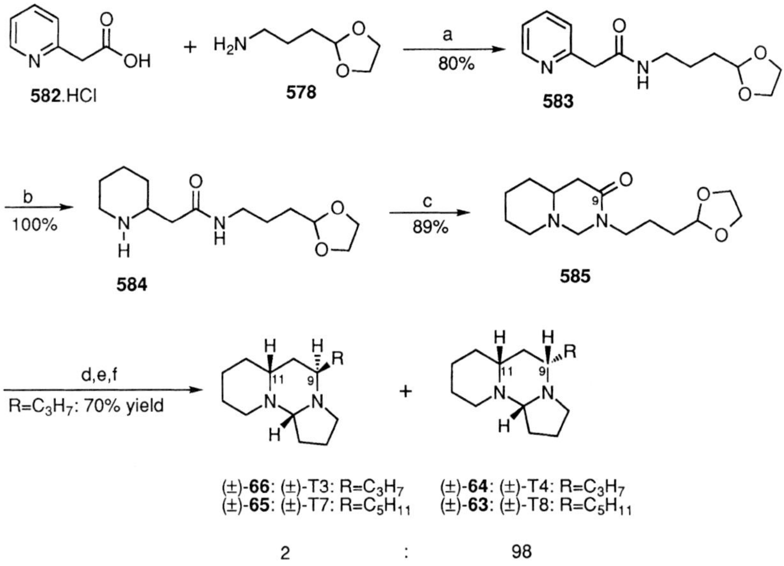

(a) DCC, DMAP, CH$_2$Cl$_2$, Et$_3$N; (b) 5% Rh/Al$_2$O$_3$, H$_2$, MeOH; (c) CH$_2$O, KOH cat., THF; (d) TMEDA, for **64** and **66**: propylmagnesium chloride, for **63** and **65**: pentylmagnesium chloride; (e) reducing agent, then HCl; (f) KOH.

Scheme 74. Synthesis of (±)-T4 [(±)-**64**] and (±)-T8 [(±)-**63**] *(50)*

The most recent synthesis of racemic tetraponerines is the synthesis of (±)-T8 described by BARLUENGA and co-workers in 1994 *(51)* who achieved a high yield synthesis totally diastereoselective in favor of (±)-

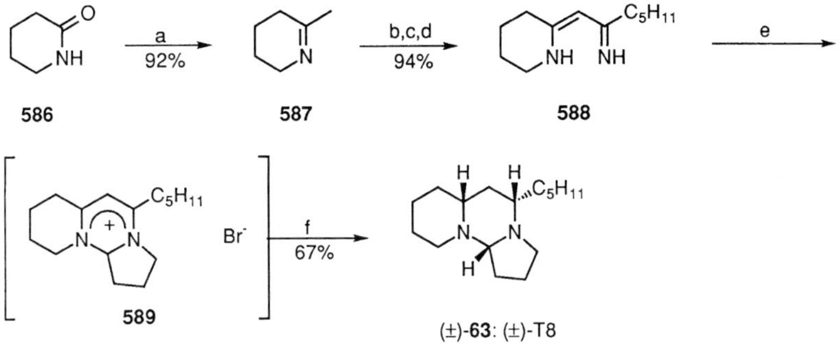

(a) MeLi, ether; (b) LDA; (c) C$_5$H$_{11}$CN; (d) NaOH; (e) 4-bromobutanal, THF, Na$_2$SO$_4$; (f) NaBH$_4$, THF, MeOH.

Scheme 75. Synthesis of (±)-T8 [(±)-**63**] *(51)*

T8 [(±)-**63**]. The tetraponerine skeleton was constructed by a tandem cyclization/reduction of azadiene **588** which in turn was readily available from δ-valerolactam **586** via imine **587** (Scheme 75). Slow addition of a solution of hexanenitrile to the metalated imine **587** resulted in the formation of pure azadiene **588** in nearly quantitative yield. (±)-T8 was then obtained in a one-pot reaction by sequential condensation of azadiene **588** with 4-bromobutanal followed by reduction with sodium borohydride. (±)-T8 was isolated as a single stereoisomer in 58% overall yield from **586**.

B. Syntheses of Nonracemic Tetraponerines

Three asymmetric syntheses of tetraponerines have been described up to now. The first to be dealt with is one reported by MACOURS et al. in 1995 (52). (+)- and (−)-T8, (+)- and (−)-T7 were synthesized in six steps and 27% overall yield from chiral acetylenic sulfoxide (+)-**592** via a cycloaddition reaction with 3,4,5,6-tetrahydropyridine-1-oxide (Scheme 76). Chiral acetylenic sulfoxide (S)-(+)-**592** was prepared in 72% yield by addition of acetylenic Grignard reagent **590** to (S)-menthyl-p-toluenesulfinate **591**. The key step was the cycloaddition reaction of sulfoxide (S)-**592** and 3,4,5,6-tetrahydropyridine-1-oxide, itself produced by HgO oxidation of N-hydroxypiperidine. The cycloaddition afforded an approximately 1:1 mixture of Δ^4-isoxazolines (R,S)-**593** and (S,S)-**593**, a result which confirmed the low asymmetric induction power of chiral sulfoxide groups in this type of cycloaddition reactions. After flash chromatography on silica gel, the two Δ^4-isoxazolines were transformed into β-aminoketones (R)- and (S)-**594**, respectively, by hydrogenolysis with H_2/PtO_2 in MeOH in the presence of citric acid. Next, the amino group of (R)-**594** was protected by reaction with benzylchloroformate to afford (R)-**595**. The last steps of the synthesis are based on the synthesis of racemic tetraponerines described earlier by JONES (49) (see Scheme 73). Unstable imine (R)-**596**, prepared from carbamate (R)-**595** by treatment with 4,4-diethoxybutylamine, was immediately reduced with $NaBH_4$ to yield a mixture of (R,S)-**597** and (R,R)-**597** in a temperature-dependent ratio (R,S/R,R = 4:6 at room temperature and 2:8 at −78°C). Cleavage of the carbamate group of the unseparable diastereoisomeric mixture of (R,S)-**597** and (R,R)-**597** and treatment of the resulting diaminoacetals first with HCl, then with NaOH to effect the cyclization, afforded a mixture of the expected tricyclic derivatives (+)-T8 and (+)-T7 (11% and 16% overall yield for (+)-T8 and (+)-T7, respectively). (−)-T8 and

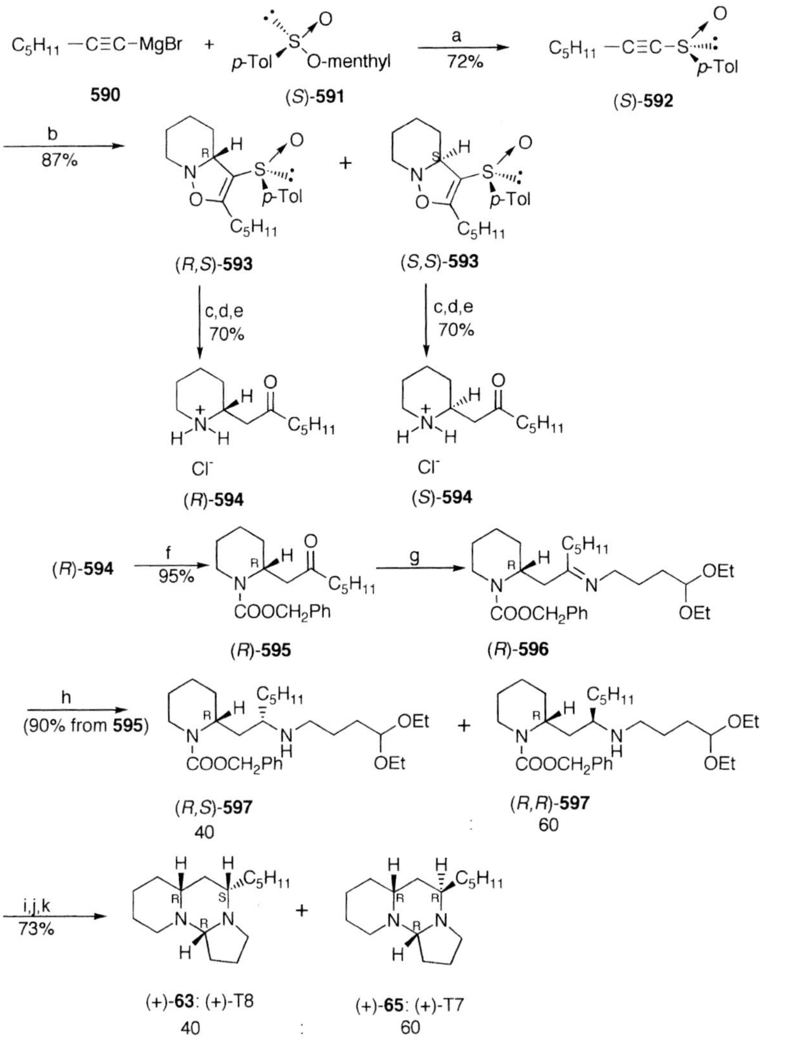

(a) ether-toluene; (b) 3,4,5,6-tetrahydropyridine-1-oxide, CHCl$_3$; (c) SiO$_2$ chromatography; (d) H$_2$, PtO$_2$, citric acid, MeOH; (e) elution on Sephadex DEAE A-25; (f) ClCOOCH$_2$Ph, K$_2$CO$_3$; (g) 4,4-diethoxybutylamine, Amberlyst A-15, 3A molecular sieves; (h) NaBH$_4$, MeOH; (i) H$_2$, Pd-C, MeOH; (j) HCl 3%; (k) NaOH to pH=8.2.

Scheme 76. Synthesis of (+)-T7 [(+)-**65**] and (+)-T8 [(+)-**63**] (*52*)

(−)-T7 were obtained in the same manner using (S)-β-aminoketone (S)-**594**.

Two other asymmetric syntheses of natural tetraponerine have been reported by HUSSON and co-workers in the 90's (*46, 135*). They use as

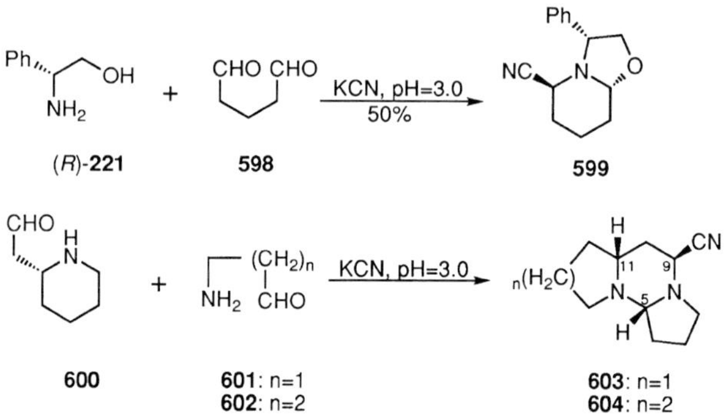

Scheme 77. The CN(R,S) method for the asymmetric synthesis
of the (+)-tetraponerines (46, 135)

starting material (−)-2-cyano-6-oxazolopiperidine **599**, which is a 1,4-dihydropyridine equivalent obtained by a Robinson-Schöpf type condensation between glutaraldehyde and (−)-phenylglycinol [(R)-**221**], in the presence of KCN (Scheme 77). Cyano aminals **603** and **604** are the key intermediates in these two syntheses; they are prepared by cross-condensation of two different aminoaldehydes, **600** and either **601** or **602**, via an enamine-iminium species in a scheme which parallels the previously reported preparation of the chiral 2-cyano-6-oxazolopiperidine **599**. This approach can be considered as an extension of the CN(R,S) method not only because of the analogy between the syntheses of **599** and **603** (**604**), but also because of the possible transformation of **603** and **604** into the tetraponerines with the required stereochemistry. Indeed it has been shown that replacement of the nitrile function of **599** by an alkyl chain can be directed to produce either equatorial or axial α-alkylpiperidine derivatives. Chiral aminoaldehyde **600** was prepared with the desired configuration starting from **599**. Finally, the relative configuration of the tetraponerines depends on the reaction conditions chosen for the C-9 alkylation step of **603** (**604**).

The first synthesis reported by HUSSON et al. based on the CN(R,S) method is that of natural (+)-T8 which was accomplished in six steps and 34% yield from **599** (Scheme 78) (46). The anion of chiral 2-cyano-6-oxazolopiperidine **599** was alkylated with 2-(bromomethyl)-1,3-dioxolane in the presence of HMPA to give **605** in 84% yield. Reduction of **605** with NaBH$_4$ in ethanol resulted in totally stereoselective preparation of acetal (2R)-**606** in quantitative yield. Hydrogenolysis of the N-benzylic bond and condensation of the aminoaldehyde resulting from acid cleavage of the acetal of (2R)-**607** with 4-aminobutanal diethyl

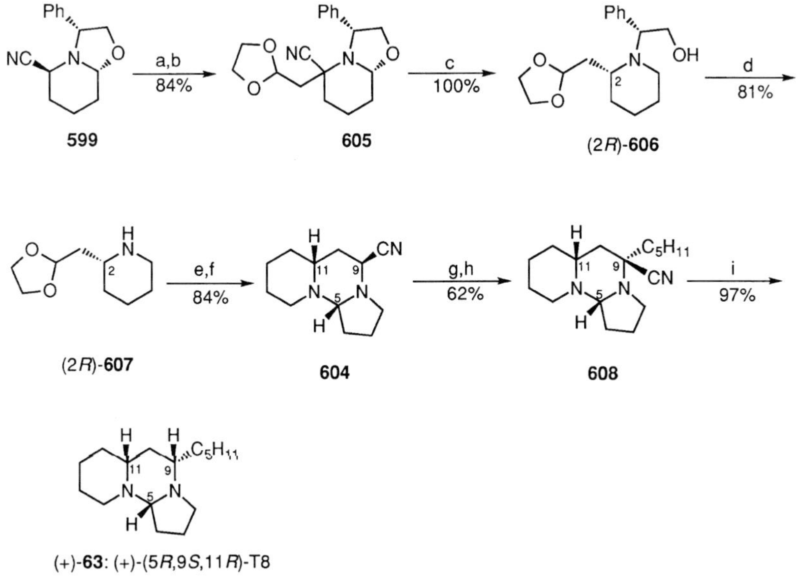

(a) LDA-HMPA, THF; (b) 2-(bromomethyl)-1,3-dioxolane; (c) NaBH$_4$, EtOH; (d) H$_2$, Pd/C, MeOH; (e) HCl; (f) 4-aminobutanal diethyl acetal, KCN, pH=2-3; (g) LDA-HMPA, THF; (h) Br-(CH$_2$)$_4$-CH$_3$; (i) Na, NH$_3$ liq.

Scheme 78. Asymmetric synthesis of (+)-T8 [(+)-**63**] (*46*)

acetal led to the most stable tricyclic cyano aminal **604** in 84% yield. NMR analysis confirmed that the protons on C-5 and C-11 were both axial and that the cyano group occupied the axial position. Introduction of the pentyl side chain in the proper equatorial configuration was carried out by alkylation with *n*-pentyl bromide after deprotonation by LDA-HMPA to yield **608**. The cyano group of **608** was removed in a highly stereoselective manner using Na in liquid ammonia to give (+)-T8 in 97% yield.

The strategy HUSSON *et al.* used for the asymmetric synthesis of (+)-T8 in Scheme 78 allowed them to synthesize all eight tetraponerines (+)-T1 to (+)-T8 in a convergent and concise approach (Scheme 79) (*135*). The asymmetric synthesis of (+)-T3, (+)-T4, (+)-T7 and (+)-T8 started with 2-cyano-6-oxazolopiperidine **599** which was converted into cyano aminal **604** in four steps and 64% overall yield as previously described in Scheme 78. Direct nucleophilic alkylation of cyano aminal **604** with propylmagnesium bromide led to introduction of the side chain in the axial position to give tetraponerine (+)-T3 [(+)-**66**] in a 77% yield and 97% de. Alkylation of **604** with pentylmagnesium bromide

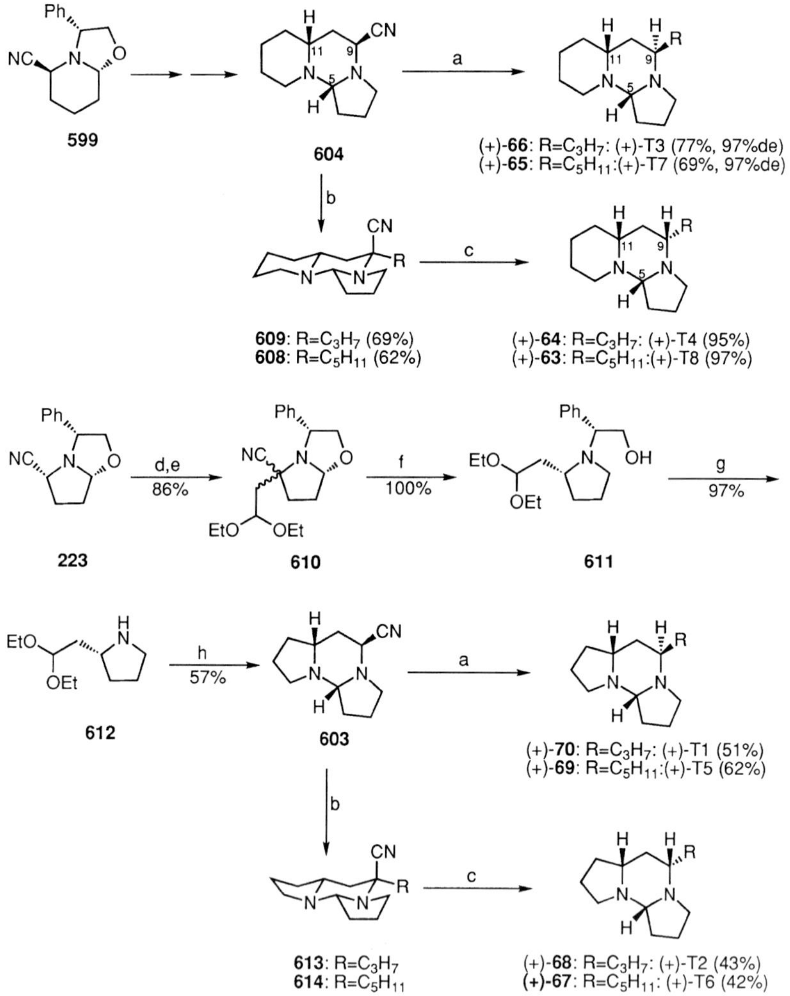

Scheme 79. Asymmetric synthesis of (+)-T1 [(+)-**70**] to (+)-T8 [(+)-**63**] (*135*)

(a) ether, for **64,66,68,70**: propylmagnesium bromide, for **63,65,67,69**: pentylmagnesium bromide; (b) LDA, for **64,66,68,70**: propyl magnesium bromide, for **63,65,67,69**: pentylmagnesium bromide; (c) Na/NH$_3$; (d) LDA, HMPA; (e) bromoethanal diethyl acetal; (f) NaBH$_4$; (g) H$_2$, Pd/C; (h) 4-aminobutanal diethyl acetal, 10% HCl, KCN.

under the same conditions gave (+)-T7 [(±)-**65**] in a 69% yield. On the other hand, introduction of the alkyl chain in the equatorial position necessitates two steps. Alkylation of cyanoaminal **604** was achieved *via* its anion to give **608** as a single isomer. Stereoselective decyanation of

608 using Na in liquid ammonia yielded exclusively the all-*cis* (+)-T8 [(±)-**63**]. Using a similar procedure tetraponerine (+)-T4 [(±)-**64**] was prepared from **604** in a 65% yield for the two steps.

For the asymmetric synthesis of (+)-T1, (+)-T2, (+)-T5 and (+)-T6, the same strategy was applied starting from 2-cyano-5-oxazolopyrrolidine **223** (Scheme 79). This compound, which has received less attention than the piperidine analog, can also be easily prepared and submitted to highly diastereoselective alkylation. Thus, **223** was diastereoselectively transformed into the pyrrolidine acetal derivative **612** in three steps and 83% overall yield. Cross-condensation of **612** with aminobutanal provided aminonitrile **603** as a single diastereoisomer upon quenching with cyanide ion. Aminonitrile **603** is the precursor of the four remaining tetraponerines (+)-T1 [(+)-**70**], (+)-T2 [(+)-**68**], (+)-T5 [(+)-**69**], and (+)-T6 [(+)-**67**], using the methodology described above to introduce the alkyl group in the correct configuration.

3. Nonalkaloidal Compounds

As already stated earlier this chapter is limited to the discussion of defensive compounds produced in the two glands linked to the sting, namely the poison and the Dufour glands. As we have seen in the preceeding sections, the majority of low molecular weight defensive compounds produced in the poison gland of ants are alkaloids. The obvious exception to this generalization if formic acid which seems to be a diagnostic character for the subfamily Formicinae. Formic acid is produced in the poison gland and transferred to a reservoir, where it accumulates as a very concentrated aqueous solution (50–60%). Although formicine ants are devoid of a functional sting, such highly acidic solutions provide them with a very efficient defensive weapon. In the genus *Formica*, for example, the workers discharge a stream of formic acid mixed with undecane originating from the Dufour gland. The acid serves as a toxin to repel invaders, whereas undecane serves not only as a pheromone to alert other workers, but also as a wetting agent which promotes the spreading of formic acid onto the cuticle of the enemy (6, 7). Aspects of formic acid biosynthesis will be discussed in section 4.

Most of the nonalkaloidal defensive compounds isolated so far from ants originate from the Dufour gland. The primary functions of the Dufour gland are thought to be defense and communication (6, *136*). The most common constituents of this gland are linear odd-numbered alkanes and alkenes which may or not be accompanied by even-

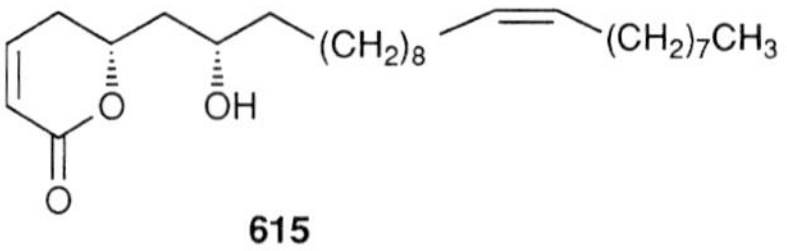

Fig. 8. Compounds isolated from the Dufour gland of *Tetramorium aculeatum*

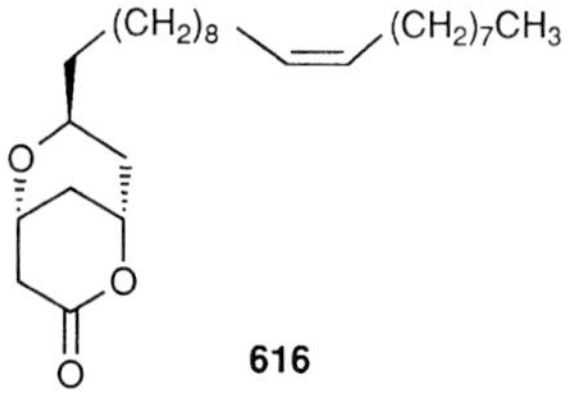

numbered and branched-chain hydrocarbons. Simple oxygenated compounds (esters, ketones) may also be present as well as sesquiterpenoids. These compounds play various biological roles (*e.g.* alarm pheromone, trail pheromone, home-range marking pheromone, wetting agents to enhance the toxicity of the poison gland product *etc.*). However, very few compounds which are toxic *per se* have been isolated up to now from the Dufour gland of ants and we will discuss them below.

Tetramorium aculeatum is an African ant notorious for its aggressiveness and for causing severe skin irritation by biting and stinging. In Congo, it is named the "urticating ant". Dissection of worker ants revealed that they possess a hypertrophied Dufour gland, from which (6R*)-[(2S*)-2-hydroxyheneicos-12-enyl]-5,6-dihydro-2H-pyran-2-one (**615**) as major and (1S*,5R*,7S*)-7-(nonadec-10-enyl)-2,6-dioxabicyclo[3.3.1]nonan-3-one (**616**) as minor constituent have been isolated (Fig. 8) (*137*). In solution, **615** is slowly transformed into **616** on standing, thus suggesting that the former is the compound biosynthesized in the gland. In common with other unsaturated lactones which are well known skin irritants, (*e.g.* massoilactone), **615** can be viewed as an efficient Michael acceptor and thus could be responsible for the urticating properties of this ant.

Ants of the genus *Crematogaster* are characterized by a unique defensive mechanism that requires the cooperation of the Dufour and poison glands in the production of both topical poison and alarm pheromone (*138–140*). These ants are able to raise their abdomen forward and over the thorax and head. Thus, the abdominal tip can be pointed in nearly all directions. Moreover, the sting has a spatulate tip which is not a suitable injection device. In many *Crematogaster* species, the venom is emitted as a froth that accumulates on the spatulate portion and at the basis of the sting, from where it can easily be applied to the integument of other insects. Up to now, the venoms of the three European species of *Crematogaster* ants, three species from Papua New Guinea and two species from Brazil have been studied. In all cases, the

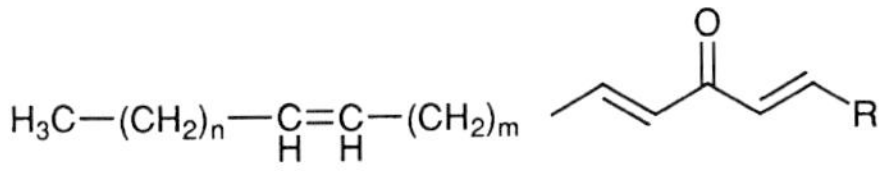

$H_3C-(CH_2)_n-\underset{H}{\overset{H}{C}}=\underset{}{\overset{}{C}}-(CH_2)_m$ —CH=CH—C(=O)—CH=CH—R

	n	m	R=CH$_2$OAc	R=CHO	R=COOH
C$_{19}$	3	7	617a	619a	620a
	5	5	617b	619b	620b
	7	3	617c	619c	620c
C$_{21}$	3	9	617d	619d	620d
	5	7	617e	619e	620e
	7	5	617f	619f	620f
C$_{23}$	3	11	617g	619g	620g
	5	9	617h	619h	620h
	7	7	617i	619i	620i
C$_{25}$	3	13	617j	619j	620j
	5	11	617k	619k	620k
	7	9	617l	621l	620l

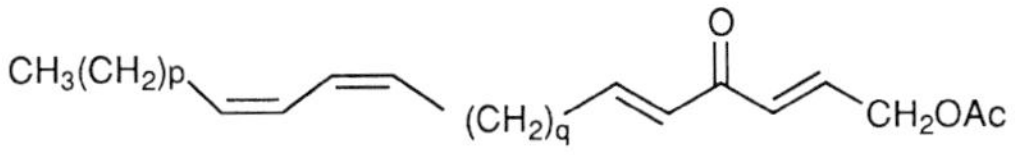

$CH_3(CH_2)_p$—CH=CH—CH=CH—$(CH_2)_q$—CH=CH—C(=O)—CH=CH—CH_2OAc

618a   p = 5   q = 5
618b   p = 3   q = 7

Fig. 9. Compounds isolated from the Dufour gland of European and Neo-Guinean *Crematogaster* ants

venoms consist of long chain unsaturated compounds bearing peculiar terminal functional groups, such as a cross-conjugated dienone, or a furan ring.

The major components of the Dufour gland of the European species *Crematogaster scutellaris* were found by NMR and GC/MS studies to be (2*E*,5*E*,12*Z*)-4-oxo-heneicosa-2,5,12-trien-1-ol acetate (**617d**), its Δ^{14} and Δ^{16} double bond isomers (**617e** and **617f**), and the corresponding (*Z*,*Z*)-dienes (**618a** and **618b**) (Fig. 9) (*138–140*). The location of the isolated double bond in the chain followed from ozonolysis and epoxidation experiments whereas its *Z* geometry was established by $^1H/^{13}C$ NMR heterocoupling experiments. The location of the (*Z*,*Z*)-diene in **618a** and **618b** is tentative. All these compounds derive from an acetylated 21 carbon long chain. Minor derivatives based on an homologous C$_{23}$ chain are also present (**617g–i**), as well as traces of acetylated C$_{19}$ homologs tentatively identified as **617a–c**. The venoms of the two other European species, *C. auberti* and *C. sordidula*, have also been studied. *C. auberti* is similar to *C. scutellaris* in producing mainly **617d–f**, together with the same higher and lower homologs, but it lacks the dienic derivatives **618**. *C. sordidula* essentially contains the acetylated C_{19} compound **617a–c**, accompanied by acetylated C_{17} homologs.

All these derivatives (*e.g.* **617a–i**), which share the same functional group, namely an (*E,E*)-cross-conjugated dienone linked to a primary acetate function, are stored in the Dufour gland of these ants. When the venom is emitted, these compounds are transformed into highly electrophilic 4-oxo-2,5-dienals (**619a–i**) by the action of two enzymes, an esterase and an oxygen-dependent alcohol oxidase, which are both stored in the poison gland. Thus, this elegant mechanism allows the ants to store venom precursors (**617a–i**) of relatively low toxicity, the production of the true toxins (**619a–i**) being triggered during the simultaneous emission of both the Dufour and the poison glands constituents (*139*). The presence of an esterase and of an oxygen-dependent oxidase in the poison gland was substantiated by *in vitro* experiments in phosphate buffer, using the poison gland as enzyme source. Acetic acid which is freed during this process and is easily detected by the human nose was identified in venom samples by GC of its *t*-butyldimethylsilyl derivative. It reinforces the efficiency of the defensive mechanism by acting as an alarm pheromone. Biological testing has indeed demonstrated that the toxicity of the native Dufour gland constituents (**617a–i**) is markedly lower than that of the 4-oxo-2,5-dienals-enriched secretion. It should also be mentioned that the 4-oxo-2,5-dienal derivatives are exceedingly unstable and that, once produced, they are quickly oxidized or rearranged into an array of compounds, among which the corresponding carboxylic acids (**620**) (Fig. 9), the lactols (**621**), and the α-angelica lactones (**622**) have been identified (Fig. 10) (*138*). The latter probably arise from addition of the carboxyl

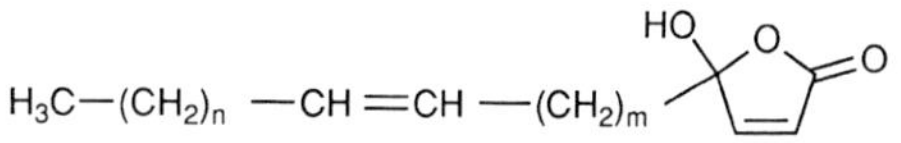

$H_3C-(CH_2)_n -CH=CH-(CH_2)_m$

621

$H_3C-(CH_2)_n -CH=CH-(CH_2)_m$

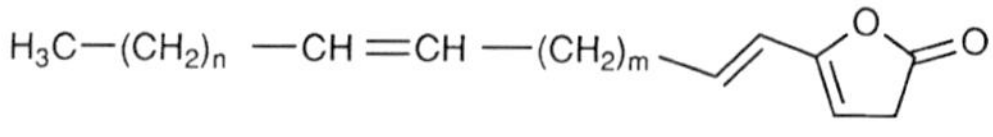

622

Fig. 10. Cyclized compounds isolated from the Dufour gland of *Crematogaster* ants

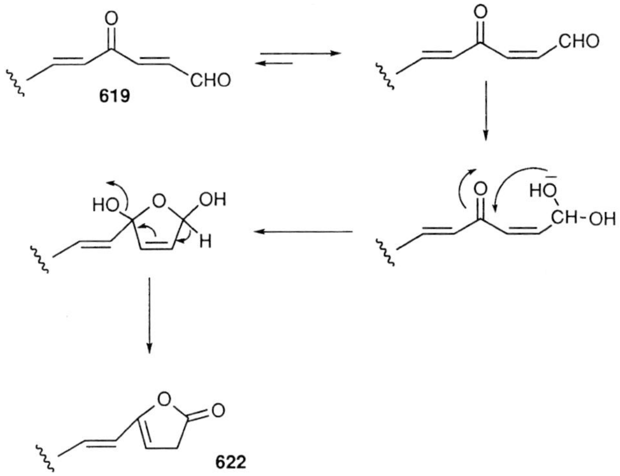

Scheme 80. Possible origin of α-angelica lactones **622** from 4-oxo-dienals **619** (*139*)

group onto the C-4 ketone carbonyl, after isomerization of the *trans* conjugated double bond to the *cis* isomer, whereas the formation of the α-angelica lactones may be rationalized as shown in Scheme 80 (*139*).

The investigation of *Crematogaster* ant venoms was pursued by the study of three as yet unidentified species originating from Papua New Guinea (*141*). *Crematogaster* sp. 1 was shown to contain, besides the 4-oxo-2,5-dienals (**619g–l**), a series of derivatives, whose structures were determined to be **623a–f**, on the basis of a complete 1D and 2D NMR study at 600 MHz (Fig. 11). Thus, in the latter compounds, the usual cross-conjugated dienone is replaced by a furan ring conjugated to an *E* double bond. It has been shown, at least in some species from Papua New Guinea, that the furans are native compounds in the Dufour gland (unpublished results). Biosynthetically, it is not known if these compounds arise from 1,4-dione precursors by a Paal-Knorr-type cyclization, as shown in Scheme 81 or from the α-angelica lactones (**622**) by reduction of the lactone carbonyl followed by loss of water. In this latter study (*141*), the position of the isolated double bond in the chain was established by dimethyldisulfide treatment, followed by linked scan MS/MS analysis of the resulting mixture of adducts. The results showed that, as in the case of the European species, there are always three position isomers of the isolated double bond for each chain length. Moreover, the positions of this double bond are always the same with respect to the terminal methyl group, namely ω^5, ω^7 and ω^9. This observation could have interesting biosynthetic implications.

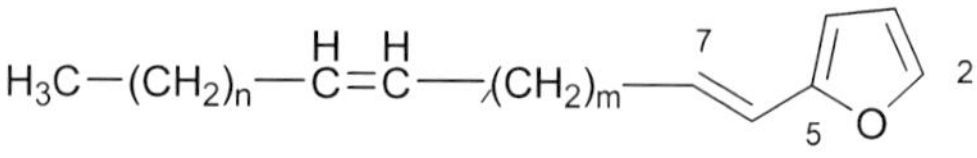

$H_3C-(CH_2)_n-\overset{H}{C}=\overset{H}{C}-(CH_2)_m-$ [furan, positions 7, 5, 2, O]

	n	m
623a	3	11
623b	5	9
623c	7	7
623d	3	13
623e	5	11
623f	7	9

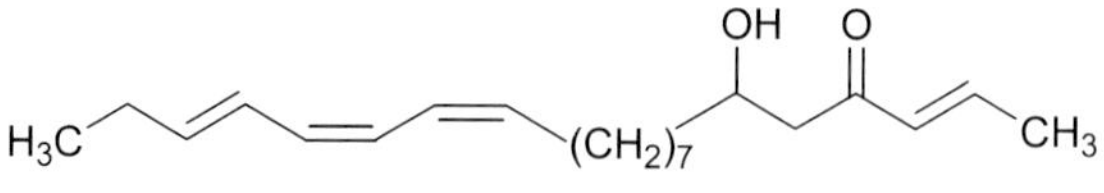

624

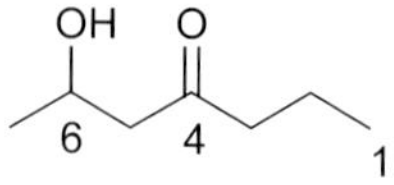

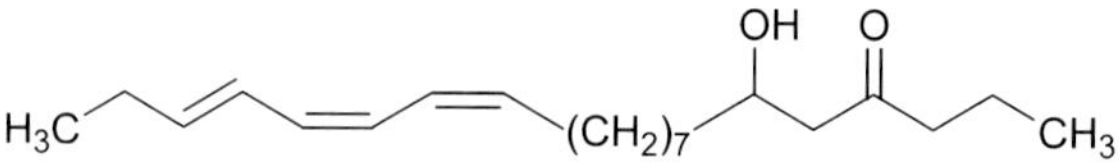

625

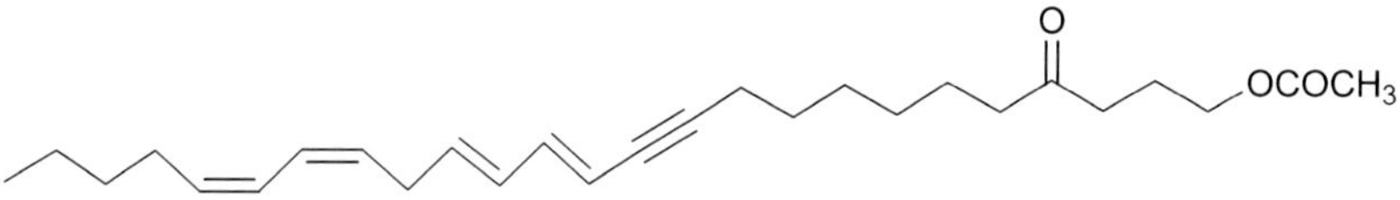

626

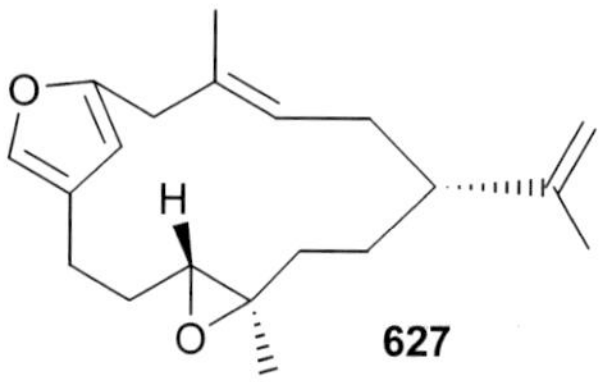

627

Fig. 11. Compounds isolated from the Dufour gland of Neo-Guinean and Brazilian *Crematogaster* ants

In contrast to all the other species of *Crematogaster* studied till now, the venom of *C.* sp. 2 from Papua New Guinea did not contain mixtures of homologous compounds. Two derivatives, **624** and **625**, characterized

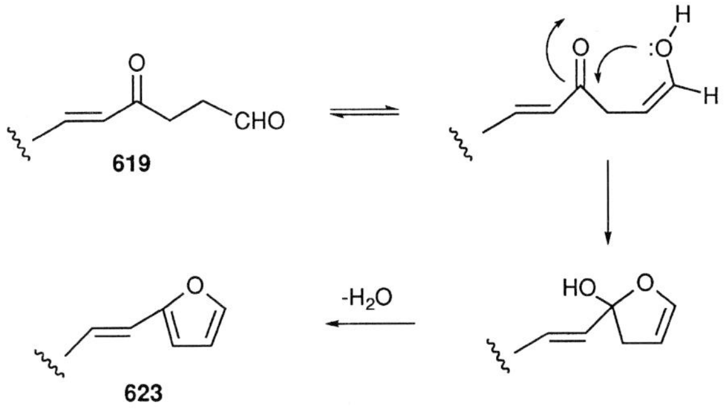

*Scheme 81. Possible biosynthetic origin of furans **623** from 4-oxo-dienals **619** (141)*

by the presence of a conjugated triene on one end of the chain and by an 1,3-hydroxyketone at the other end were isolated from this ant. These structures depart somewhat from the common theme observed up to now and could constitute biogenetic intermediates *en route* to the cross-conjugated dienone system. The venom of *C.* sp. 3 was shown to contain 4-oxo-2,5-dienyl acetates (**617g–l**) similar to those already reported from other species. It should be pointed out that in *C.* species 1 and 3 from Papua New Guinea, the major derivatives are based on C_{25} chains, instead of C_{19}, C_{21} or C_{23} in the European species. This could constitute a useful taxonomic marker for these species.

Quite recently, an investigation of several *Crematogaster* species from Brazil has been undertaken. So far several compounds have been isolated and their structures determined. The first was identified as (13*E*,15*E*,18*Z*,20*Z*)-1-hydroxypentacosa-13,15,18,20-tetraen-11-yn-4-one 1-acetate (**626**) (Fig. 11) (*142*). This is the first compound of this type to be isolated from ants. Its exact function is not yet established, but its occurrence in the venom suggests that, as in the other long chain derivatives from *Crematogaster* ants, it should have toxic properties.

The other compounds isolated from Brazilian *Crematogaster* ants are cembrane diterpenes, such as **627** (Fig. 11), which proved to be toxic towards other ants (*143*). This is the first report of cembrane diterpenes as defensive compounds in ants.

4. Biosynthesis

Considering the number and variety of ant alkaloids already reported, it is intriguing that so far only the biosynthesis of the tetraponerines and the solenopsins has been studied. As discussed in Section 2.1.5, the

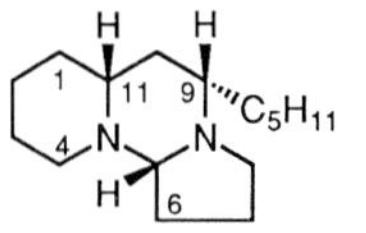

T8 (**63**)

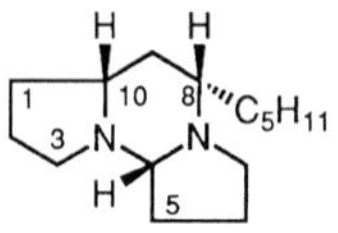

T6 (**67**)

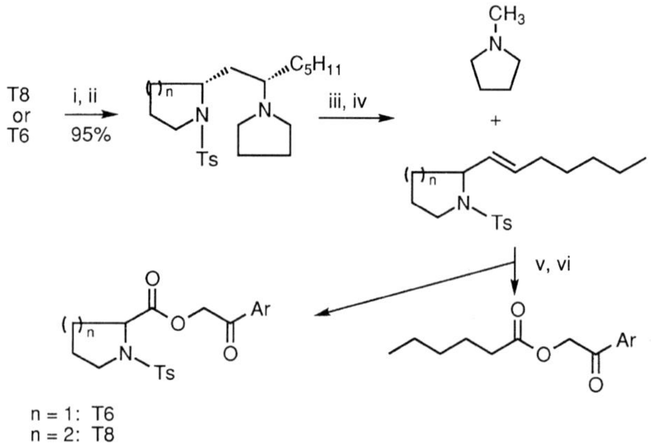

Reagents and conditions: i. H_2, PtO_2, EtOH, HCl; ii. TsCl, pyridine; iii. CH_3I, CH_3CN; iv. OH^-; v. 160 °C, 3h; vi. $KMnO_4$, $NaIO_4$, 24h; vii. *p*-$BrC_6H_4CH_2Br$, CH_3CN, reflux, 1h (Ts = *p*-$CH_3C_6H_4SO_2$; Ar = 4-BrC_6H_4).

Scheme 82. Degradation scheme of labeled tetraponerines T6 and T8 (*144*, *145*)

tetraponerines, which are produced in the poison gland of *Tetraponera* sp. ants, belong to two different structural types with the decahydropyrido[1,2-c]pyrrolo[1′,2′-a]pyrimidine and the decahydrodipyrrolo[1,2-a:1′,2′-c]pyrimidine skeletons, exemplified by T8 (**63**) and T6 (**67**), respectively (see structures **63–70**, Fig. 6). Administration of sodium [1-^{14}C] and [2-^{14}C]acetate, L-[U-^{14}C]glutamic acid, γ-amino[U-^{14}C]butanoic acid, L-[U-^{14}C]ornithine hydrochloride and [1,4-^{14}C]putrescine dihydrochloride to *Tetraponera* ants followed by chemical degradation (Scheme 82) of labeled T8 (**63**), led to the conclusion that this compound has a mixed biogenetic origin. The pyrrolidine ring derives from glutamic acid *via* L-ornithine and putrescine, whereas the remaining twelve carbons are derived from a polyacetate chain formed from six acetate units (Scheme 83) (*144*). By contrast, the same incorporation experiments followed by degradation of the labeled T6 (**67**) showed that putrescine serves as precursor to both pyrrolidine rings of this alkaloid with the remaining carbon atoms coming from a seven carbon moiety of polyacetate origin (Scheme 83)

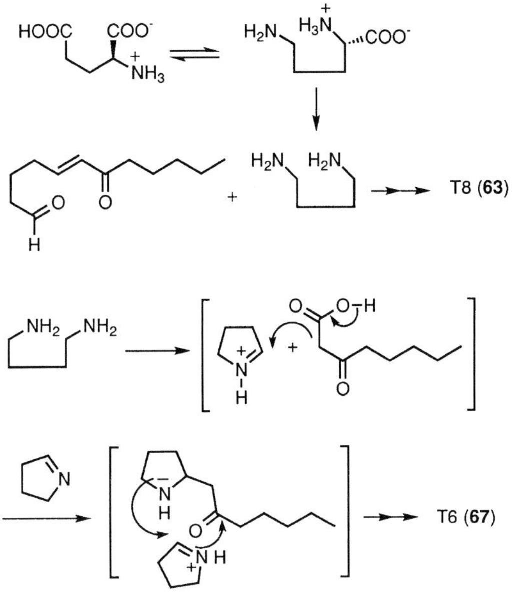

Scheme 83. Biosynthetic pathways of tetraponerines T6 and T8 (*144, 145*)

(*145*). These studies demonstrated that two closely related alkaloid skeletons produced by one and the same organism arise by two different biosynthetic pathways.

The biosynthesis of *cis*- and *trans*-solenopsins A (*cis*- and *trans*-**628**) was also studied by tracer experiments with *Solenopsis geminata*. Feeding 3,000 to 4,000 worker ants with [1-^{14}C]- and [2-^{14}C]acetate yielded a mixture of radioactive *cis*- *trans*-solenopsins A, which was diluted with "cold" synthetic material and subjected to the chemical degradation depicted in Scheme 84. The distribution of label in the degradation compounds clearly supports the hypothesis that the solenopsins A are formed from an eighteen carbon polyacetate chain resulting from the condensation of acetyl-coenzyme A with eight subsequent units of malonyl-coenzyme A (*146*). The hypothetical 18-carbon polyacetate chain could then afford *cis*- and *trans*-solenopsin A by either of two pathways, routes A and B depicted in Scheme 85. Which is the correct one is not yet known.

Among non-alkaloidal compounds, only the biosynthesis of formic acid has been studied so far, using intact poison glands dissected from workers of the ant *Camponotus pennsylvanicus* (*147*). The results

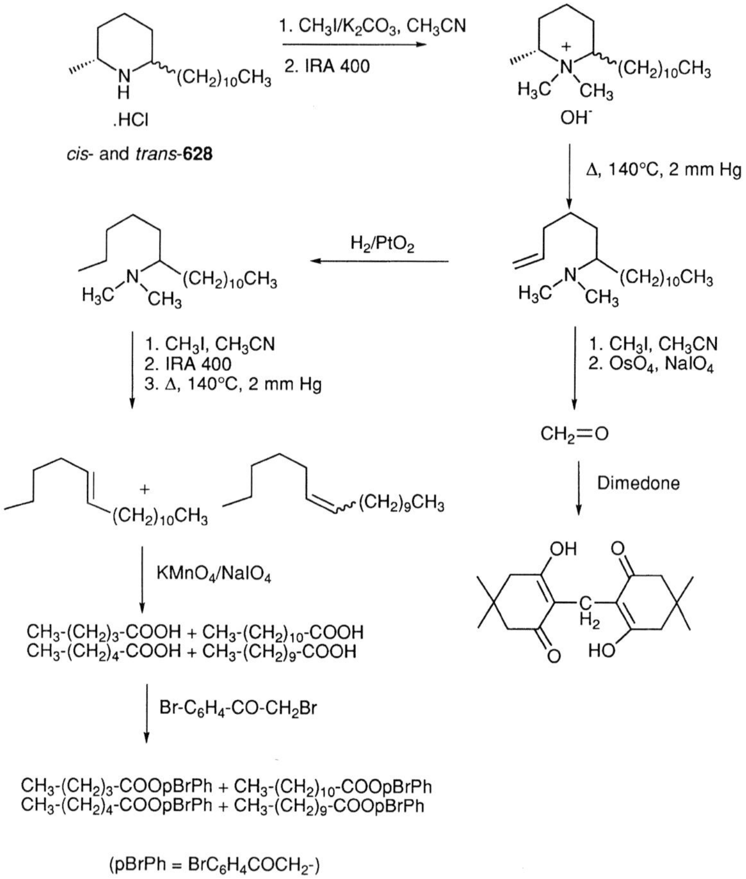

Scheme 84. Degradation scheme of labeled solenopsins A (*146*)

demonstrate that the pathway of formic acid production in ants is not different from that already reported for bacterial and mammalian cells. Thus, serine serves as major C-1 donor to formic acid through several tetrahydrofolate intermediates. The last intermediate in this biosynthetic pathway, 10-formyltetrahydrofolate, is hydrolyzed to the final product, formic acid. Since the equilibrium between formic acid and 10-formyltetrahydrofolate is 1 : 20, only small amounts of formic acid are produced in the glandular cells. Formic acid can thus only accumulate through transfer to a second compartment (the poison gland reservoir), which is insulated by a cuticular intima (*147*). Thus, the originality of this system is the morphological compartimentalization of the gland for

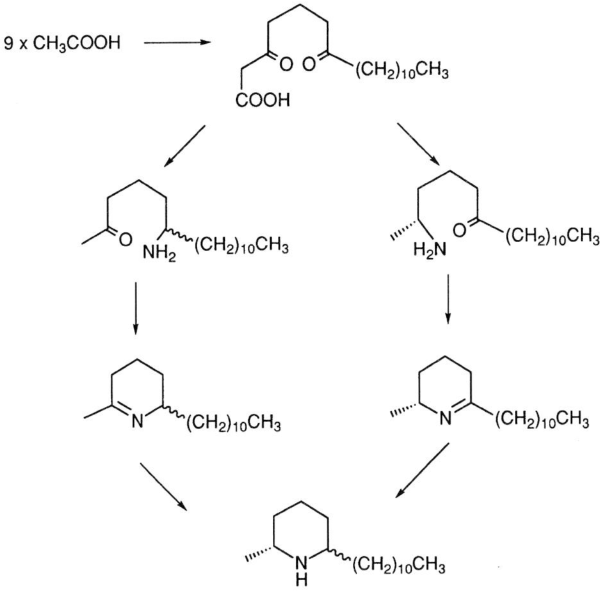

Scheme 85. Possible biosynthetic pathways to solenopsins A *(146)*

accumulating large amounts of this cytotoxic compound without harming the producing organism.

References

1. Buschinger A, Maschwitz U (1984) Defensive Behavior and Defensive Mechanisms in Ants. In: Hermann HR (ed) Social Insects. Praeger, New York, pp 95–149
2. Hölldobler B, Wilson EO (1990) The Ants. Springer, Berlin
3. Maschwitz U (1975) Old and New Chemical Weapons in Ants. In: Pheromones and Defensive Secretions in Social Insects. Proceedings of the IUSSI Symposium, Dijon, pp 41–45
4. Schmidt JO (1970) Hymenopteran Venoms: Striving toward the Ultimate Defense against Vertebrates. In: Evans DL, Schmidt JO (eds) Insect defenses. Adaptative Mechanisms and Strategies of Prey and Predators. State University of New York Press, Albany, pp 387–419
5. Schmidt JO (1986) Chemistry, Pharmacology, and Chemical Ecology of Ant Venoms. In: Piek T (ed) Venoms of the Hymenoptera. Biochemical, Pharmacological and Behavioural Aspects. London, Academic Press, pp 425–508
6. Attygalle AB, Morgan ED (1984) Chemicals from the Glands of Ants Chem Soc Rev **13**: 245

7. Blum MS (1981) Chemical Defenses of Arthropods. Academic Press, New York
8. Numata A, Ibuka T (1987) Alkaloids from Ants and Other Insects. In: Brossi A (ed) The Alkaloids, Vol. 31. Academic Press, San Diego, p 193
9. Braekman JC, Daloze D (1990) Chemical Defense in Ants. In: Atta ur Rahman (ed) Studies in Natural Products Chemistry, Vol. 6. Elsevier, Amsterdam, p 421
10. Jones TH, Blum MS, Robertson HG (1990) Novel Dialkylpiperidines in the Venom of the Ant *Monomorium delagoense*. J Nat Prod **53**: 429
11. Jones TH, Torres JA, Spande TF, Garraffo HM, Blum MS, Snelling RR (1996) Chemistry of Venom Alkaloids in some *Solenopsis (Diplorhoptrum)* Species from Puerto Rico. J Chem Ecol **22**: 1221
12. Gorman JST, Jones TH, Spande TF, Snelling RR, Torres JA, Garraffo HM (1998) 3-Hexyl-5-methylindolizidine Isomers from Thief Ants, *Solenopsis (Diplorhoptrum)* Species. J Chem Ecol **24**: 933
13. Attygalle AB, Xu SC, McCormick KD, Meinwald J, Blankespoor CL, Eisner T (1993) Alkaloids of the Mexican Bean Beetle, *Epilachna varivestis* (Coccinellidae). Tetrahedron **49**: 9333
14. Tarawa JN, Blokhin A, Foderaro TA, Stermitz FR, Hope H (1993) Toxic Piperidine Alkaloids from Pine and Spruce Trees. New Structures and a Biosynthesis Hypothesis. J Org Chem **58**: 4813
15. Hill RK, Chan TH (1965) Magnetic Non-Equivalence of Methylene Protons in Dissymetric Benzylamines. Tetrahedron **21**: 2015
16. Garraffo HM, Simon LD, Daly JW, Spande TF (1994) *Cis*- and *Trans*-Configurations of α,α'-Disubstituted Piperidines and Pyrrolidines by GC-FTIR: Application to Decahydroquinoline Stereochemistry. Tetrahedron **50**: 11329
17. Leclercq S, Thirionet I, Broeders F, Daloze D, Van der Meer R, Braekman JC (1994) Absolute Configuration of the Solenopsins, Venom Alkaloids of the Fire Ants. Tetrahedron **50**: 8465
18. Wheeler JW, Olubajo O, Storm CB, Duffield RM (1981) Anabaseine: Venom Alkaloid of *Aphaenogaster* Ants. Science **211**: 1051
19. Attygalle AB, Kern F, Huang Q, Meinwald J (1998) Trail Pheromone of the Myrmicine Ant *Aphaenogaster rudis* (Hymenoptera: Formicidae). Naturwissensch **85**: 38
20. Jackson BD, Wright PJ, Morgan ED (1989) 3-Ethyl-2,5-Dimethylpyrazine, a Component of the Trail Pheromone of the Ant *Messor bouvieri*. Experientia **45**: 487
21. Brand JM, Mpuru SP (1993) Dufour's Gland and Poison Gland Chemistry of the Myrmicine Ant, *Messor capensis*. J Chem Ecol **19**: 1315
22. Coll M, Hefetz A, Lloyd HA (1987) Adnexal Glands Chemistry of *Messor ebeninus* Forel (Formicidae: Myrmicinae). Z Naturforsch **46c**: 1027
23. Van der Meer RK, Morel LM (1995) Ant Queens Deposit Pheromones and Antimicrobial Agents on Eggs. Naturwissensch **82**: 93
24. Van der Meer RK, Lofgren CS (1988) Use of Chemical Characters in Defining Populations of Fire Ants, *Solenopsis saevissima* Complex, (Hymenoptera: Formicidae). The Florida Entomologist **71**: 323
25. Jones TH, Blum MS, Andersen AN, Fales HM, Escoubas P (1988) Novel 2-Ethyl-5-Alkylpyrrolidines in the Venom of the Australian Ant of the Genus *Monomorium*. J Chem Ecol **14**: 35
26. Bacos D, Basselier JJ, Celerier JP, Lange C, Marx E, Lhommet G, Escoubas P, Lemaire M, Clement JL (1988) Ant Venom Alkaloids from *Monomorium* Species: Natural Insecticides. Tetrahedron Lett **29**: 3061

27. Jones TH, Stahli SM, Don AW, Blum MS (1988) Chemotaxonomic Implications of the Venom Chemistry of some *Monomorium antarcticum* Populations. J Chem Ecol **14**: 2197
28. Jones TH, Laddago A, Don AW, Blum MS (1990) A Novel (5*E*,9*Z*)-Dialkylindolizidine from the Ant *Monomorium smithii*. J Nat Prod **53**: 375
29. Jones TH, Blum MS, Fales HM, Brando CRF, Lattke J (1991) Chemistry of Venom Alkaloids in the Ant Genus *Megalomyrmex*. J Chem Ecol **17**: 1897
30. Jones TH, De Vries PJ, Escoubas P (1991) Chemistry of Venom Alkaloids in the Ant *Megalomyrmex foreli* (Myrmicinae) from Costa Rica. J Chem Ecol **17**: 2507
31. Reder E, Veith HJ, Buschinger, A (1995) Novel Alkaloids from the Poison Glands of Ants Leptothoracini. Helv Chim Acta **78**: 73
32. Koob R, Rudolph C, Veith HJ (1997) The Absolute Configuration of 3-Methylpyrrolidine Alkaloids from Poison Glands of Ants Leptothoracini (Myrmicinae). Helv Chim Acta **80**: 267
33. Buschinger A (1972) Giftdrüsensekret als Sexualpheromon bei der Ameise *Harpagoxenus sublaevis*. Naturwissensch **59**: 313
34. Jones TH, Blum MS, Fales HM, Thompson CR (1980) (5*Z*,8*E*)-3-Heptyl-5-Methylpyrrolizidine from a Thief Ant. J Org Chem **45**: 4778
35. Jones TH, Highet RJ, Don AW, Blum MS (1986) Alkaloids of the Ant *Chelaner antarcticus*. J Org Chem **51**: 2712
36. Ritter FJ, Rothgans IEM, Talman E, Verwiel PEJ, Stein F (1973) 5-Methyl-3-Butyl-Octahydroindolizine, a Novel Type of Pheromone Attractive to Pharaoh's Ants (*Monomorium pharaonis*). Experientia **29**: 530
37. Jones TH, Highet RJ, Blum MS, Fales HM (1984) (5*Z*,9*Z*)-3-Alkyl-5-Methylindolizidines from *Solenopsis* (*Diphorhoptrum*) Species. J Chem Ecol **10**: 1233
38. Francke W, Schröder F, Walter F, Sinnwell V, Baumann H, Kaib M (1995) New Alkaloids from Ants: Identification and Synthesis of (3*R*,5*S*,9*R*)-3-Butyl-5-(1-Oxopropyl)Indolizidine and (3*R*,5*R*,9*R*)-3-Butyl-5-(1-Oxopropyl)Indolizidine, Constituents of the Poison Gland Secretion in *Myrmicaria eumenoides* (Hymenoptera, Formicidae). Liebigs Ann 965
39. Schröder F, Franke S, Francke W, Baumann H, Kaib M, Pasteels JM, Daloze D (1996) A New Family of Tricyclic Alkaloids from *Myrmicaria* Ants. Tetrahedron, **52**: 13539
40. Schröder F, Francke W (1998) Synthesis of Myrmicarin 217, a Pyrrolo[2,1,5-cd]Indolizine from Ants. Tetrahedron **54**: 5259
41. Schröder F, Sinnwell V, Baumann H, Kaib M (1996) Myrmicarin 430A: A New Heptacyclic Alkaloids from *Myrmicaria* Ants. Chem Commun 2139
42. Schröder F, Sinnwell V, Baumann H, Kaib M, Francke W (1997) Myrmicarin 663: A New Decacyclic Alkaloid from Ants. Angew Chem Int Ed Engl **36**: 77
43. Braekman JC, Daloze D, Pasteels JM, Van Hecke P, Declercq JP, Sinnwell V, Francke W (1987) Tetraponerine-8, an Alkaloidal Contact Poison in a Neo-Guinean Pseudomyrmecine Ant, *Tetraponera* sp. Z Naturforsch **42c**: 627
44. Merlin P, Braekman JC, Daloze D, Pasteels JM (1988) Tetraponerines, Toxic Alkaloids in the Venom of the Neo-Guinean Pseudomyrmecine Ant, *Tetraponera* sp. J Chem Ecol **14**: 517
45. Merlin P (1990) PhD Thesis, University of Brussels
46. Yue C, Royer J, Husson HP (1990) The First Enantioselective Synthesis of Natural (+)-Tetraponerine-8: A New Extension of the CN(R,S) Method to an Uncommon Skeleton. J Org Chem **55**: 1140
47. Merlin P, Braekman JC, Daloze D (1988) Stereoselective Synthesis of (±)-Tetraponerine-8, a Defense Alkaloid of the Ant *Tetraponera* sp. Tetrahedron Lett **29**: 1691

48. Merlin P, Braekman JC, Daloze D (1991) Stereoselective Total Synthesis of (±)-Tetraponerine-8. Tetrahedron **47**: 3805
49. Jones TH (1990) A Short Tetraponerine Synthesis. Tetrahedron Lett **31**: 1535
50. Jones TH (1990) A Stereoselective Synthesis of the (9Z,11Z) Tetraponerines T4 and T8. Tetrahedron Lett **31**: 4543
51. Barluenga J, Tomas M, Kouznetsov V, Rubio E (1994) An Extremely Short Stereoselective Synthesis of (±)-Tetraponerine-8. J Org Chem **59**: 3699
52. Macours P, Braekman JC, Daloze D (1995) Concise Asymmetric Syntheses of (+)- and (−)- Tetraponerine-8, (+)- and (−)-Tetraponerine-7, and their Ethyl Homologues. A Correction of the Structures of Tetraponerine-3 and -7. Tetrahedron **51**: 1415
53. Devijver C, Macours P, Braekman JC, Daloze D, Pasteels JM (1995) Short Syntheses of (±)-Tetraponerines-5 and -6. The Structures of Tetraponerines-1 and -2, and a Revision of the Structures of (+)-Tetraponerines-5 and -6. Tetrahedron **51**: 10913
54. Morgan ED, Hölldobler B, Vaisar T, Jackson BD (1992) Contents of Poison Apparatus and their Relation to Trail-Following in the Ant *Daceton armigerum*. J Chem Ecol **18**: 2161
55. Janssen E, Bestmann HJ, Hölldobler B, Kern F (1995) *N,N*-Dimethyluracil and Actinidine, Two Pheromones of the Ponerine Ant *Megaponera foetens*. J Chem Ecol **21**: 1947
56. Jones TH, Torres JA, Snelling RR, Spande TF (1996) Primary Tetradecenyl Amines from the Ant *Monomorium floricola*. J Nat Prod **59**: 801
57. Brophy JJ, Clezy PS, Leung CWF, Robertson PL (1993) Secondary Amines Isolated from Venom Gland of Dolichoderine Ant, *Technomyrmex albipes*. J Chem Ecol **19**: 2183
58. Leclercq S, Daloze D, Braekman JC (1996) Synthesis of the Fire Ant Alkaloids, Solenopsins. A Review. Organic Preparations and Procedures Int **28**: 499
59. Chakalamannil S, Wang Y (1997) An Enantioselective Route to *Trans*-2,6-Disubstituted Piperidines. Tetrahedron **53**: 11203
60. Adams DR, Carruthers W, Williams MJ, Crowley PJ (1989) Synthesis of *trans*-2,6-Dialkylpiperidines by Intramolecular Amidomercuriation and by 1,3-Cycloaddition of Alkenes to 2-Methyl-2,3,4,5-tetrahydropyridine Oxide. J Chem Soc Perkin Trans I, 1507
61. Escoubas P, Fales HM, Andersen AN, Blum MS, Jones TH (1988) Novel 2-Ethyl-5-alkylpyrrolidines in the Venom of an Australian Ant of the Genus *Monomorium*. J Chem Ecol **14**: 35
62. Blum MS, Don AW, Stahly SM, Jones TH (1988) Chemotaxanomic Implications of the Venom Chemistry of Some *Monomorium antarcticum*. J Chem Ecol **14**: 2197
63. Fales HM, Blum MS, Jones TH (1979) Synthesis of Unsymmetrical 2,5-di-*n*-Alkylpyrrolidines: 2-Hexyl-5-pentylpyrrolidine from the Thief Ant *Solenopsis molesta*, *Solenopsis texana*, and its Homologues. Tetrahedron Lett **12**: 1031
64. Tufariello JJ, Puglis JM (1986) The α-α'-Dialkylation of Cyclic Amines. The Synthesis of *Solenopsis* Ant Venoms. Tetrahedron Lett **27**: 1489
65. Renko ZD, Schink HE, Bäckvall JE (1990) A Stereocontrolled Organopalladium Route to 2,5-Disubstituted Pyrrolidine Derivatives. Application to the Synthesis of a Venom Alkaloid of the Ant Species *Monomorium latinode*. J Org Chem **55**: 826
66. Okukado N, Van Horn DE, Baba S, Takahashi T, Negishi E (1970) A Novel Stereoselective Synthesis of 1,3-Dienes from Alkynes *via* the Addition of Cuprous Chloride to Vinylalanes. J Am Chem Soc **92**: 6678

67. Dumas F, D'Angelo J (1992) A New Route to *Trans*-2,5-Dialkylpyrrolidines. Tetrahedron Lett **38**: 2005
68. Fouquet G, Schlosser M (1974) Improved Carbon-Carbon Linking by Controlled Copper Catalysis. Angew Chem Int Edit **13**: 82
69. Meyers AI, Edwards PD, Bailey TR, Jagdmann GE (1985) α-Amino Carbanions. Preparation, Metalation, and Alkylation of Enamidines. Synthesis of Piperidine and Pyrrolidine Natural Products and Homologation of Carbonyl Compounds. J Org Chem **50**: 1019
70. Bacos D, Celerier JP, Marx E, Rosset S, Lhommet G (1990) Stereoselective Synthesis and Stereochemical Determination of 2,5-Dialkylpyrrolidines and 2,6-Dialkylpiperidines. J Het Chem **27**: 1387
71. Miyashita M, Awen BZE, Yoshikoshi A (1990) Acyl Nitronates in Organic Synthesis. An Expeditious Synthesis of 2,5-Dialkylpyrrolidines and 2,5-Dialkylpyrrolines Including Ant Venom Alkaloids. Chem Lett 238
72. Veith J, Collas M, Zimmer R (1997) Simple Synthesis of both Enantiomers of 3-Methyl-*N*-(3-methylbutyl)pyrrolidine. Liebigs Ann 391
73. Takahata H, Takehara H, Ohkubo N, Momose T (1990) An Efficient Route to Chiral *Trans*-2,5-dialkylpyrrolidines *via* Stereoselective Intramolecular Amidomercuration. Tetrahedron Asym **1**: 561 (1990).
74. Schlessinger RH, Iwanowicz EJ (1987) The Synthesis of either (+)- or (−)-*Trans*-2,5-dimethylpyrrolidine. Tetrahedron Lett **28**: 2083
75. Wistrand LG, Skrinjar M (1991) Chirospecific Synthesis of *Trans*-2,5-disubstituted Pyrrolidines *via* Stereoselective Addition of Organocopper Reagents to *N*-Acyliminium ions. Tetrahedron **47**: 573
76. Shono T, Matsumura Y, Tsubata K, Sugihara Y, Yamane S, Kanazawa T, Aoki T (1982) Electroorganic Chemistry, 60. Electroorganic Synthesis of Enamides and Enecarbamates and their Utilization in Organic Synthesis. J Am Chem Soc **104**: 6697
77. Rapoport H, Shiosaki K (1985) α-Amino-acids as Chiral Educts for Asymmetric Products. Chirospecific Syntheses of the 5-Butyl-2-heptylpyrrolidines from Glutamic Acid. J Org Chem **50**: 1229
78. Rosset S, Celerier JP, Lhommet G (1991) Enantioselective Syntheses of *Monomorium minutum* Ant Venom Alkaloids: (5*R*)-2-(5-Hexenyl)-5-nonyl-3,4-dihydro-2H-pyrrole and (2*R*,5*R*)-2-(5-Hexenyl)-5-nonylpyrrolidine from (*S*)-Pyroglutamic Acid. Tetrahedron Lett **32**: 7521
79. Arseniyadis S, Huang PQ, Piveteau D, Husson HP (1988) Asymmetric Synthesis XII: Stereocontroled Electrophilic-Nucleophilic α,α'-Substitution of the Pyrrolidine Ring. Tetrahedron **44**: 2457
80. Huang PQ, Arseniyadis S, Husson HP (1987) Asymmetric Synthesis X: A Chiral Pyrrolidine Synthon for a New Approach to the Synthesis of Alkaloids. Tetrahedron Lett **28**: 547
81. Grierson DS, Royer J, Guerrier L, Husson HP (1986) Asymmetric synthesis 6. Practical Synthesis of (+)-Solenopsin A. J Org Chem **51**: 4475
82. Machinaga N, Kibayashi C (1991) Enantioselective Total Synthesis of (+)- and (−)-Pyrrolidine 197B, a New Class of Alkaloids from the Dendrobatid Poison Frog: Assignment of the Absolute Configuration. J Org Chem **56**: 1386
83. Mendoza R, Netzel DA, Sonnet PE (1979) ^{13}C NMR Assignments of Selected Octahydroindolizines. J Het Chem 1041
84. Lathbury D, Gallagher T (1986) Stereoselective Synthesis of Pyrrolizidine Alkaloids *via* Substituted Nitrones. J Chem Soc Chem Commun 1017

85. Vavrecka M, Janowitz A, Hesse M (1991) Transformation of 4-Nitroalkane-1,7-diones into Pyrrolizidine. Tetrahedron Lett **32**: 5543
86. Provot O, Lhommet G, Célérier JP (1998) Diastereoselective Synthesis of *Cis*-3- and 3,5-Alkylpyrrolizidines. J Het Chem **35**: 371
87. Takahata H, Bandoh H, Momose T (1991) Chirospecific Synthesis of an Ant Venom Alkaloid (5Z,8E)-3-Heptyl-5-methylpyrrolizidine. Tetrahedron Asymm **2**: 351
88. Provot O, Célérier JP, Petit H, Lhommet G (1992) Synthesis of Ant Venom Alkaloids from Chiral β-Enamino Lactones: (3R,5R,8S)-3-Heptyl-5-methylpyrrolizidine. J Org Chem **57**: 2163
89. Grandjean C, Rosset S, Célérier JP, Lhommet G (1993) A New Synthesis of Ant Venom Alkaloid: (3S,5R,8S)-3-Heptyl-5-methylpyrrolizidine. Tetrahedron Lett **34**: 4517
90. Guerrier L, Royer J, Grierson DS, Husson HP (1983) Chiral 1,4-Dihydropyridine Equivalents: A New Approach to the Asymmetric Synthesis of Alkaloids. The Enantiospecific Synthesis of (+)- and (−)- Coniine and -Dihydropinidine. J Am Chem Soc **105**: 7754
91. Arseniyadis S, Huang PQ, Husson HP (1988) Asymmetric Synthesis XIV: A Short and Efficient Synthesis of 3,5-Disubstituted Pyrrolizidine Alkaloids *via* the CN(R,S) Method. Tetrahedron Lett **29**: 1391
92. Takahata H, Bandoh H, Momose T (1992) A Short, Chirospecific Synthesis of the Ant Alkaloid (3R,5S,8S)-3,5-Dialkylpyrrolizidines. J Org Chem **57**: 4401
93. Iida H, Watanabe Y, Kibayashi C (1986) A Stereoselective Synthesis of the Ant Trail Pheromone (±)-Monomorine I. Tetrahedron Lett **27**: 5513
94. Iida H, Wananabe Y, Kibayashi C (1989) Total Synthesis of (±)-Dihydropinidine, (±)-Monomorine I, and (±)-Indolizidine 223 AB (Gephyrotoxin 223 AB) by Intramolecular Nitroso Diels-Alder Reaction. J Org Chem **54**: 4088
95. McGrane PL, Livinghouse T (1992) Synthetic Applications of Group IV Metal Imido Complex-Alkyne [2+2] Cycloadditions. A Concise Total Synthesis of (±)-Monomorine. J Org Chem **57**: 1323
96. Stevens RV, Lee AWM (1982) Studies on the Stereochemistry of Nucleophilic Additions to Tetrahydropyridinium Salts. A Stereospecific Total Synthesis of (±)-Monomorine I. J Chem Soc Chem Commun 102
97. Nagasaka T, Kato H, Hayashi H, Shioda M, Hikasa H, Hamaguchi F (1990): Stereoselective Formal Synthesis of (±)-Monomorine I from 6-Methyl-2-piperidinone. Heterocycles **30**: 561
98. Ohta T, Hosoi A, Kimura T, Nozoe S (1987) Direct Chain Elongation of *N*-Carbamoylpyroglutamate. An Efficient Synthesis of (−)-Pyrrolidine-2,5-dicarboxylic Acid. Chem Lett 2091
99. Ohta T, Hosoi A, Nozoe S (1988) Stereoselective Hydroxylation of *N*-Carbamoyl- L-pyroglutamate. Synthesis of (−)-Bulgecitine. Tetrahedron Lett **29**: 329
100. Ohta T, Hosoi A, Kimura T, Nozoe S (1988) Chirospecific Synthesis of (+)-PS-5 from L-Glutamic Acid. Tetrahedron Lett **29**: 4305
101. Yamaguchi R, Hata E, Matsuki T, Kawanishi M (1987) An Efficient Regio- and Stereoselective Synthesis of (±)-Monomorine I *via* the Highly Regioselective α-Alkynylation of a 1-Acylpyridinium Salt. J Org Chem **52**: 2094
102. Castano AM, Cuerva JM, Echavarren AM (1994) A Concise Synthesis of (±)-Monomorine I by Way of a Palladium-Catalyzed Reductive Coupling. Tetrahedron Lett **34**: 7435

103. Sheils CJ, Gray SM, Conard JL, Shawe TT (1994) Iterative Reductive Alkylation Approach to Alkaloids: A Synthesis of (±)-Monomorine I and its C-3 Epimer. J Org Chem **59**: 5841
104. Somfai P, Jarevang T, Lindström UM, Svensson A (1997) Bicyclo[3.3.1]nonanes as Synthetic Intermediates. Part 20. Asymmetric Synthesis of the Indolizidine Alkaloids Monomorine I and Indolizidine 223 AB. Acta Chem Scand **51**: 1024
105. Zaslona A, Tang Q, Jefford CW (1989) A Short, Simple Synthesis of (±)-Monomorine. Helv Chim Acta **72**: 1749
106. Zeller E, Grierson DS (1991) Reactions of α-Aminonitriles under Dissolving Metal Conditions: a Concise Synthesis of (±)-Monomorine I. Synlett **12**: 878
107. Vavrecka M, Hesse M (1991): Synthese von Monomorine I, einem Spurpheromon der Pharao-Ameise (*Monomorium pharaonis*). Helv Chim Acta **74**: 438
108. Mori M, Masanori H, Sato Y (1998) Atmospheric Nitrogen Fixation. Short-step Synthesis of Monomorine I. J Org Chem **63**: 4832
109. Tang Q, Zaslona A, Jefford CW (1991): Short, Enantioselective Syntheses of (−)-Indolizidine 167B and (+)-Monomorine. J Am Chem Soc **113**: 3513
110. Angle SR, Breitenbucher JG (1993) A General Route for the Synthesis of Enantiopure Indolizidine Alkaloids from α-Amino Acids. Total Synthesis of (+)-Monomorine. Tetrahedron Lett **34**: 3985
111. Ibuka T, Habashita H, Otaka A, Fujii N, Oguchi Y, Nobutaka F, Uyehara T, Yamamoto Y (1991) A Highly Stereoselective Synthesis of (E)-Alkene Dipeptide Isosteres *via* Organocopper–Lewis Acid Mediated Reaction. J Org Chem **56**: 4370
112. Takahata H, Bandoh H, Momose T (1993) A Short, Practical Synthesis of the Ant Venom Alkaloid, Three ($3R,5S,8aS$)-3-Alkyl-5-methylindolizidines. Tetrahedron **49**: 11205
113. Sienkiewicz K, Thornton SR, Jefford CW (1994) The Enantiospecific Synthesis of (−)-Monomorine from L-Glutamic Ester. Tetrahedron Lett **35**: 4759
114. Saliou C, Fleurant A, Célérier JP, Lhommet G (1991) Total Synthesis of (+)-Monomorine I from Chiral Cyclic β-Enamino Ester. Tetrahedron Lett **32**: 3365
115. Smith AL, Williams SF, Holmes AB (1988) Stereoselective Synthesis of (±)-Indolizidines 167B, 205A, and 207A. Enantioselective Synthesis of (−)-Indolizidine 209B. J Am Chem Soc **110**: 8696
116. Kang TS, Chung CK, Lee E (1996) Radical Cyclization of β-Aminoacrylates: Expedient Synthesis of (+)-Monomorine I and (+)-Indolizidine 195B. Bull Korean Chem Soc **17**: 212
117. Iida H, Yamazaki N, Kibayashi C (1986) α-Chelation Controlled Nucleophilic Addition to Chiral α,β-Dialkoxycarbonyl Compounds. Diastereoselective Preparation of L-Xylo and L-Lyxo Triols. J Org Chem **51**: 3769
118. Yamazaki N, Kibayashi C (1988) Enantioselective Total Synthesis of (+)-Monomorine I. Tetrahedron Lett **29**: 5767
119. Ito M, Kibayashi C (1990) An Alternative Enantioselective Total Synthesis of (+)-Monomorine I. Tetrahedron Lett **31**: 5065
120. Ito M, Kibayashi C (1991) Total Synthesis of (+)-Monomorine I *via* Nitrone Cycloaddition Route. Tetrahedron **47**: 9329
121. Fujita F, Nakai H, Kobayashi S, Inoue K, Nojima S, Ohno M (1982): An Efficient and Stereoselective Synthesis of Platelet-Activating Factors and the Enantiomers from D- and L-Tartaric Acids. Tetrahedron Lett **23**: 3507
122. Houk KN, Moses SR, Wu YD, Rondan NG, Jäger V, Schohe R, Fronczek FR (1984) Stereoselective Nitrile Oxide Cycloadditions to Chiral Allyl Ethers and Alcohols. The "Inside Alkoxy" Effect. J Am Chem Soc **106**: 3880

123. Berry MB, Craig D, Jones PS, Rowlands GJ (1977) 5-Endo-trig Cyclization in Heterocycle Synthesis: Enantiospecific Synthesis of (+)-Monomorine I. Chem Comm 2141
124. Osborn HMI, Sweeney JB, Howson W (1994) The Synthesis and Reactivity of N-Diphenylphosphinyl Aziridines. Synlett 145
125. Artis DR, Cho I, Jaime-Figueroa S, Muchowski JM (1994) Oxidative Radical Cyclization of (ω-iodoalkyl)Indoles and Pyrroles. Synthesis of (−)-Monomorine and Three Diastereoisomers. J Org Chem **59**: 2456
126. Shirai R, Tanaka M, Koga K (1986) Enantioselective Deprotonation by Chiral Lithium Amide Bases: Asymmetric Synthesis of Trimethylsilyl Enol Ethers from 4-Alkylcylohexanones. J Am Chem Soc **108**: 543
127. Toyooka N, Hirai Y, Momose T (1990) Total Synthesis of (+)-Monomorine I via Asymmetric α-Ketonic Cleavage of 8-Azabicyclo[3.2.1.]octan-3-one. Chem Pharm Bull **38**: 2072
128. Higashiyama K, Nakahata K, Takahashi H (1994) Asymmetric Synthesis of (+)-Monomorine I by Way of a Diastereoselective Reaction of 1,3-Oxazolidine with a Grignard Reagent. J Chem Soc Perkin Trans 1, 351
129. Hase TA, Ourila A, Holmberg C (1981) A Short Route to Pyrenophorin and Vermiculine. J Org Chem **46**: 3137
130. Chu GH, Solladié G (1996) Total Synthesis of (+)-Indolizidine 195B and (+)-Monomorine. Tetrahedron Lett **37**: 111
131. Munchhof MJ, Meyers AI (1995) Novel Asymmetric Route to Chiral, Nonracemic Cis-2,6-Disubstituted Piperidines. Synthesis of (+)-Pinidinone and (+)-Monomorine. J Am Chem Soc **117**: 5399
132. Eschenmoser A (1970) Roads to Corrins. Quart Rev **24**: 366
133. Takahata H, Bandoh H, Momose T (1996) An Asymmetric Synthesis of the Ant Venom Alkaloid (3S,3S,8aR)-3-Butyl-5-(4-pentenyl)indolizidine via the Sharpless Asymmetric Dihydroxylation. Heterocycles **42**: 39
134. Macdonald TL (1980) Indolizidine Alkaloid Synthesis. Preparation of the Pharaoh Ant Trail Pheromone and Gephyrotoxin 223 Stereoisomers. J Org Chem **450**: 193
135. Yue C, Gauthier I, Royer J, Husson HP (1996) Concise and Stereoselective Syntheses of the Eight Natural and Defense Alkaloids (+)-Tetraponerine-1 to (+)-Tetraponerine-8 According to the CN(R,S) Strategy. J Org Chem **61**: 4949
136. Blum MS, Hermann HR (1978) In: Bettini S (ed) Venoms and Venom Apparatuses of the Formicidae: Myrmeciinae, Ponerinae, Dorylinae, Pseudomyrmecinae, Myrmicinae and Formicinae. Springer, Berlin, p 801
137. Merlin P, Braekman JC, Daloze D, Pasteels JM, Dejean A (1992) New δ-Lactones from the Dufour's Gland of the Urticating Ant *Tetramorium aculeatum*. Experientia **48**: 111
138. Daloze D, Braekman JC, Vanhecke P, Boeve JL, Pasteels JM (1987) Long Chain Electrophilic Contact Poisons in the Dufour's Gland of the Ant *Crematogaster scutellaris*. Can J Chem **65**: 432
139. Pasteels JM, Daloze D, Boeve JL (1989) Aldehydic Contact Poisons of the Ant *Crematogaster scutellaris* (Hymenoptera: Myrmicinae): Enzyme-Mediated Production from Acetate Precursors. J Chem Ecol **15**: 1501
140. Daloze D, Kaisin M, Detrain C, Pasteels JM (1991) Chemical Defence in the Three European Species of *Crematogaster* Ants. Experientia **47**: 1082
141. Leclercq S, Daloze D, Braekman JC, Kaisin M, Detrain C, De Biseau JC, Pasteels JM (1997) Venom Constituents of Three Species of *Crematogaster* Ants from Papua New Guinea. J Nat Prod **60**: 1148

142. Daloze D, De Biseau JC, Leclercq S, Braekman JC, Quinet Y, Pasteels JM (1998) (13*E*,15*E*,18*Z*,20*Z*)-1-Hydroxypentacosa-13,15,18,20-tetraen-11-yn-4-one 1-Acetate, from the Venom of a Brazilian *Crematogaster* Ant. Tetrahedron Lett **39**: 4671
143. Leclerq S, Braekman JC, Daloze D, Luhmer M, De Biseau JC, Pasteels JM, Quinet Y (unpublished results)
144. Renson B, Merlin P, Daloze D, Braekman JC (1994) Biosynthesis of Tetraponerine 8, a Defence Alkaloid of the Ant *Tetraponera* sp. Can J Chem **72**: 105
145. Devijver C, Braekman JC, Daloze D, Pasteels JM (1997) The Biosynthesis of Tetraponerine 6: Evidence that Different Pathways are Operating in the Biosynthesis of the Two Tetraponerine Skeletons. J Chem Soc Chem Comm 661
146. Leclercq S, Braekman JC, Daloze D, Pasteels JM, Van der Meer RK (1996). Biosynthesis of the Solenopsins, Venom Alkaloids of the Fire Ants. Naturwissenschaften **83**: 222
147. Hefetz A, Blum MS (1978): Biosynthesis and Accumulation of Formic Acid in the Poison Gland of the Carpenter Ant *Camponotus pennsylvanicus*. Science **201**: 454

(Received April 12, 1999)

Author Index

Page numbers printed in *italics* refer to References

Adams, D.R. *224*
Akao, T. *110*
Alazard, J.P. *111*
Andersen, A.N. *222*, *224*
Angle, S.R. 174, 176, 177, *227*
Aoki, T. *225*
Aoyagi, Y. *111*
Arai, Y. *111*
Arseniyadis, S. *225*, *226*
Artis, D.R. *228*
Attygalle, A.B. 117, *221*, *222*
Awen, B.Z.E. *225*

Baba, S. *224*
Bacher, A. *107*
Bäckvall, J.E. 133, *224*
Bacos, D. *222*, *225*
Baghdikian, B. *110*
Bailey, T.R. *225*
Balansard, G. *110*
Ban, C. *111*
Bandoh, H. *226–228*
Barluenga, J. 205, *224*
Bärtels, C. *108*, *112*
Basselier, J.J. *222*
Baumann, H. *223*
Beal, J.L. *107*
Belotti, D. *110*
Benkrief, R. *110*
Berdini, R. *109*
Berkowitz, W.F. *108*
Bernini, R. *114*
Berry, M.B. *228*
Bestmann, H.J. *224*
Bhaduri, A.P. *114*
Bianco, A. 92, *107*, *109*, *110*, *112*, *114*
Blankespoor, C.L. *222*
Blokhin, A. *222*

Blum, M.S. 117, 160, *222–224*, *228*, *229*
Bobbitt, J.M. *107*
Boeve, J.L. *228*
Bonadies, F. *114*
Bonin, M. *111*
Bonini, C. *114*
Boros, C. *107*, *110*
Bowers, D.M. *107*
Braekman, J.C. 118, 120, *222–224*, *228*, *229*
Brand, J.M. *222*
Brando, C.R.F. *223*
Braun, G. *113*
Brayer, J.L. *111*
Breinholt, J. *108*
Breitenbucher, J.G. 174, 176, 177, *227*
Briggs, L.H. *108*
Broeders, F. *222*
Brophy, J.J. *224*
Brownell, J. *114*
Buschinger, A. 116, *221*, *223*

Cachet, X. *109*
Carnevale, G. *114*
Carruthers, W. *224*
Castano, A.M. *226*
Célérier, J.P. *222*, *225–227*
Cerichelli, G. *112*
Chakalamannil, S. 128, *224*
Chan, T.H. *109*, *222*
Chase, M.W. *107*
Chen, C. *110*
Chen, D. *111*
Chi, Y. *111*
Chia, W. *110*
Cho, I. *228*
Choudhry, S. *108*
Chu, G.H. 188, *228*

Chung, C.K.   *227*
Cid, M.M.   *111, 112*
Classon, B.   *114*
Clement, J.L.   *222*
Clezy, P.S.   *224*
Coll, M.   *222*
Collas, M.   *225*
Conard, J.L.   *227*
Contin, A.   *107*
Cordell, G.A.   *107, 112*
Cossy, J.   *110, 111*
Craig, D.   183, *228*
Crowley, P.J.   *224*
Cuerva, J.M.   *226*

Daloze, D.   118, 120, *222–224, 228, 229*
Daluge, S.   *114*
Daly, J.W.   *222*
Damtoft, S.   *108, 109*
D'Angelo, J.   134, *225*
D'Annibale, A.   *114*
Davini, E.   *114*
De Biseau, J.C.   *228, 229*
Debrauwer, L.   *110*
Declercq, J.P.   *223*
Deguin, B.   *109*
Dejean, A.   *228*
Demuth, H.   *108*
Detrain, C.   *228*
Devijver, C.   203, *224, 229*
De Vries, P.J.   *223*
Dietz, U.   *113*
Di Fabio, R.   *114*
Don, A.W.   *223, 224*
Duffield, R.M.   *222*
Dumas, F.   134, *225*
Dumenil, G.   *110*

Echavarren, A.M.   167, *226*
Edwards, P.D.   *225*
Eggnauer, U.   *111*
Eichinger, D.   *107*
Eisenreich, W.   *107*
Eisner, T.   *222*
El-Naggar, L.J.   *107*
El-Sedawy, A.I.   *110*
Eschenmoser, A.   *228*
Escoubas, P.   *222–224*
Escudero-Hernandez, M.L.   *111*

Fales, H.M.   *222–224*
Faure, R.   *110*
Felicio, J.D.   *111*
Fennen, J.   *112*
Fiumana, A.   *111*
Fleurant, A.   *227*
Foderaro, T.A.   *222*
Fouquet, G.   *225*
Francke, W.   194, 196, *223*
Franke, S.   *223*
Franzyk, H.   *107–109, 114*
Frederiksen, L.B.   *108*
Frederiksen, S.M.   *107, 109, 110*
Fronczek, F.R.   *227*
Fujie, I.   *108*
Fujihira, S.   *111*
Fujii, N.   *227*
Fujikawa, S.   *112*
Fujita, F.   *227*
Fujita, T.   *108, 109*
Fukui, Y.   *112*

Gallagher, T.   152, *225*
Garnier, J.   *113*
Garraffo, H.M.   119, *222*
Gauthier, I.   *228*
Ge, Y.   *109, 112, 113*
Gethöffer, H.   *113*
Ghisalberti, E.L.   *107*
Gopinath, P.M.   *108*
Gorman, J.S.T.   *222*
Gournelis, D.   *110*
Grandjean, C.   *226*
Gray, S.M.   *227*
Grayer, R.J.   *107*
Grierson, D.S.   170, *225–227*
Gubbiotti, A.   *114*
Guerrier, L.   *225, 226*
Guiraud-Dauriac, H.   *110*
Guiso, M.   *109, 112*

Habashita, H.   *227*
Hadi, H.A.   *111*
Hamaguchi, F.   166, *226*
Han, Q.   *108, 109*
Haremsa, S.   *108, 113*
Harris, G.H.   *110*
Hase, T.A.   *228*
Hashimoto, F.   *111*
Hata, E.   *226*

Hattori, M. *110*
Hayashi, H. *226*
Hayashi, M. *114*
Hefetz, A. *222, 229*
Hermann, H.R. *228*
Hesse, M. 152, 171, *226, 227*
Higashiyama, K. *228*
Highet, R.J. *223*
Hikasa, H. *226*
Hill, R.K. *222*
Hirai, Y. *228*
Hiraishi, A. *108, 109*
Hölldobler, B. 116, *221, 224*
Holmberg, C. *228*
Holmes, A.B. *227*
Honda, G. *109*
Honda, T. *111*
Hope, H. *222*
Hosoi, A. *226*
Hotellier, F. *110*
Houk, K.N. 183, *227*
Howson, W. *228*
Hrabie, J. *108*
Hua, M. *114*
Huang, P.Q. *225, 226*
Huang, Q. *222*
Huber, W. *113*
Huber-Patz, U. *108, 113*
Husson, H. *111*
Husson, H.P. 146, 158, 207–209, *223, 225, 226, 228*

Iatridou, H. *113*
Iavarone, C. *114*
Ibuka, T. 117, 118, 120, 163, *222, 227*
Iida, H. *226, 227*
Iitaka, Y. *111*
Ikeda, Y. *114*
Ikumoto, T. *112*
Inariyama, T. *111*
Inoiue, K. *110*
Inoue, K. *108, 110, 112, 227*
Inouye, H. 36, *106, 108–110, 112*
Irngartinger, H. *108, 113*
Ishiguro, K. *114*
Isoe, S. 82, *107–109, 112, 113*
Ito, K. *114*
Ito, M. 180, *227*
Itô, S. *112*
Iwahashi, Y. *107*

Iwanowicz, E.J. 140, *225*
Iwashita, T. *112*
Iyer, R.I. *108*

Jackson, B.D. *222, 224*
Jacobson, E.J. *111*
Jagdmann, G.E. *225*
Jäger, V. *227*
Jaggy, H. *108*
Jahn, R. *108*
Jaime-Figueroa, S. *228*
Janowitz, A. *226*
Janssen, E. *224*
Janssens, A. *113*
Jarevang, T. *227*
Jefford, C.W. 170, 173, 177, *227*
Jensen, S.R. 22, *106–109, 114*
Jessen, C.U. *108*
Jewett, B. *109*
Jones, K. *111*
Jones, P.S. *228*
Jones, T.H. 118, 120, 125, 129–131, 150, 151, 159, 162, 190, 193, 203, 204, 206, *222–224*
Junior, P. *107*

Kaib, M. *223*
Kaisin, M. *228*
Kam, T. *111*
Kametani, T. *111*
Kanazawa, T. *225*
Kan-Fan, C. *111*
Kang, T.S. *227*
Katano, N. *110*
Kataoka, K. *112*
Kato, H. *226*
Katsumura, S. *112, 113*
Kawakami, K. *114*
Kawamura, I. *112*
Kawanishi, M. *226*
Kawata, Y. *110*
Kern, F. *222, 224*
Kibayashi, C. 147, 163, 180, *225–227*
Kigawa, M. *109, 113*
Kimura, T. *226*
Kirk, O. *108, 109*
Knudsen, T.B. *108*
Kobashi, K. *110*
Kobayashi, H. *111*
Kobayashi, S. *227*

Koch, M. *109*, *110*
Kocsis, A. *112*
Koga, K. *112*, 184–186, *228*
Koleva, I.I. *108*
Komura, H. *112*
Kondo, S. *112*, *113*
Koob, R. *223*
Kouznetsov, V. *224*
Krajsovszky, G. *112*
Krull, R.E. *107*
Kurihara, T. *111*
Kuwajima, H. *110*
Kvarnström, I. *114*

Labarre, S. *110*
Laddago, A. *223*
Lange, C. *222*
Lathbury, D. 152, *225*
Lattke, J. *223*
Lavelle, G.C. *114*
Leblanc, C. *110*, *111*
Leboff, A. *111*
Leclercq, S. 119, *222*, *224*, *228*, *229*
Lee, A.W.M. 166, *226*
Lee, E. 180, *227*
Lee, F. *114*
Lefeber, A.W.M. *107*
Lemaire, M. *222*
Le Men, J. *110*
Lernhardt, U. *113*
Leung, C.W.F. *224*
Lhommet, G. 136, 144, 149, 153, 156, 178, *222*, *225*–*227*
Li, J. *111*
Lichtenthäler, J. *108*
Lindström, U.M. *227*
Lins, A.P. *111*
Livinghouse, T. 165, *226*
Lloyd, H.A. *222*
Lo Baido, G. *112*
Lofgren, C.S. *222*
Lozanova, A.V. *111*
Luhmer, M. *229*

Macdonald, T.L. 196, *228*
Machinaga, N. 147, *225*
Macours, P. 206, *224*
Maklouf, K. *109*
Marini, E. *109*
Marubayashi, N. *111*

Marx, E. *222*, *225*
Masanori, H. *227*
Maschwitz, U. 116, *221*
Mataloni, F. *114*
Mathad, V.T. *114*
Mathuram, V. *108*
Matsuki, T. *226*
Matsumoto, T. *109*
Matsumura, Y. *225*
Matuda, Y. *111*
Maurer, W. *109*
Mazzei, R.A. *112*, *114*
McCormick, K.D. *222*
McGrane, P.L. 165, *226*
Meinwald, J. *222*
Melzer, E. *108*
Mendoza, R. *225*
Merlin, P. 200, *223*, *224*, *228*, *229*
Meyers, A.I. 135, 189, *225*, *228*
Michel, S. *110*
Mitsuhashi, H. *109*, *113*
Miyashita, M. *225*
Mohammad-Ali, A.K. *109*
Momose, T. 140, 155, 161, 176, 184, 193, *225*–*228*
Morel, L.M. *222*
Morgan, E.D. 117, *221*, *222*, *224*
Mori, M. *227*
Morimoto, Y. *109*
Moritome, N. *112*
Moses, S.R. *227*
Mpuru, S.P. *222*
Muchowski, J.M. 184, *228*
Munchhof, M.J. 189, *228*
Murai, F. 15, *108*
Muto, N. *114*

Naccarato, G. *110*
Nagasaka, T. *226*
Nagino, C. *112*
Nakahata, K. *228*
Nakai, H. *227*
Nakajima, H. *108*
Nakamura, Y. *109*
Nakatani, K. *108*, *109*, *112*, *113*
Namba, T. *110*
Naruse, N. *113*
Naruto, M. *113*
Negishi, E. 133, *224*
Netzel, D.A. *225*

Neuberger, K. *108*, *113*
N'Guyen, A. *110*
Nicholls, G.A. *108*
Nicoletti, M. *109*
Nielsen, B.J. *108*, *109*
Nishioka, T. *109*
Nixdorf, M. *113*
Nobutaka, F. *227*
Nogushi, H. *111*
Nohara, T. *111*
Nojima, S. *227*
Nomoto, K. *112*
Nose, T. *114*
Nozoe, S. 166, *226*
Numata, A. 117, 118, 120, 163, *222*

Odagaki, Y. *112*
Oeser, T. *113*
Oguchi, Y. *227*
Oguri, M. *112*
Ohkubo, N. *225*
Ohno, K. *113*
Ohno, M. *227*
Ohta, A. *111*
Ohta, T. *226*
Okinawa, M. *109*
Okukado, N. *224*
Ollivier, E. *110*
Olsen, C.E. *109*
Olubajo, O. *222*
Ono, M. *108*
Ooi, A. *110*
Oppolzer, W. *111*
Osborn, H.M.I. *228*
Oster, B. *113*
Otaka, A. *227*
Otsuka, H. *109*
Ourila, A. *228*
Ozaki, F. *112*

Pandey, C.P. *114*
Passacantilli, P. *109*, *110*
Pasteels, J.M. *223*, *224*, *228*, *229*
Patnaik, G.K. *114*
Peng, C. *110*
Petit, H. *226*
Pierce, D. *108*
Piveteau, D. *225*
Podányi, B. *112*
Poisson, J. *113*

Pombo-Villar, E. *111*, *112*
Provot, O. *226*
Puglis, J.M. 132, *224*
Pusset, J. *110*

Qualls, J. *114*
Quinet, Y. *229*
Quirion, J. *111*

Raj, K. *114*
Ranarivelo, Y. *110*
Rapoport, H. 142, *225*
Rasmussen, J.H. *107*, *114*
Rathelot, P. *110*
Reder, E. *223*
Reifenstahl, U. *113*
Renko, Z.D. *224*
Renson, B. *229*
Righi, G. *109*, *110*
Rimpler, H. *107*
Ritter, F.J. *223*
Robertson, H.G. *222*
Robertson, P.L. *224*
Rodewald, H. *108*, *113*
Rolland, Y. *113*
Rondan, N.G. *227*
Rosset, S. *225*, *226*
Rothgans, I.E.M. *223*
Rowlands, G.J. *228*
Royer, J. *223*, *225*, *226*, *228*
Rubio, E. *224*
Rudolph, C. *223*

Sakaguchi, K. *113*
Saliou, C. *227*
Sampathkumar, P.S. *108*
Samuelsson, B. *114*
Sankawa, U. *111*
Sasson, I. *108*
Sato, Y. *227*
Scarpati, M.L. *110*
Schacht, M. *109*
Schick, H. *108*, *113*
Schink, H.E. *224*
Schlessinger, R.H. 140, *225*
Schlosser, M. 135, *225*
Schmidbauer, S.B. *109*
Schmidt, J.O. 117, *221*
Schohe, R. *227*
Schröder, F. *223*

Schwartz, A. *112*
Segebarth, K.P. *107*
Sévenet, T. *110, 111*
Sezik, E. *109*
Shannon, W.M. *114*
Shawe, T.T. *167, 227*
Sheils, C.J. *227*
Shen, C. *110*
Shen, Y. *110*
Shiao, M. *110*
Shimano, K. *109, 113*
Shingu, T. *112*
Shioda, M. *226*
Shiosaki, K. *142, 225*
Shirai, R. *228*
Shirakata, N. *112*
Shono, T. *142, 225*
Shu, Y.Z. *110*
Sienkiewicz, K. *227*
Simmonds, M.S.J. *107*
Simon, L.D. *222*
Sinnwell, V. *223*
Skaltsounis, A.L. *110*
Skrinjar, M. *141, 142, 225*
Smith, A.L. *227*
Snelling, R.R. *222, 224*
Solladié, G. *188, 228*
Somfai, P. *168, 227*
Sonnet, P.E. *225*
Spande, T.F. *222, 224*
Stahly, S.M. *223, 224*
Stammler, H. *113*
Stein, F. *223*
Stepanov, A.V. *111*
Stermitz, F.R. *107, 109, 110, 222*
Stevens, R.V. *166, 226*
Stevenson, T. *111*
Storm, C.B. *222*
Strunz, G.M. *109*
Sugihara, Y. *225*
Suzuki, Y. *111*
Svensson, A. *227*
Sweeney, J.B. *184, 228*
Swiatek, L. *107*
Szabó, L.F. *112*
Szabó-Pusztay, K. *112*

Tabata, M. *109*
Tagaki, S. *114*
Tagawa, M. *15, 108*

Takahashi, H. *186, 228*
Takahashi, T. *224*
Takahata, H. *192, 225–228*
Takeda, Y. *109, 112*
Takehara, H. *225*
Takeuchi, H. *113*
Talman, E. *223*
Tamaoka, Y. *112*
Tanahashi, T. *110*
Tanaka, K. *109*
Tanaka, M. *109, 113, 228*
Tang, Q. *227*
Tarawa, J.N. *222*
Tatsuki, S. *112*
Thal, C. *111*
Thale, Z. *109*
Thirionet, I. *222*
Thomas, A.W. *109*
Thompson, C.R. *223*
Thornton, S.R. *227*
Tietze, L.F. *108, 112*
Tillequin, F. *109, 110*
Tixidre, A. *113*
Tobita, S. *109*
Tomas, M. *224*
Torres, J.A. *222, 224*
Touyama, R. *112*
Toyooka, N. *228*
Trogolo, C. *114*
Tsubata, K. *225*
Tsuchida, S. *111*
Tsujino, M. *114*
Tsunoda, T. *112*
Tufariello, J.J. *132, 224*

Ueda, S. *107*
Uyehara, T. *227*

Vaisar, T. *224*
Van der Eycken, E. *113*
Van der Heijden, R. *107*
Van der Meer, R.K. *222, 229*
Vandewalle, M. *113*
Van Hecke, P. *223, 228*
Van Horn, D.E. *224*
Vavrecka, M. *226, 227*
Veith, H.J. *223*
Veith, J. *138, 225*
Verpoorte, R. *107*
Verwiel, P.E.J. *223*

Veselovsky, V.V. *111*
Vince, R. *114*

Wachtmeister, J. *114*
Wakamatsu, T. *109, 113*
Walter, F. *223*
Wang, Y. 128, *224*
Watanabe, Y. *226*
Weber, H.P. *111*
Wei, C. *111*
Weinges, K. *108, 109, 113*
Weislow, O.S. *114*
Weiss, J. *113*
Wheeler, J.W. 119, *222*
Wiesner, K. *110*
Wildman, W.C. *110*
Williams, M.J. *224*
Williams, S.F. *227*
Wilson, E.O. 116, *221*
Wistrand, L.G. 141, 142, *225*
Wright, P.J. *222*
Wu, Y.D. *227*
Wysokinska, H. *107*

Xu, S.C. *222*

Yaginuma, S. *114*
Yamaguchi, R. 166, *226*
Yamaki, M. *114*
Yamamoto, H. *110, 112*
Yamamoto, K. *112*
Yamamoto, Y. 174, *227*
Yamane, S. *225*
Yamashota, M. *111*
Yamazaki, N. 180, *227*
Yan, W. *111*
Yatsuzuka, M. *112*
Ye, Y. *112*
Yesilada, E. *109*
Yoganathan, K. *111*
Yokoi, T. *112*
Yoshida, T. *109*
Yoshikoshi, A. 137, 150, *225*
Yue, C. 203, *223, 228*

Zaslona, A. *227*
Zeller, E. 170, *227*
Zenk, M.H. *107*
Ziegler, H.J. *108, 109, 113*
Zimmer, R. *225*

Subject Index

[1-^{14}C]Acetate 219
[2-^{14}C]Acetate 219
Acetic acid 11, 15, 63, 83, 98, 129, 137, 144, 150, 162, 214
Acetic anhydride 162, 181
2-Acetonaphthone 44
Acetone 10, 13
Acetonitrile 55
2-Acetylbutyrolactone 156
Acetylenic sulfoxide 206
Acidic ethanol 178
Actinidiaceae 26, 49
Actinidia polygama 26
Actinidine 55–57, 128
(−)-Actinidine 8, 49, 58
Adoxoside 16
epi-Adoxoside 16
Aglucone derivatives 87
Ajmalicine 8
Alangiside 8
D-Alanine 155, 161
L-Alanine 129, 155, 156, 173–176, 193, 195
N-Cbz-L-Alanine methyl ester 128
Alarm pheromones 128, 212, 214
Alboside A 15
Alboside B 15
Alkaloidal glucosides 7
Alkaloidal iridoids 7
Alkaloids 117
Alkenals 131
1-Alken-3-ones 131
Alkenylpiperidines 120
N-Alkylated-3-methylpyrrolidines 138
7-*O*-Alkyl-10-hydroxyoleoside secoiridoid glucosides 40
2-Alkyl-6-methylpiperidines 118, 119
N-Alkyl-3-methylpyrrolidines 122
Alkylpyrazines 128
(+)-Allamandicin 33

(+)-*epi*-Allamandicin 33
(+)-*iso*-Allamandicin 33
Allamandin 6
Allylmagnesium chloride 185, 186
Aluminium oxide 195
α-Amino-acids 173
4-Aminobutanal diethyl acetal 208–210
γ-Amino[U-^{14}C]butanoic acid 218
2-Aminoethanethiol 68
(*R*)-4-Aminopentanoic acid 184, 185
2-(3-Aminopropyl)-1,3-dioxolane 204
Ammonium acetate 47, 49, 51, 131, 159
Ammonium chloride 53
Ammonium formate 93, 142
Anabaseine 119, 120
Anabasine 119, 120
Antifeedant activity 28, 91
Antirrhide 9
Antirrhinolide 28
Antirrhinoside 4, 9, 24, 25, 28, 51, 60, 64, 66, 100
epi-Antirrhinoside 24
Antirrhinoside tetrabenzoate 100
Antirrhinum majus 9, 28
Antitumor activity 103
Antiviral activity 98, 102
Aphaenogaster fulva 119, 120
Aphaenogaster rudis 119, 120
Aphaenogaster tennesseensis 119, 120
Apocynaceae 59
Aqueous acid 47
Aqueous sodium periodate 103
Arbortristoside A 105
Asperuloside 12, 13, 36
Asperuloside aglycone 1-*O*-TBDMS ether 33, 35, 36
Asperulosidic acids 36
Aucubaceae 12
Aucuba japonica 12
8β-Aucuban tetraacetate 27

Aucubigenin 7, 53, 98
Aucubin 4, 11, 12, 21–23, 25, 53, 91, 92, 94–97
6-*epi*-Aucubin 23
Aucubin hexaacetate 21, 27–29, 91–93
Aucubinine A 53
Aucubinine B 52–54
Aucubovir 98
Austrodimerine 48
Azabicyclo[4.3.0]nonane alkaloids 59, 61, 64
3-Azabicyclo[4.3.0]nonane alkaloids 66
8-Azabicyclo[3.2.1]octan-3-one 185
Aza-[2,3]-Wittig rearrangement 168, 169
Azobisisobutyronitrile 30

Baeyer-Villiger oxidation 36, 44
Bakankosin 8, 67, 68
Baldrinal 29, 30, 84
Barium manganate 30
Bartsioside tetraacetate 16
Beatrine A 52
Beatrine B 52
Belleau's reagent 189
Benzoyl chloride 175
Benzoyl peroxide 128
6-*O*-Benzoylshanzhiside methyl ester 47
Benzylamine hydrochloride 65, 67
Benzyl bromide 202
Benzylchloroformate 153, 157, 166, 190, 206
3-Benzyl-5-(2-hydroxyethyl)-4-methylthiazolium chloride 191
Benzylmercaptan 87
p-(Benzyloxy)benzoyl chloride 28
(*S*)-*N*-(Benzyloxycarbonyl)-1-methyl-5-hexenylamine 176
(*R*)-*N*-Benzylphenylglycinol 187
Bicyclic cyclopentanoid piperidine alkaloids 45
Bicyclic cyclopentanoid piperidines 60, 63
Bicyclic piperidines 66
Bignoniaceae 49, 59
Biological activity 82, 102, 141
2,3-Bipyridyl 119
Birch reduction 84
Bisiridoids 4, 14
N,*O*-Bis(trimethylsilyl)acetamide 58
N-Boc-L-alanine ethyl ester 174

9-Borabicyclo[3.3.1]nonane 61
Boric acid 137, 144
Boschnaloside 18
Boschniakine 47
Bovine serum albumin 105
4-Bromobutanal 205, 206
Bromoethanal diethyl acetal 210
2-(2-Bromoethyl)-1,3-dioxolane 130
2-(Bromomethyl)-1,3-dioxolane 208, 209
5-Bromopentane 130
1-Bromo-4-pentene 144
5-Bromopentylbenzylether 196
α-Bromophenylacetate 175
3-Bromopropanal 146
3-Bromopyridine 56
N-Bromosuccinimide 83
n-Butanol 11, 12
1-Butene 133
(2*S*,5*S*)-*trans*-5-Butyl-2-alkylpyrrolidines 140
Butyl bromide 196
tert-Butyl bromoacetate 169
tert-Butyldimethylsilyl chloride 193
tert-Butyl formamidine 135
2-Butyl-5-(1*E*,3*E*)-heptadienyl-5-pyrroline 122
2-Butyl-5-((*E*)-1-heptenyl)-5-pyrroline 122
trans-2-Butyl-5-heptylpyrrolidine 132
(2*R*,5*R*)-*trans*-5-Butyl-2-heptylpyrrolidine 141
(2*S*,5*S*)-*trans*-5-Butyl-2-heptylpyrrolidine 141
(2*R*,5*R*)-*trans*-5-Butyl-2-(5-hexenyl)-pyrrolidine 141
(5*E*,8*E*)-3-Butyl-5-hexylpyrrolizidine 152, 159
n-Butyllithium 136
t-Butyllithium 136
tert-Butyllithium 135, 136
n-Butylmagnesium bromide 189, 190
(2*S*)-2-Butyloxirane 197
3-Butyl-5(4-penten-1-yl)indolizidine 190
(2*S*,5*S*)-*trans*-2-Butyl-5-pentylpyrrolidine 147, 148
(2*S*,5*S*)-*trans*-5-Butyl-2-pentylpyrrolidine 141
2-Butylpyrrolemagnesium bromide 192

Subject Index

Caffeic esters 11
Camponotus pennsylvanicus 219
Cantleyine 49
Capensioside 16
Caprifoliaceae 13, 40
Carbocyclic iridoids 3, 4, 99
Carbovir 98
4-Carboxylated iridoids 15
(−)-Carvone monoepoxide 61
Catalpol 4, 12, 13, 22, 87, 88, 91, 101, 102
Catalpol hexaacetate 89, 91
Cembrane diterpenes 217
Cerbinal 7, 29, 30, 84
Charcoal 11, 178
Charcoal method 11, 13
Chelaner antarcticum 120, 122
Chelaner sp. 160, 162
Chemotaxonomic markers 8
Chiral 1,3-oxazolidines 186, 187
Chiral sulfoxides 188
4-Chloro-1-butanol 190
Chloroform 11, 13, 28, 136
m-Chloroperbenzoic acid 19, 21, 28, 68
m-Chloroperoxybenzoic acid 132
6-Chloropurine 101
N-Chlorosuccinimide 62, 200, 201
Chugaev reaction 17
Citric acid 206, 207
(−)-(*S*)-Citronellol 58
Claisen rearrangement 174, 175
Collins oxidation 164
Complex indole alkaloids 7
Copper(II) bromide 91
Cornin 18, 49
Corninine 49
Cornin tetraacetate 18
8-*O*-*p*-Coumaroylharpagide 51, 54
Crematogaster auberti 214
Crematogaster scutellaris 213, 214
Crematogaster sordidula 214
Crematogaster sp. 117, 212–217
Cyanomethylenetrimethyl-
  phosphorane 64
(−)-2-Cyano-6-oxazolopiperidine 208
2-Cyano-5-oxazolopyrrolidine 146, 155, 211
Cyclopentanoid amide 64
Cyclopentanoid compounds 9
Cyclopentanoids 87, 98

Cyclopentano-isoxazolidines 58
(+)-Cyclosarkomycin 89–91
Cytotoxic activity 221

Decarboxylated iridoids 15
9-Decenal 160
Dec-1-en-3-one 153
7,8-Dehydro-6β,10-dihydroxy-11-*nor*-
  iridomyrmecin 25
(−)-*N*-Demethyl-δ-skytanthine 62, 63
10-Deoxyaucubin 21
10-Deoxycatalpol 21
Deoxygeniposide 16, 17, 36, 103
Deoxygeniposide tetraacetate 19, 37
Deoxyloganin 18, 19, 41, 103
8-*epi*-Deoxyloganin 18, 103
Deoxyloganin aglucone 26
Deoxyrhexifoline 46
(−)-Deoxyrhexifoline 49
3,5-Dialkenylpyrrolizidines 160
2,5-Dialkyl-2-hydroxypyrrolidines 150
2,6-Dialkylpiperidines 118, 128, 129
2,6-Dialkylpyridines 129
2,5-Dialkylpyrrolidines 118, 120, 136–138, 149, 150
trans-2,5-Dialkylpyrrolidines 122, 123
(2*R*,5*R*)-*trans*-2,5-Dialkyl-
  pyrrolidines 141
3,5-Dialkylpyrrolidines 123
Dialkylpyrrolines 150
2,5-Dialkylpyrrolines 150
3,5-Dialkylpyrrolizidines 122, 152
N,*N*-Diallyl-2-oxocyclopentane-
  carboxamide 55
Diazabicyclo[5.4.0]undec-7-ene 44
Diazoketones 170
Diazomethane 184
3,4-(Di-*O*-benzyl)dopaol 36
2,3-Dichloro-5,6-dicyano-1,4-
  benzoquinone 30
Dichloromethane 171
Dicyclohexylcarbodiimide 34
6,10-Dideoxyaucubin 16, 21
5,7-Dideoxycynanchoside 99
5,7-Dideoxy-3,4-dihydro-
  cynanchoside 87
Didrovaltrate 30, 31
Diels-Alder cycloaddition 163, 164
1,1-Diethoxy-4-aminobutane 202, 203
4,4-Diethoxybutylamine 206, 207

Diethyl chlorophosphate 151, 171
Diethylcyanophosphonate 171
Diethyl ether 11
Diethyl L-glutamate hydrochloride 177
Diethyl L-tartrate 180, 181, 182
L-Diethyl tartrate 180
(5E,8Z)-3,5-Di(5-hexenyl)pyrrolizidine 161
(5Z,8E)-3,5-Di(5-hexenyl)pyrrolizidine 161
Dihydrojasminine 48
(S)-4,5-Dihydro-3-methyl-2(3H)furanone 138
8,10-Dihydrosweroside 76, 77
8,10-Dihydrosweroside aglucone 76, 78
5,6β-Dihydroxyadoxoside 23
ortho-Dihydroxylated phenolics 11
Diisobutylaluminum hydride 26, 128
1,4-Diketones 131, 150
Dimethoxytetrahydrofuran 184
2,5-Dimethoxytetrahydrofuran 174, 177
N,N-Dimethylbarbituric acid 73
N,N-Dimethylbutyramide 203
Dimethylformamide 24
Dimethyl malonate 93
Dimethyl (R)-2-methylsuccinate 138
Dimethyl(2-oxohexyl)phosphonate 169
2,2-Dimethyl-1,3-propantriol 30
Dimethylsulfate 136, 156, 178
Dioxan 93
Diphenyl diselenide 135
Diphenyl disulfide 30
Diphenylphosphinic chloride 184
Diplorhoptrum ssp. 119, 125
2,5-Disubstituted pyrrolidines 131
Dolichoderines 117
Doronomyrmex goesswaldi 122

Entomopathogenic fungi 119
(+)-Epidihydrotecomanine 62, 64
Epoxydecaloside 4
Epoxysecologanins 41, 42
8,10-Epoxysecologanins 39, 40
Epoxysecoxyloganins 41
8,10-Epoxysecoxyloganins 39, 40
Epoxyswerosides 42
8,10-Epoxyswerosides 40
Erythrocentaurin 52

Eschweiler-Clarke procedure 64
Ethanol 10, 12, 130, 208
Ethanolic triethyl orthoformate 91
Ethyl acetate 11, 13
(2R,5R)-trans-5-Ethyl-2-heptylpyrrolidine 141
Ethyl iodide 136
Ethyl lithium 195, 196
3-Ethyl-5-methylindolizidine 124
Ethyl-3-oxoheptanoate 172
trans-2-Ethyl-5-(12-tridecen-1-yl)pyrrolidine 120
2-Ethyl-5-undecylpyrrolines 150
trans-2-Ethyl-5-undecylpyrrolidine 120
γ-Ethynyl bromide 56
Eucommial 98
Euphroside 47
Euphrosine 47
Evans rearrangement 31, 41

Favorskii-type rearrangement 61
Flash column chromatography 12
Flavonoids 11
Fontanesioside 14
Formaldehyde 146, 204
Formic acid 117, 211, 219, 220
Formica sp. 211
Formicinae 116, 211
3-Formyloxy-4-hydroxy derivatives 89
10-Formyltetrahydrofolate 220
Forsythide dimethyl ester 14, 15
Fructose 52
Fulvoharpagides 69
Fulvoiridoids 69
Fulvolamiols 69
2-Furyllithium 162

Gardenine 54
Gardenoside 10, 16, 36, 53, 54
Gardenoside aglycone 33
Gardenoside aglycone 1-O-TBDMS ether 33, 35
Gardenoside hexaacetate 17
Gardiol 54
Gardoside 4
Gardoside methyl ester 20, 21
Garjasmine 33, 35, 36
Genipa americana 10
Genipin 7, 29–31, 33, 51, 54, 70–72, 82–85

Genipin monosilyl ether 41
Genipin 1-*O*-TBDMS ether 32
Genipinine 54
Geniposide 12, 13, 16, 26, 51, 53, 54
Geniposidic acid 10
Geniposidic acid tetraacetate 16
Gentianine 8, 52
Gentiopicral 52
Gentiopicroside 5
Gentiopicroside aglycone 1-*O*-TBDMS ether 44
7-*O*-Gentisoylsecologanol 5
Geranyl tolyl sulfone 82, 83
β-Glucosidase 47, 49–52, 54, 66, 78, 79, 97, 102
Glucosidic monoterpenoid alkaloids 45
5-Glucosylantirrhinoside 9
Glutamic acid 218
D-Glutamic acid 179, 180
L-Glutamic acid 177
L-[U-^{14}C]Glutamic acid 218
Glutaraldehyde 208
Glycine 70, 71
β-Glycosidases 6
Grignard reaction 56, 155, 157, 159, 164, 180
Grignard reagent 147, 187, 206

Halitunal 82, 84, 86
Harpagide 47, 48, 51, 54, 69, 70
Harpagophytum procumbens 51, 52
Harpagophytum sp. 54
Harpagoside 51, 54
Harpagoxenus sublaevis 122
16-Heptadecene-2,5,8-trione 160
6-Heptenal 161
6-Hepten-2,5-dione 160
1-Hepten-3-one 171, 194
Heptyl bromide 146
($\pm$)-*trans*-2-Heptyl-5-ethylpyrrolidine 134, 135
Heptylmagnesium bromide 155
(5*Z*,8*E*)-3-Heptyl-5-methylpyrrolizidine 122
Hexahydropyrrolo[2,1,5-cd]-indolizines 126
Hexanenitrile 206
(*S*,*S*)-1,2,5,6-Hexanetetraol 147
2-Hexanone 166
(*E*)-2-Hexenal 152

Hex-5-enal 184
(2*R*,5*R*)-*trans*-2-(5-Hexenyl)-5-nonylpyrrolidine 144, 145
n-Hexylidenetriphenylphosphorane 141
3-Hexyl-5-methylindolizidine 193
(5*Z*,9*Z*)-3-Hexyl-5-methylindolizidine 124, 193
trans-2-Hexyl-5-pentylpyrrolidine 132, 133
Homoiridoids 73
Horner-Wittig olefination 176
Hydrochloric acid 51, 52, 79, 153, 199
Hydrogen peroxide 21
Hydroxamic acid 163
10-Hydroxycornin 23
7-Hydroxydehydroskytanthine 59
5-(2'-Hydroxyethyl)-4-methyl-3-benzylthiazolium chloride 132, 159, 160, 162, 163
10-Hydroxyhastatoside 23
(6*R**)-[(2*S**)-2-Hydroxyheneicos-12-enyl]-5,6-dihydro-2H-pyran-2-one 212
Hydroxyincarvilline 59
5-Hydroxymethylisochroman-1-one 52
2-Hydroxyoctanoic acid 142
10-Hydroxyoleosides 42
(13*E*,15*E*,18*Z*,20*Z*)-1-Hydroxpentacosa-13,15,18,20-tetraen-11-yn-4-one 1-acetate 217
1-Hydroxypiperidine 200
N-Hydroxypiperidine 206
5-Hydroxysecologanol 15
5-Hydroxyskytanthine 59
4-Hydroxytecomanine 59
Hymenoptera 116
(−)-Hypnophilin 89, 90

Incarvillateine 59
Incarvillea alkaloids 59, 60
Incarvillea sinensis 59
Incarvilline 8, 59
Incarvine C 59
($\pm$)-Indolizidine 195B 170
Indolizidine alkaloids 119
Indolizidines 124, 125, 163
Inhibitory activity 119
Insecticidal activity 118, 120
Intestinal bacteria 52–54
(*E*)-1-Iodo-1-hexene 133

6-Iodopurine tetrabutylammonium salt   101, 102
7-Iodo-11-trimethylsilyl ester   68
Iridodial   7, 26
Iridodial glucoside   18
epi-Iridodial glucoside   18
(+)-α-Iridodiol   61
Iridoid acids   14
Iridoid esters   13, 14
Iridoid glucosides   9, 12, 16–18, 22, 23, 25, 36, 43, 45–51, 65, 69, 87, 98, 99, 103
Iridoid glycosides   3, 5, 11, 13, 105, 106
Iridoid lactones   15
Iridoid methyl esters   14
Iridoid-related alkaloids   8
Iridoids   2, 7–15, 24, 45, 52, 69, 87, 92, 98, 103, 104
C_8-Iridoids   4
C_9-Iridoids   4
C_{10}-Iridoids   4
Iridolactones   27
Iridomyrmecin   7
(+)-Iridomyrmecin   26
Iridomyrmex sp.   118
Isobutyric anhydride   98
Isodihydronepetalactone   26
Isoeucommiol   97
Isolinarioside   24
N-Isopentyl-3-hydroxymethylpyrrolidine   138
N-Isopentyl-2-phenylethylamine   119
Isoplectrodorine   46
Isopropanol   13
5,6-*O*-Isopropylidene-3,4-dihydroantirrhinoside   28
Isovaleric acid   31, 32
Isoxazolones   74, 75

Jasminin   5
Jasminine   47
Jasmolactone B   40
Jasmolactone D   40
Jasmolactones   40, 42
Jones oxidation   36, 38–40, 62, 104, 152, 193
Jones reagent   38–40

Ketologanin   19, 105, 106
Ketopyrrolidines   168

Kinabalurine A   8, 59
Kinabalurine B   59
Kinabalurine C   59
Kinabalurine D   59
Kinabalurine E   59
Kinabalurine F   59
Kingiside   5, 40
8-*epi*-Kingiside   40, 48
Kingiside aglycone   7
Kingiside aglycone silyl ethers   45
Kingiside aglycone 1-*O*-TBDMS ether   44
8-*epi*-Kingiside aglycone 1-*O*-TBDMS ether   44
Kingiside tetraacetate   38, 39
8-*epi*-Kingiside tetraacetate   38, 39
Kopsia alkaloids   60
Kopsia sp.   59

Lamiaceae   13
Lamiol   69, 70
Lawesson's reagent   144
Lead tetraacetate   37, 42
Leptothoracini sp.   122, 138
Leptothorax acervorum   122
Leptothorax muscorum   122
Ligstroside   48
(+)-Limonene   125
Linarioloside   21
Linarioloside pentaacetate   93
Linarioside   9
Lipophilic compounds   117
Lithio(phenylsulfonyl)-methane   184
Lithium aluminum hydride   22, 89
Lithium azide   147
Lithium dimethylcuprate   89
Lithium dipentenyl cuprate   145
Lithium dipropyl cuprate   178
7-*epi*-Loganic acid   19
Loganin   4, 19, 41, 49, 105
8-*epi*-Loganin   20, 49
Lonicera japonica   40, 43
Loxylostosidine   68
2,6-Lutidine   85, 130
3,4-Lutidine   57

Male attracting activity   122
D-Mannitol   140, 147, 148
Marine diterpenoids   82
Massoilactone   212

Subject Index

Medium pressure liquid chromatography 12
Megalomyrmex foreli 122, 151
Megalomyrmex goeldii 121
Megalomyrmex leoninus 121
Megalomyrmex modestus 123, 159
Megalomyrmex sp. 121–123
Meldrum's acid 73, 136, 137, 149, 157, 158, 178
Menthiafolin 5
Mercuric acetate 140
Mercury(II) acetate 94, 97
Messor bouvieri 120
Messor capensis 120
Messor ebeninus 120
Methanesulfonyl chloride 147
Methanol 10, 12, 26, 95, 150
Methanol racemigerine 51
Methanolic bromine 79
Methanolic sodium methoxide 101
Methoxymethylenetriphenylphosphonium chloride 88
(−)-3′-Methoxyspecionin 91, 92
3-Methyl-5-alkenylpyrrolizidines 160
2-Methyl-6-alkyl piperidines 128
cis-2-Methyl-6-alkyl piperidines 119
trans-2-Methyl-6-alkyl piperidines 120
Methylamine 77
Methylamine hydrochloride 71
Methyl γ-aminobutyrate 104
N-Methylbakankosin 77
Methyl α-bromoacetate 189
3-Methylbutylamine 138
4-Methylcyclohexane-1,3-dione 55
Methylgluco-oleoside 5
Methyl iodide 171
Methyl jasmonate 88
(+)-Methyl jasmonate 87–89
(−)-Methyl jasmonate 87
Methylmagnesium bromide 56, 147
Methylmagnesiumcarbonate 202
Methyl 5-methylnicotinate 55
N-Methylmorpholine 174
N-Methylmorpholine-*N*-oxide 32, 39
cis-2-Methyl-6-((Z)-4-nonenyl)piperidine 119
trans-2-Methyl-6-((Z)-4-nonenyl)piperidine 119
cis-2-Methyl-6-((4Z)-nonenyl)piperidine 130

(±)-*cis*-2-Methyl-6-((4Z)-nonenyl)piperidine 131
trans-2-Methyl-6-((4Z)-nonenyl)piperidine 130
2-Methyl-6-((Z)-4-nonenyl)pyridine 130
(5Z,8E)-3-Methyl-5-(8-nonenyl)pyrrolizidine 160
cis-2-Methyl-6-nonylpiperidine 119, 125
trans-2-Methyl-6-nonylpiperidine 119
6-Methylpiperidin-2-one 166, 167
6-Methylpyridine-2-carboxaldehyde 197
3-Methylpyrrolidine 138
(*R*)-5-Methylpyrrolidinone 156
3-Methyltetrahydrofuran-2-one 138
cis-2-Methyl-6-undecylpiperidine 119
Methylvinylketone 151–153
Michael addition 32
Michael reaction 153
(+)-Mitsugashiwa-lactone 27
Mitsunobu inversion 19, 23
Mitsunobu reaction 180, 202
Mitsunobu-type reaction 64
Monochloroacetic acid 23
Monomorine I 124, 163, 184
(+)-Monomorine I 174, 175, 176, 178–184, 186–190, 193
(−)-Monomorine I 177, 184, 185
(±)-Monomorine I 163–174
(+)-(3*R*,5*S*,8*S*)-Monomorine I 174–176, 178, 179, 181–183, 186–188, 190
(+)-(3*R*,5*S*,9*S*)-Monomorine I 173, 176, 178
(−)-(3*S*,5*R*,8*R*)-Monomorine I 177, 185
(−)-(3*S*,5*R*,9*R*)-Monomorine I 178
Monomorium antarcticum 120, 122, 123
Monomorium delagoense 118, 129
Monomorium floricola 128
Monomorium pharaonis 124
Monomorium smithii 121, 123, 125
Monomorium sp. 118–122, 150, 160
Monoterpenoid piperidines 7
Morroniside 5
Morroniside aglycone 1-*O*-TBDMS ether 44
8-*epi*-Morroniside aglycone 1-*O*-TBDMS ether 44
epi-Muralioside 24
Myrmicaria eumenoides 125, 194, 200
Myrmicaria opaciventris 125, 126, 200
Myrmicaria sp. 125

Myrmicaria striata 126, 200
Myrmicarin 213A 126
Myrmicarin 213B 126
Myrmicarin 215A 126
Myrmicarin 215B 126
Myrmicarin 215C 126
Myrmicarin 217 126, 198–200
Myrmicarin 237A 125, 194, 196, 197
(−)-Myrmicarin 237A 194, 198
Myrmicarin 237B 125, 194, 196, 197
(+)-Myrmicarin 237B 194, 198
Myrmicarin 430A 126
Myrmicarin 663 126
Myrmicarins 125, 126
Myrmicinae 118, 119, 121, 122, 124, 125

Nepeta sp. 22
(+)-Nepetalactone 7, 26
Nepetanudoside 22
Nepetanudoside tetraacetate 22
Neplanocin B 102
Neplanocin C 102
Nitrobenzene 55
2-Nitrocyclohexanone 171
6-Nitrohexanoic acid 195
Nitromethane 153
5-Nitropentadecane-2,8-dione 152
1-Nitropentane 151
(1S^*,5R^*,7S^*)-7-(Nonadec-10-enyl)-2,6-dioxabicyclo[3.3.1]nonan-3-one 212
1-Nonen-3-one 159
(5E,8Z)-3-(1-Non-8-enyl)-5-((E)-1-prop-1-enyl) pyrrolizidine 122, 162
Non-glycosidic iridoids 6, 7
Nonracemic pyrrolidines 139
Nonylmagnesium bromide 145
L-Norleucine 140, 192, 193
D-Norleucinol 183, 184
p-Nucleophiles 142
Nyctanthes arbor-tristis 11

Oleaceae 48
Oleoside dimethyl ester 14, 15
Oleuropein 5, 36, 38, 48
Optical activity 47, 62, 82, 84, 86, 147, 155
L-Ornithine 218
L-Ornithine hydrochloride 203
L-[U-^{14}C]Ornithine hydrochloride 218
Orthocarpus sp. 47

Osmanthus austrocaledonica 47, 48
Osmium tetroxide 32, 36, 39
Oxerine 56, 57
(−)-Oxerine 47, 48
2-Oxo-3-azabicyclo-[4.3.0]-nonan-6-ols 55
4-Oxo-2,5-dienals 214, 215
4-Oxo-2,5-dienyl acetates 217
(2E,5E,12Z)-4-Oxo-heneicosa-2,5,12-trien-1-ol acetate 213
4-Oxooctanal 159
3-Oxooctanoic acid 203

Paraformaldehyde 165
Parikh-Doering oxidation 155
Patrinoside 6
Penstemide 10
Penstemide aglycone 6, 32
Penstemon serrulatus 10
Penstemon ssp. 24
1,5-Pentanediol 196
1-Pentenylmagnesium bromide 192
4-Pentenylmagnesium bromide 191, 192
Pent-4-enylmagnesium bromide 187
n-Pentyl bromide 209
Pentylmagnesium bromide 204, 209, 210
n-Pentylmagnesium bromide 135
Pentylmagnesium chloride 205
1-Pentyn-3-one 180
Peroxyformic acid 89
Petiodial 82
(+)-Petiodial 83
(−)-Petiodial 82
Pheidole sp. 118
(R)-1-Phenylethylamine 62
(S)-1-Phenylethylamine 134
2-(1-Phenylethyl)-2-azabicyclo[2.2.1]hept-5-ene 62
N-(S)-1-Phenylethylcarboxyamide 64
(R)-Phenylglycinol 158
(−)-Phenylglycinol 146, 158, 208
(1)-Phenylmethylamine 135
Phenylselenol 136
Phenylsulfide 58
Phosphorus oxychloride 33
2-Picolyllithium 203
Picrorhiza kurrooa 13
(+)-Pipecolic acid 127
Piperidine 75, 76, 78, 93

Subject Index

Piperidine monoterpene alkaloids 59
Piperidines 118, 128
cis-Piperidines 130
trans-Piperidines 130
Plantarenaloside 4, 47
Plectrodorine 46
Plectronia odorata 46
Plumeiride aglycone 1-*O*-TBDMS ether 33
Plumeria iridoids 6, 32–34
(+)-Plumericin 33
(+)-*iso*-Plumericin 33
Plumieride 6
Potassium acetate 31
Potassium *tert*-butoxide 184
Potassium cyanide 23
Potassium fluoride 58
Potassium hydride 34
Potassium monopersulfate triple salt 31
Potassium selenocyanate 162
L-Proline 141
(*S*)-Proline 141, 142
Propanal 131, 150
Propylamine 77
n-Propylamine 75–77
Propylene oxide 162, 163
Propyl lithium 142
Propylmagnesium bromide 180, 209, 210
Propylmagnesium chloride 205
Propyltriphenylphosphonium chloride 88
Prostaglandin analogues 87
Prostaglandins 88, 92
Pseudoalkaloids 7
Pummerer rearrangement 189
Putrescine 218
[1,4-^{14}C]Putrescine dihydrochloride 218
Pyrazolones 74, 75
Pyridine 36
Pyridine monoterpene alkaloids 7, 45
Pyridine monoterpenes 46
Pyridines 118
Pyridinium bromide 24
Pyridinium chloride 24
Pyridinium chlorochromate 26, 106
Pyridinium dichromate 26, 88, 193, 194, 196, 197, 199
Pyridinium halides 24
2-Pyridylacetic acid hydrochloride 204
(*S*)-Pyroglutamic acid 142–145, 149, 155–157, 178

Pyrrolidine alkaloids 121, 133
Pyrrolidines 120, 122, 131, 140
cis-Pyrrolidines 137
trans-Pyrrolidines 137
(2*R*,5*R*)-*trans*-Pyrrolidines 144
Δ^1-Pyrroline 203
1-Pyrroline 1-oxide 132
Pyrrolines 120, 122
1-Pyrrolines 120
Pyrrolizidines 122, 123
Pyrrolo[2,1,5-cd]indolizines 126

Quinoline 16, 17

Racemigerine 51
Radiatoside 4
Ramberg-Bäcklund ring-contraction 89
Randioside 4, 17
Rehmaglutin A 7
Repellent activity 118
(+)-Rhexifoline 8
Rhodium(II) acetate 173
Ruthenium(VIII) oxide 28, 89

Samarium(II) iodide 63
(+)-Sarkomycin 89
Sarracenin 7
(−)-Sarracenin 44
8-*epi*-Sarracenin 44
Sarracenins 45
Scabrosidol 4
Scandoside hexaacetate 22
Scrophulariaceae 9, 13, 28, 47
Scutellaria sp. 13
Scutellaria subvelutina 13
Scutellarioside I 13, 99
Secoiridoid aglycone silyl ethers 41, 43
Secoiridoid glucosides 39
Secoiridoids 3, 5, 15, 36, 45, 79
Secologanin 5, 7, 12, 13, 41, 67–69, 73, 75, 77
Secologanin aglycone 1-*O*-TBDMS ether 42
Secologanin dimethyl acetal 79
Secologanin tetraacetate 39
Secologanol 15
Secoxyloganin 48
L-Selectride 180
Serrulatoloside 10
Sesquiterpene derivatives 117

Sesquiterpenoids 212
Shanzhiside methyl ester 36, 47
Silica gel 12, 15, 134, 142, 188, 197, 203, 206
Silver(I) oxide 62
α-Skytanthine 59
(+)-α-Skytanthine 8, 60, 61, 64, 65
(−)-α-Skytanthine 62, 64
β-Skytanthine 59
δ-Skytanthine 59
(+)-δ-Skytanthine 60, 61
Skytanthus alkaloids 59, 60
Skytanthus sp. 59
Sodium[1-^{14}C]acetate 218
Sodium[2-^{14}C]acetate 218
Sodium azide 168, 189
Sodium bicarbonate 47, 87
Sodium bis-(2-methoxyethoxy)aluminium hydride 133
Sodium borohydride 23, 26, 36, 38, 42, 49, 94, 99, 100, 106, 137, 144, 157, 206
Sodium bromide 176
Sodium carbonate 48
Sodium chloride 56
Sodium cyanoborohydride 65, 67, 100, 130, 168, 177
Sodium hydride 137, 144, 184
Sodium hydroxide 24, 47
Sodium methoxide 32, 36, 44
Sodium naphthalenide 184
Sodium periodate 17, 36, 37, 87
Sodium sulfide 89
cis-Solenopsin A 219–221
trans-Solenopsin A 219–221
(+)-Solenopsin A 128, 129
trans-(+)-Solenopsin A 128, 129
Solenopsins 128, 146
Solenopsis conjurata 124
Solenopsis diplorhoptrum 193
Solenopsis fugax 133
Solenopsis geminata 119, 219
Solenopsis invicta 119
Solenopsis molesta 133
Solenopsis punctaticeps 141
Solenopsis sp. 118–121, 124, 125, 130, 150, 160
Solenopsis texana 133
Solenopsis xenovenum 122, 152
Soxhlet method 13
(−)-Specionin 7, 28, 29, 90

Succinic anhydride 192
Succinimide 200–202
Sulfuric acid 13
Swern oxidation 63, 83, 141, 161, 169, 176, 180, 185, 188, 198
Sweroside 15, 74–76
Sweroside aglycone 1-*O*-TBDMS ether 42
Sweroside tetraacetate 40, 68
Swertiamarin 15, 52, 53
Symphoricarpos albus 13

Tandem-Knoevenagel-hetero-Diels-Alder reaction 73–75
Tarennoside 10
Tartaric acid 173
L-Tartaric acid 180
Technomyrmex albipes 128
Tecoma alkaloids 59, 60
Tecoma stans 49, 59
(+)-Tecomanine 61, 62
(−)-Tecomanine 59
(−)-Tecostidine 49
1-Tetradecen-3-one 131, 150
(Z)-7-Tetradecenylamine 128
(Z)-9-Tetradecenylamine 128
Tetrahydrocantleyine 49
Tetrahydrofuran 34
3,4,5,6-Tetrahydropyridine-1-oxide 206, 207
Tetramorium aculeatum 212
Tetraponera sp. 126, 127, 200, 218
Tetraponerine-1 126, 127
(+)-Tetraponerine-1 209–211
Tetraponerine-2 126, 127
(+)-Tetraponerine-2 209–211
Tetraponerine-3 126, 127, 203
(+)-Tetraponerine-3 209, 210
(±)-Tetraponerine-3 203, 204
Tetraponerine-4 126, 127, 203
(+)-Tetraponerine-4 209–211
(±)-Tetraponerine-4 203, 204, 206
Tetraponerine-5 126, 127
(+)-Tetraponerine-5 209–211
(±)-Tetraponerine-5 202, 203
Tetraponerine-6 126, 127, 218, 219
(+)-Tetraponerine-6 209–211
(±)-Tetraponerine-6 202, 203
Tetraponerine-7 126, 127, 203
(+)-Tetraponerine-7 206, 207, 209, 210

(−)-Tetraponerine-7   206, 207
(±)-Tetraponerine-7   204
Tetraponerine-8   126, 127, 203, 218, 219
(+)-Tetraponerine-8   206–211
(−)-Tetraponerine-8   206, 207
(±)-Tetraponerine-8   200, 201, 204–206
Tetraponerines   127, 200
(+)-Tetraponerines   208
Thallium(III) nitrate   26
1,1′-Thiocarbonyldiimidazole   30, 86
Thionyl chloride   36, 147
Thunbergioside   4
Toluene   85
p-Toluenesulfonic acid   28
Tosyl azide   55
Tosyl chloride   197
Toxic activity   214, 217
Trail pheromones   128, 212
Tributylphosphine   30, 151, 153
Tributyltin hydride   56, 63, 180
Tricyclic alkaloids   69
Triethylamine   55, 58, 131, 142, 147, 160, 175, 184, 189
Triethylsilane   32, 33
Trifluoroacetic acid   32, 91
Triisopropylsilyltrifluoromethane-sulfonate   175
Trimethylamine-*N*-oxide   36
Trimethylphosphite   189
Trimethylsilyl iodide   68
(Trimethylsilyl)tributylstannane   41
Trioxadamantane compounds   79–81
Trioxadamantane epimers   79
Triphenylphosphine   142
Triphenylphosphine oxide   101
Triquinane sesquiterpenes   89
Trityl pyridinium tetrafluoroborate   104
Trityl tetrafluoroborate   32
Tryptophan   7
Tungsten(VI) oxide   21

Udoteatrial   82
Udoteatrial hydrate   84
ent-Udoteatrial hydrate   85
Unedoside   4
Uracil   97, 98

Vacuum liquid chromatography   12
Valerianaceae   49
Valeriana iridoids   6
Valeriana-type iridoids   10, 31
Valtrate   6
6-*O*-Vanilloylcatalpol   90
6-*O*-Vanilloylcatalpol hexaacetate   92
Verbascoside   11
Vilka's reagent   135
Vilsmeier formylation   22
Vincoside   8
Vinyl sulfides   58
Vorbrüggen procedure   98

Wacker oxidation   180
Wadsworth-Emmons reaction   152
Weinreb's procedure   193
Wittig-Horner reaction   155
Wittig olefination   56, 161, 169
Wittig reaction   88, 128, 130, 141, 186
Wolff-Kishner reduction   192
Wolff rearrangement   55

Xenovenine   152, 153, 155, 161
(±)-Xenovenine   152–154
(3*R*,5*S*,8*R*)-Xenovenine   155
(3*S*,5*R*,8*S*)-Xenovenine   155–158
Xylene   55
Xylostosidine   68

Zemplén deacetylation   18
Zinc borohydride   180
Zinc iodide   177

SpringerChemistry

Fortschritte der Chemie organischer Naturstoffe
Progress in the Chemistry of Organic Natural Products

Founded by L. Zechmeister
Editors: W. Herz, H. Falk, G. W. Kirby, R. E. Moore, and C. Tamm

Volume 78

1999. VIII, 168 pages. 2 figures.
Hardcover DM 250,–, öS 1750,–
Special price for subscribers to the series DM 225,–, öS 1575,–
3-211-83311-0

Contents
- Brassinosteroids (G. Adam, J. Schmidt, B. Schneider)
- Chemistry of the Neem Tree (Azadirachta indica A. Juss.)
  (A. Akhila, K. Rani)

Volume 77

1999. VII, 187 pages. 3 partly coloured figures.
Hardcover DM 250,–, öS 1750,–
Special price for subscribers to the series DM 225,–, öS 1575,–
3-211-83264-5

Contents
- Secondary Metabolites and the Control of Some Blue Stain
  and Decay Fungi (W. A. Ayer, L. S. Trifonov)
- Condensed Tannins
  (D. Ferreira, E. V. Brandt, J. Coetzee, E. Malan)
- Constituents of *Lactarius* (Mushrooms)
  (W. M. Daniewski, G. Vidari)

For further informations please visit our homepage: **www.springer.at**

All prices are recommended retail prices

SpringerWienNewYork

A-1201 Wien, Sachsenplatz 4–6, P.O.Box 89, Fax +43.1.330 24 26, *e-mail:* books@springer.at, **www.springer.at**
D-69126 Heidelberg, Haberstraße 7, Fax +49.6221.345-229, e-mail: orders@springer.de
USA, Secaucus, NJ 07096-2485, P.O. Box 2485, Fax +1.201.348-4505, e-mail: orders@springer-ny.com
EBS, Japan, Tokyo 113, 3–13, Hongo 3-chome, Bunkyo-ku, Fax +81.3.38 18 08 64, e-mail: orders@svt-ebs.co.jp

SpringerChemistry

Fortschritte der Chemie organischer Naturstoffe
Progress in the Chemistry of Organic Natural Products

Founded by L. Zechmeister
Editors: W. Herz, H. Falk, G. W. Kirby, R. E. Moore, and C. Tamm

Volume 76

1999. VII, 211 pages. 57 partly coloured figures.
Hardcover DM 250,–, öS 1750,–
Special price for subscribers to the series DM 225,–, öS 1575,–
3-211-83165-7

Contents
- Nitric Oxide: Physiological Roles, Biosynthesis and Medical Uses (D. R. Adams, M. Brochwicz-Lewinski, A. R. Butler)

Volume 75

1998. VII, 226 pages. 26 figures.
Hardcover DM 250,–, öS 1750,–
Special price for subscribers to the series DM 225,–, öS 1575,–
3-211-83053-7

Contents
- Cyclopeptide Alkaloids
  (D.C. Gournelis, G.G. Laskaris, R. Verpoorte)
- Naturally Occurring 6-Substituted 5,6-Dihydro-Alpha-Pyrones
  (L.A. Collett, M.T. Davies-Coleman, D.E.A. Rivett)

For further informations please visit our homepage: **www.springer.at**

All prices are recommended retail prices

SpringerWienNewYork

A-1201 Wien, Sachsenplatz 4–6, P.O.Box 89, Fax +43.1.330 24 26, e-mail: books@springer.at, **www.springer.at**
D-69126 Heidelberg, Haberstraße 7, Fax +49.6221.345-229, e-mail: orders@springer.de
USA, Secaucus, NJ 07096-2485, P.O. Box 2485, Fax +1.201.348-4505, e-mail: orders@springer-ny.com
EBS, Japan, Tokyo 113, 3–13, Hongo 3-chome, Bunkyo-ku, Fax +81.3.38 18 08 64, e-mail: orders@svt-ebs.co.jp

SpringerChemistry

Mikrochimica Acta
Micro and Trace Analysis

Managing Editor: W. Wegscheider
and an International Editorial Board

Presenting the latest results from all areas of analytical chemistry, "Mikrochimica Acta" is a journal of some tradition, published regularly since 1923, when it was founded by Nobel Prize winner F. Pregl. It has pioneered the present trend in analytical chemistry. In contrast to many of the highly specialized journals, "Mikrochimica Acta" covers a variety of topics, such as

Materials Science • Microsensors • Microchemistry • Elemental Analysis • Organic Analysis • Trace Analysis • Enrichment Techniques • Surface Characterization • Chemometrics • Molecular and Atomic Spectroscopic Techniques • Computer Applications in Analysis • Chromatographic Analysis • Electrochemical Analysis • Sampling Methods • Standard Reference Materials and Methods • Analysis in Biotechnology

Special attention is given to emerging new technologies, new techniques, and important trends in analytical chemistry

Subscription information

2000. Vols. 134–136 (4 issues each) Title No. 604
ISSN 0026-3672 (print version), ISSN 1436-5073 (electronic version)
DM 2442.–, öS 17178.– plus carriage charges
approx. US $ 1,517.00 including carriage charges

View table of contents and abstracts online at: **www.springer.at/mikro**

 This journal is included in the program:
"LINK – Springer Print Journals Go Electronic"
ISSN (electronic edition): 1436-5073

A-1201 Wien, Sachsenplatz 4–6, P.O.Box 89, Fax +43.1.330 24 26, e-mail: journals@springer.at, **www.springer.at**
D-69126 Heidelberg, Haberstraße 7, Fax +49.6221.345-229, e-mail: orders@springer.de
USA, Secaucus, NJ 07096-2485, P.O. Box 2485, Fax +1.201.348-4505, e-mail: orders@springer-ny.com
EBS, Japan, Tokyo 113, 3–13, Hongo 3-chome, Bunkyo-ku, Fax +81.3.38 18 08 64, e-mail: orders@svt-ebs.co.jp

SpringerChemistry

Amino
Acids

Editors-in-Chief: G. Lubec, G. C. Barrett, and C. MacLeod
and an International Editorial Board

"Amino Acids" publishes contributions from all fields of amino acid research: analysis, separation, synthesis, biosynthesis, cross linking amino acids, racemization/enantiomers, modification of amino acids as phosphorylation, methylation, acetylation, glycosylation and nonenzymatic glycosylation, new roles for amino acids in physiology and pathophysiology, biology, amino acid analogues and derivatives, polyamines, radiated amino acids, peptides, stable isotopes and isotopes of amino acids. Applications in medicine, food chemistry, nutrition, gastroenterology, nephrology, neurochemistry, pharmacology, excitatory amino acids are just some topics to be listed. We also encourage the submission of papers of interdisciplinary borderlines.

Subscription information

2000. Vols. 18+19 (4 issues each) Title No. 726
ISSN 0939-4451 (print version), ISSN 1438-2199 (electronic version)
DM 1308.–, öS 9202.– plus carriage charges
approx. US $ 823.00 including carriage charges

Special subscription price available for members of the International Society for Amino Acids Research: US $ 138.00 + US $ 54.00 carriage charges
(Orders have to be sent directly to Springer-Verlag Wien)

View table of contents and abstracts online at:
    www.springer.at/amino_acids

This journal is included in the program:
"LINK – Springer Print Journals Go Electronic".
ISSN (electronic version): 1438-2199

A-1201 Wien, Sachsenplatz 4–6, P.O.Box 89, Fax +43.1.330 24 26, e-mail: journals@springer.at, **www.springer.at**
D-69126 Heidelberg, Haberstraße 7, Fax +49.6221.345-229, e-mail: orders@springer.de
USA, Secaucus, NJ 07096-2485, P.O. Box 2485, Fax +1.201.348-4505, e-mail: orders@springer-ny.com
EBS, Japan, Tokyo 113, 3–13, Hongo 3-chome, Bunkyo-ku, Fax +81.3.38 18 08 64, e-mail: orders@svt-ebs.co.jp

Springer-Verlag
and the Environment

WE AT SPRINGER-VERLAG FIRMLY BELIEVE THAT AN international science publisher has a special obligation to the environment, and our corporate policies consistently reflect this conviction.

WE ALSO EXPECT OUR BUSINESS PARTNERS – PRINTERS, paper mills, packaging manufacturers, etc. – to commit themselves to using environmentally friendly materials and production processes.

THE PAPER IN THIS BOOK IS MADE FROM NO-CHLORINE pulp and is acid free, in conformance with international standards for paper permanency.